This unique volume provides a comprehensive survey of our understanding of the Universe based on the exact solutions of the theory of relativity. More precisely, it describes those models that fit with astronomical observations of galaxy clusters, cosmic voids and other key features of our Universe.

This authoritative account achieves two important goals. Firstly, it collects together all independently derived cosmological solutions from the birth of relativity in 1915 to the present day, and clearly shows how they are inter-related. Secondly, it presents a coherent overview of the physical properties of these inhomogeneous models. It demonstrates, for instance, that the formation of voids and the interaction of the cosmic microwave background radiation with matter in the Universe can be explained by exact solutions of the Einstein equations, without the need for approximations.

This book will be of particular interest to graduates and researchers in gravity, relativity and theoretical cosmology, as well as historians of science.

T0254509

INHOMOGENEOUS COSMOLOGICAL MODELS

Inhomogeneous Cosmological Models

ANDRZEJ KRASIŃSKI

N. Copernicus Astronomical Center and School of Sciences,
Polish Academy of Sciences, Warsaw

CAMBRIDGE
UNIVERSITY PRESS

CAMBRIDGE UNIVERSITY PRESS
Cambridge, New York, Melbourne, Madrid, Cape Town, Singapore, São Paulo

Cambridge University Press
The Edinburgh Building, Cambridge CB2 2RU, UK

Published in the United States of America by Cambridge University Press, New York

www.cambridge.org
Information on this title: www.cambridge.org/9780521481809

© Cambridge University Press 1997

First published 1997
This digitally printed first paperback version (with corrections) 2006

A catalogue record for this publication is available from the British Library

Library of Congress Cataloguing in Publication data

Krasiński, Andrzej.
Inhomogeneous cosmological models / Andrzej Krasiński.
p.   cm.
Includes bibliographical references.
ISBN 0 521 48180 5 (hardcover)
1. Cosmology – Mathematical models.   2. Relativity (Physics) – Mathematical models.   I. Title.
QB981.K76 1997
523.1′01′5118–dc20 95-23549 CIP

ISBN-13 978-0-521-48180-9 hardback
ISBN-10 0-521-48180-5 hardback

ISBN-13 978-0-521-03017-5 paperback
ISBN-10 0-521-03017-X paperback

Contents

Illustrations

NOTE: A "Box" is a magnified piece of a "Figure", but they do not differ at first sight.

Preface

This work is a review of inhomogeneous cosmological exact solutions of the Einstein equations and their properties. A solution is, by definition, cosmological if it can reproduce any metric of the Friedmann (1922, 1924)–Lemaître (1927, 1931)–Robertson (1929, 1933)–Walker (1935) class (abbreviated FLRW) by taking limiting values of arbitrary constants or functions. The book is intended to attain two objectives: (*i*) to list all independently derived cosmological solutions and reveal all interconnections between them; (*ii*) to collect in one text all discussions of the physical properties of our Universe based on inhomogeneous models. Objective (*i*) is attained by showing that the more than 300 independently published solutions can be derived by limiting transitions from about 60 parent solutions. The solutions are arranged in relatively few disjoint families, and a few of them have been rediscovered up to 20 times. In attaining objective (*ii*) it is shown that exact inhomogeneous solutions can describe several features of the Universe in agreement with observations. Examples are the presence of voids and high-density membranes in the distribution of mass or the fluctuations of temperature of the microwave background radiation. Among the effects predicted are: (*i*) cosmological expansion of planetary orbits (unmeasureably small, but nonzero); (*ii*) prevention of the Big Bang singularity by an arbitrarily small charge anywhere in the Universe. In addition, papers that have discussed averaging the small-scale Einstein equations to obtain large-scale cosmological field equations have been reviewed. It is hoped that this work will prevent further duplication of effort already spent, and that it will channel the research activity in those directions that have not yet been explored.

The reader is assumed to be familiar with Einstein's relativity theory at a level somewhat higher than elementary. For those readers who need an introduction, the book by Stephani (1990) is recommended.

This review was initially intended to be just an appendix to my paper on the Barnes models (Krasiński 1989). When I began compiling it in 1987, I expected to end up, in a couple of months, with about 20 reviewed and classified papers. A year later I decided to make the review a separate paper. Another year later, compiling the review became my main job. The work finally lasted over six years and resulted in a text suitable for a book rather than a paper. The book includes more than 700 papers chosen from more than 2300 that I had to look up or read (see Section 1.1 for the criteria of choice). With the work already finished, I wish to emphasize that Warsaw was probably the best place in the world to do it. The long tradition of nearly equally good access to the scientific communities both in the West and in the East resulted in a large and well-balanced representation of literature from both those parts of the world in the Warsaw libraries. Each single library here is incomplete, but taken together they comprise one of the most splendid collections in the world of books and journals on physics, mathematics and astronomy. Familiarity with the Russian language, rather uncommon outside Eastern Europe, has been essential in the attempt to make the review complete. Finally, while projects lasting six years are not easily tolerated by any employers, elsewhere they seem not to be tolerated at all.

The book would not have come into being without the kind assistance that I received from several persons and institutions. Their contributions are described in the Acknowledgements. I thank them all for all they did to support this project.

A. Krasiński
Warsaw 1995

Acknowledgements

The research for this book relied heavily on the author's access to several libraries. Apart from my home libraries at the Institute of Theoretical Physics of the Warsaw University and at the N. Copernicus Astronomical Center of the Polish Academy of Sciences I have made substantial use of the collections of libraries at the following institutions (listed here in chronological order):

1. Department of Nuclear and Theoretical Physics, University of Pavia (access made possible by the courtesy of M. Carfora, visit financed by Gruppo Nazionale per la Fisica Matematica, Consiglio Nazionale delle Ricerche).
2. Raman Research Institute, Bangalore (courtesy of C. V. Vishveshwara).
3. Tata Institute of Fundamental Research, Bombay (courtesy of J. V. Narlikar).
4. Mathematics and Physics Library, University of Colorado at Boulder (courtesy of H. G. Ellis).
5. Linda Hall Library, Kansas City (courtesy of J. Urani, visit financed by the Department of Physics, University of Missouri at Kansas City).
6. Department of Physics and Astronomy, University of Missouri at Columbia (courtesy of B. Mashhoon).
7. Institute of Theoretical Physics, Stockholm University (courtesy of B. Laurent and K. Rosquist).
8. Mathematical Institute, Polish Academy of Sciences.
9. Institute of Physics, Polish Academy of Sciences.
10. Uji Research Center, Kyoto University (courtesy of K. Tomita, visit financed by the Japanese Society for the Promotion of Science).

(Unless stated otherwise, the foreign visits were financed by the host institutions.) I thank my hosts for their hospitality and kind guidance.

The journals and books that I could not find by myself were traced for me around Poland by the staff of the Central Catalogue of the National Library in Warsaw. Whenever this search failed, Mrs Wanda Dorociak from the International Loans Section of the National Library traced them worldwide, and then obtained the appropriate copies for me. Many gaps in the bibliography were closed in this way. I am very grateful for the kind help of all the persons involved in this activity.

I am grateful to: H. Knutsen for a copy of his large bibliography on spherically symmetric solutions (Knutsen 1983d) which became the seed for this book; to B. Iyer for sending me several copies from Indian journals that were not accessible elsewhere; to J. Urani for exploiting on my behalf the inexhaustible resources of the

Linda Hall Library; to G. Hall, A. Kembhavi, M. Goller and H. Knutsen for sending me xerox copies of various articles on request; to V. P. Frolov, I. S. Shikin, I. Dymnikova and D. Yakovlev for their help in obtaining copies from Russia; to S. Bażański for giving me access to his collection of rare conference proceedings; and to M. A. H. MacCallum, K. Tomita and, most of all, V. P. Frolov for their help in completing the bibliographical data for American translations from Russian literature. I am also grateful (in chronological order) to H. Stephani, J. Wainwright, J. Pullin, S. W. Goode, F. Occhionero, R. N. Henriksen, J. M. M. Senovilla, T. Papacostas, W. B. Bonnor, A. Chamorro, C. Hellaby, A. Coley, U. Brauer, J. Ehlers, J. B. Griffiths, J. P. S. Lemos, J. Novotny, A. Zecca, M. B. Ribeiro, R. M. Zalaletdinov, B. Mashhoon, J. Krishna Rao and P. S. Joshi for helpful comments, and to all those who responded positively to my reprint requests (too numerous to be listed).

Large numbers of papers for this review were copied in Pavia, in Stockholm and in Uji. I deeply appreciate the generosity of my host institutions there with respect to the use of their copying facilities.

Some of the calculations for this book were performed using the algebraic program Ortocartan. I am grateful to M. Demiański for providing funds from his grant to buy an Atari Mega STE computer and install it in the Copernicus Center. Without that investment, use of the program would not have been possible, and the review would be less complete than it is.

I am grateful to H. Knutsen again for a critical reading of a large part of this work and for several helpful remarks, and to A. Trautman and S. Bażański for stimulating comments during the seminars on this subject.

My special thanks are due to M. A. H. MacCallum for a very careful reading of the manuscript and many useful comments.

I thank J. Zalewski for his kind help with the software to handle the diagrams.

The final phase of the research for this book was partly supported by the Scientific Research Committee grant no 2 1242 91 01. The first draft of the book was published as a preprint and mailed to the authors of the papers reviewed. One issue of the preprint was printed and mailed at the Department of Applied Mathematics, University of Cape Town. I am very grateful to G. F. R. Ellis for arranging that support. A second issue was printed in Warsaw; its printing was supported by the aforementioned grant, and its mailing was financed by the N. Copernicus Astronomical Center, by the Department of Mathematics – University of Colorado at Boulder (support arranged by H. G. Ellis) and by K. Kuchař. I thank all the contributors for their generosity.

1
Preliminaries

1.1 The scope of this review

Ever since the 1930s, it has been conventional wisdom in cosmology that the Friedmann (1922, 1924)–Lemaître (1927, 1931)–Robertson (1929, 1933)–Walker (1935) (FLRW)[1] models describe the large-scale properties of our observed Universe faithfully. At the same time, it has been conventional wisdom in relativity theory that finding exact solutions of the Einstein equations is extremely difficult and possible only for exceptionally simple cases. Both these views were challenged repeatedly by lone rebels, but a few generations of physicists and astronomers have been educated with these conventional wisdoms solidly incorporated into their minds. As a result of this situation, a large body of literature has come into existence in which exact solutions generalizing FLRW have been derived and applied to the description of our observed Universe, but most of it remains unknown to the physics community and is not being introduced into textbooks. This book is intended to achieve the following two objectives:

1. To list all the independently derived cosmological solutions of the Einstein equations and to reveal all the interconnections between them.
2. To compile an encyclopaedia of physics in an inhomogeneous Universe by gathering together all physical conclusions drawn from such solutions.

An exact solution of the Einstein equations is termed "cosmological" if it can reproduce a FLRW metric when its arbitrary constants or functions assume certain values or limits. This requirement will be discussed in Section 1.2. The solutions are organized into a few families. Typically, within each family there is one solution from which several other independently published solutions follow as subcases. Generalizations were constructed over some of the subcases by including, for example, electric charge, magnetic field, heat-flow, viscosity or null radiation. The families are defined by invariant criteria, and several of the limiting transitions can be defined invariantly, too. This classification revealed multiple rediscoveries of a few solutions. In the cases of the spherically symmetric limit of the Stephani (1967a) Universe, and of the $\Lambda=0$ subcase of the Lemaître (1933a)–Tolman (1934) (L–T) solution, the numbers of independently published derivations both exceed 20. It is hoped that this aspect of the book will convey to the physics community the following:

---

[1] See the dictionary of frequently used abbreviations given in Appendix A.

Message 1

Although the Einstein equations tend to be complicated, several authors have coped successfully with the difficulties. We are meeting problems because of too many rather than too few solutions. Simple-minded attempts to derive a "new" solution from "natural" assumptions are likely to result in yet another rediscovery. What is really needed is the physical and geometrical interpretation of the solutions already derived. New solutions describing the still unexplored physical situations would be useful, but the unexplored area has to be carefully identified in advance.

The physical and geometrical properties were extensively worked out only for the L–T model (and, to some extent, for the Szekeres (1975a) solutions). The amount of information extracted from them is impressive, ranging from successful descriptions of various features of the Universe (voids – Sato 1984; fluctuations of temperature in the background radiation – Panek 1992), through possibly testable predictions (expansion of planetary orbits – Gautreau 1984) to striking qualitative conclusions (an uncompensated electric charge anywhere in the Universe can halt and reverse the collapse towards the final singularity – Shikin 1972a, Ivanenko, Krechet, Lapchinskii 1973; remnants of a nonsimultaneous Big Bang can be observable – Novikov 1964a). This mass of material implies the following:

Message 2

A new branch of relativistic astrophysics, using the mathematically well-founded methods of general relativity and based on exact solutions of the Einstein equations, is already in existence. One does not always have to resort to approximations in order to describe "realistic" physics.

This book was written from the point of view of relativity. The question it seeks to answer is not "What can we learn about the Universe by all means available?", but, rather, "What use do we have for relativity theory in studying the Universe?". For this reason, it does not address several problems that astrophysicists find important. Exception is made for only one such problem: the astrophysical reasons justifying the consideration of inhomogeneous models; this is reviewed in Appendix B. Certain papers discussing the influence of inhomogeneities in matter distribution on the microwave background radiation have been omitted because the model they used was constructed by patching together pieces of the FLRW and Schwarzschild spacetimes. This, the author assumed, was no advancement of the topic of exact solutions. On the other hand, the papers that discussed the same effect using the L–T model have been included.

The subject matter of the book is defined as follows:

1. The Bianchi-type homogeneous models are not included. The literature on them is enormous, and including it would double the volume of the book. The subject is in good hands, in no danger of being unappreciated (see Rosquist and Jantzen 1988), and good, if somewhat outdated, reviews are available (MacCallum

1973, 1979, 1984 and 1985, Collins and Ellis 1979). Most of the interesting physics is connected with inhomogeneities in geometry and matter distribution (for more comments on the deficiencies of the Bianchi models see Collins 1979).

2. In line with the preceding point, exact solutions in which the metric is FLRW are not included, no matter how complicated the source is. More generally, the book omits papers in which the source in the Einstein equations was generalized for a new component that has no influence on the metric. For example, Coley and Tupper (1983) considered spacetimes in which the metric was FLRW, but the velocity field was tilted with respect to the homogeneous hypersurfaces and the energy-momentum tensor contained contributions from shear viscosity and heat-conduction that cancelled the effect of tilt. For another example, Patel and Pandya (1986) considered spacetimes with a rotating fluid, null radiation and electromagnetic field with the property that the rotation and the null+electromagnetic fields cancelled each other, leaving no trace of them in the metric. The metric was the same as that in Bondarenko and Kobushkin's (1972) perfect fluid solution. See also a general discussion by Tupper (1983), explaining in detail how a perfect fluid energy-momentum tensor can be reinterpreted as that of a heat-conducting fluid in electromagnetic field. Such ambiguities in decomposing the energy-momentum tensor may turn out to be important if a physical interpretation is attached to them; so far this has not happened. Without such an interpretation, adding new and apparently unobservable components to the source seems to be in contradiction with Ockham's principle.

3. In apparent contradiction to point 1, solutions with the Kantowski-Sachs (1966) geometry are included (see Section 1.3.2 for a short description). They are simple, not too numerous and often show up as limiting cases along with FLRW, so they would have to be mentioned anyway.

4. The main topic of this book comprises those inhomogeneous solutions that can reproduce the FLRW models in a limit. This principle derives from what may be called the Cosmological Doctrine: at the largest scale of description, the FLRW models are the correct first approximation to relativistic cosmology. Hence, a more detailed model is acceptable only if it is an exact perturbation of FLRW. For a discussion on how a model may fail to have a FLRW limit – see Section 1.3.

5. Approximate solutions found within various perturbative schemes are not included even if they have a FLRW limit. Like the Bianchi models, they are numerous and under good care in other hands.

6. Numerical investigations have only been included in those cases where they help in understanding the properties of exact solutions. Purely numerical integrations of the Einstein equations have not been included.

7. In principle, all papers in which inhomogeneous generalizations of the FLRW or K–S models are discussed have been included here. However, the few cases where the discussion is short and not very illuminating, have been omitted.

8. Where necessary and possible, properties of the various solutions have been derived or verified using the algebraic computer program Ortocartan (Krasiński

1993, Krasiński and Perkowski 1981). Errors have been found in a few papers (e.g. the metric given is not a perfect fluid solution, contrary to the author's claim). Those errors have been corrected only where the correction was easy to guess. In general, papers containing errors have been omitted; although where they form the basis for an overly erroneous claim, the error is noted here without being corrected. It must be stressed that Ortocartan has only been used where either some data were missing or there was reason to suspect an error. Otherwise, this author has tended to trust the other authors.

9. Papers in which the information given is insufficient to classify the solution, while reconstruction of the missing data using Ortocartan would require excessive effort, are not mentioned.

10. Solutions of other theories of gravitation (e.g. Brans–Dicke) which become FLRW in the limit of the Einstein theory have not been included. Theories that do not include the Einstein theory as a limit were not considered.

11. Papers in which classes of metrics were investigated using the Einstein equations, but not solving them, have not as a rule been included. The only major exception is the Tabensky–Taub–Tomita family of spacetimes (see Sections 6.1-6.7) included because of the amount of attention it has received in the literature.

12. In a separate chapter, papers that deal with averaging out small-scale inhomogeneities (in order to obtain a large-scale model) are reviewed. Those papers do not usually discuss exact solutions, but the subject is of fundamental importance for interpreting inhomogeneous cosmological models and for comparing them with observations.

Papers in which solutions obeying these criteria were derived are classified in the diagrams (see the list of illustrations) according to the generality of the solutions. Figure 1.1 explains the signs used in the other diagrams. The classification reveals several repeated discoveries of the same solutions in different (or the same) coordinates. Papers in which properties of the solutions are discussed are usually mentioned or described together with the derivations. However, the papers discussing properties of the Lemaître–Tolman model (see Section 2.12) are so numerous that they are reviewed in Chapter 3 of this book, sorted by subject.

Historically, this work is not the first attempt at a review of inhomogeneous cosmological models. However, its predecessors were obviously not meant to be complete. Partial overviews may be found in the papers by MacCallum (1979, 1984 and 1985), Collins and Szafron (1979a,b), Wainwright (1979, 1981) and Knutsen (1983d); these provided many references for this book. The book by Kramer *et al.* (1980) is the most comprehensive of all the reviews, but it was written from a different point of view. It emphasizes the detection of differences between solutions, while this book will reveal inherent links between different solutions and show how many of them can be obtained by limiting transitions from relatively few (see the sentence closing our Section 1.3.1). The subject matter of this book is much narrower than that of Kramer *et al.*'s book, but it will be pursued here to a greater depth.

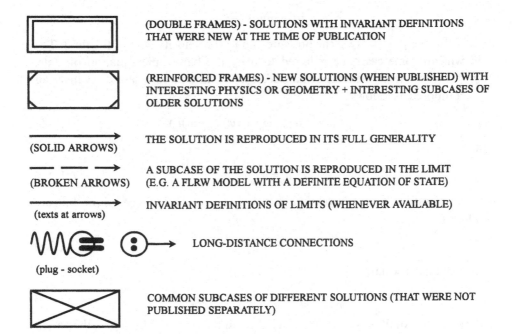

(DOUBLE FRAMES) - SOLUTIONS WITH INVARIANT DEFINITIONS
THAT WERE NEW AT THE TIME OF PUBLICATION

(REINFORCED FRAMES) - NEW SOLUTIONS (WHEN PUBLISHED) WITH
INTERESTING PHYSICS OR GEOMETRY + INTERESTING SUBCASES OF
OLDER SOLUTIONS

(SOLID ARROWS)   THE SOLUTION IS REPRODUCED IN ITS FULL GENERALITY

(BROKEN ARROWS)   A SUBCASE OF THE SOLUTION IS REPRODUCED IN THE LIMIT
(E.G. A FLRW MODEL WITH A DEFINITE EQUATION OF STATE)

(texts at arrows)   INVARIANT DEFINITIONS OF LIMITS (WHENEVER AVAILABLE)

LONG-DISTANCE CONNECTIONS

(plug - socket)

COMMON SUBCASES OF DIFFERENT SOLUTIONS (THAT WERE NOT
PUBLISHED SEPARATELY)

ABBREVIATIONS ARE EXPLAINED IN APPENDIX A

Figure 1.1. Signs used in the other diagrams.

1.2 The sources in the Einstein equations

The signature will be $(+---)$ throughout the book, and the Riemann tensor is
defined so that, for a covariant vector field k_α:

$$k_{\alpha;\beta\gamma} - k_{\alpha;\gamma\beta} = k_\rho R^\rho{}_{\alpha\beta\gamma}. \qquad (1.2.1)$$

Components of covariant derivatives will be denoted by a semicolon preceding
the appropriate subscript. Partial and ordinary derivatives will be denoted by a
comma preceding the appropriate index or the symbol of a variable.

The Einstein equations are:

$$G_{\alpha\beta} \equiv R_{\alpha\beta} - \tfrac{1}{2}g_{\alpha\beta}R = \kappa T_{\alpha\beta} + \Lambda g_{\alpha\beta}, \qquad (1.2.2)$$

$$\kappa \overset{\text{def}}{=} 8\pi G/c^4, \qquad (1.2.3)$$

where G is the gravitational constant, Λ is the cosmological constant, and the
energy-momentum tensor for the solutions considered in this book can contain the
following contributions:

$$T = T^f + T^h + T^v + T^n + T^s + T^e, \qquad (1.2.4)$$

where:

(I) T^f is the perfect fluid contribution, with:

$$T^f{}_{\alpha\beta}=(\epsilon+p)u_\alpha u_\beta-pg_{\alpha\beta}, \qquad (1.2.5)$$

ϵ being the energy-density, p the pressure and u_α the velocity field of the fluid. The fluid will, in some cases, be required to obey the barotropic equation of state, $f(\epsilon,p)=0$. The velocity field is normalized, $u_\alpha u^\alpha=1$, and its covariant derivative can be decomposed as follows:

$$u_{\alpha;\beta}=\dot{u}_\alpha u_\beta+\sigma_{\alpha\beta}+\omega_{\alpha\beta}+\tfrac{1}{3}\theta(g_{\alpha\beta}-u_\alpha u_\beta), \qquad (1.2.6)$$

where:

$$\dot{u}_\alpha=u_{\alpha;\rho}u^\rho \qquad (1.2.7)$$

is the acceleration vector, obeying $\dot{u}_\alpha u^\alpha=0$ (when $\dot{u}_\alpha=0$, the flow-lines of the fluid are geodesics);

$$\theta=u^\alpha{}_{;\alpha} \qquad (1.2.8)$$

is the expansion scalar;

$$\sigma_{\alpha\beta}=u_{(\alpha;\beta)}-\dot{u}_{(\alpha}u_{\beta)}-\tfrac{1}{3}\theta(g_{\alpha\beta}-u_\alpha u_\beta) \qquad (1.2.9)$$

is the shear tensor (it is symmetric in its indices and obeys $\sigma^\alpha{}_\alpha=0=\sigma_{\alpha\beta}u^\beta$; the parentheses around indices denote symmetrization); and

$$\omega_{\alpha\beta}=u_{[\alpha;\beta]}-\dot{u}_{[\alpha}u_{\beta]} \qquad (1.2.10)$$

is the rotation tensor (it is antisymmetric and obeys $\omega_{\alpha\beta}u^\beta=0$; the square brackets around indices denote antisymmetrization). The shear tensor is zero if and only if the shear scalar σ, defined by $\sigma^2=\tfrac{1}{2}\sigma_{\alpha\beta}\sigma^{\alpha\beta}$, is zero. The rotation tensor is zero if and only if the rotation scalar ω, defined by $\omega^2=\tfrac{1}{2}\omega_{\alpha\beta}\omega^{\alpha\beta}$, is zero. Quite often, the special perfect fluid with $p=0$ (dust matter) is considered as a source. If dust is the only source, then, necessarily, $\dot{u}^\alpha=0$.

This description of continuous media was introduced by Ehlers (1961), and then made popular by Ellis (1971) (an English translation of the Ehlers 1961 article was published recently, Ehlers 1993). A similar formalism based on arbitrary timelike congruences was introduced by Zelmanov (1959) but it did not gain wide recognition.[2]

(II) T^h is the heat-flow contribution:

$$T^h{}_{\alpha\beta}=q_\alpha u_\beta+q_\beta u_\alpha, \qquad (1.2.11)$$

where q^α is the heat-flow vector obeying $q_\alpha u^\alpha=0$. It defines the direction in which energy is transported across the flow-lines. In order to be physically meaningful, the definition (1.2.11) has to be supplemented with a few further conditions which are, unfortunately, rarely considered in papers on exact solutions. One has to assume in addition that the number-density of the fluid particles, n, is conserved:

---

[2] See "Notes added in proof" at the end of the book, Note 1.

$$(nu^\alpha)_{;\alpha} = 0, \tag{1.2.12}$$

that the first law of thermodynamics holds:

$$T\,\mathrm{d}(S/n) = \mathrm{d}(\epsilon/n) + p\,\mathrm{d}(1/n), \tag{1.2.13}$$

where T is the temperature and S is the entropy-density ((1.2.13) in fact defines S and T), and that:

$$q_\alpha = -q[(\delta_\alpha{}^\beta - u_\alpha u^\beta)T_{,\beta} + T\dot{u}_\alpha], \tag{1.2.14}$$

where q is the thermal conductivity coefficient. Equation (1.2.14) is the relativistic generalization of the Fourier law. In the nonrelativistic limit, the term $(qT\dot{u}_\alpha)$ is absent; it implies isothermal heat flow in an accelerated medium opposite to the direction of acceleration. Using (1.2.14), eq. (1.2.13) determines T and S up to rescaling, that is, up to the choice of units, and implies:

$$S^\alpha{}_{;\alpha} \geq 0, \tag{1.2.15}$$

where:

$$S^\alpha = Su^\alpha + q^\alpha/T, \tag{1.2.16}$$

see Eckart (1940) and Ehlers (1961) for proofs.

In an inhomogeneous cosmological model, spatial variations of temperature are expected. It is then natural to suppose that heat transfer between fluid particles occurs. This is the main motivation for considering cosmological solutions with heat-flow.

(III) T^v is the contribution of viscosity. The bulk viscosity term, $\zeta u^\alpha{}_{;\alpha}$ enters all equations only in the combination $(p - \zeta u^\alpha{}_{;\alpha})$, so in principle it can be assumed zero by redefining the pressure. However, the pressure thus redefined may acquire a nonzero spatial gradient even if it did not have it before. Hence, ζ provides a non-trivial correction. The shear viscosity contributes the term:

$$T^v{}_{\alpha\beta} = \eta\sigma_{\alpha\beta}, \tag{1.2.17}$$

where η is the shear viscosity coefficient. See Ehlers (1961) for more details.

(IV) T^n is the energy-momentum tensor of a null fluid:

$$T^n{}_{\alpha\beta} = \tau k_\alpha k_\beta, \tag{1.2.18}$$

where $k_\alpha k^\alpha = 0$ and the vector field k_α defines the direction of flow of the null fluid. When the null fluid is present together with a perfect fluid, then $k_\alpha u^\alpha = 1$ is assumed in addition. In the absence of any normalizing condition on k, the decomposition of T^n into k and τ is nonunique. If $T = T^n$, that is, if all the other contributions to energy-momentum vanish, then the equations of motion, $T^{\alpha\beta}{}_{;\beta} = 0$ imply that k is a geodesic vector. The energy-momentum tensor of the form of (1.2.18) may be generated by a null electromagnetic field, see below. In this case, k defines the direction of the electromagnetic wave and must obey the conditions resulting from the Maxwell equations. However, an energy-momentum tensor of the form of (1.2.18) may be produced by other fields travelling on null geodesics (e.g. neutrinos), and for each field it should obey the consequences of the appropriate field equations. Therefore, it has become the habit of authors working on the Einstein equations to ignore any limitations on k resulting from nongravitational field equations. They should be imposed when the solution is applied to a definite physical situation, but such papers are very rare.

(V)T^s is the energy-momentum tensor of a scalar field:

$$T^s{}_{\alpha\beta} = \phi_{,\alpha}\phi_{,\beta} - g_{\alpha\beta}[\tfrac{1}{2}\phi_{,\rho}\phi^{,\rho} + V(\phi)], \tag{1.2.19}$$

where ϕ is the scalar field, obeying

$$g^{\alpha\beta}\phi_{;\alpha\beta} - \frac{\partial V}{\partial\phi} = 0, \tag{1.2.20}$$

and $V(\phi)$ is the potential of self-interaction of the field. In most papers, the potential is assumed zero, and then the energy-momentum tensor (1.2.19) can be interpreted as due to a "stiff perfect fluid" with the equation of state $\epsilon = p$, where:

$$\epsilon = p = \tfrac{1}{2}\phi_{,\rho}\phi^{,\rho}, \tag{1.2.21}$$

$$u_\alpha = \phi_{,\alpha}/(\phi_{,\rho}\phi^{,\rho})^{1/2}. \tag{1.2.22}$$

This interpretation is allowed only when the gradient of the scalar field is a timelike vector – a condition usually taken for granted. The "stiff perfect fluid" is a hypothetical medium in which the velocity of sound equals the velocity of light. It is reported to have possibly existed in the early Universe. Conversely, a nonrotating "stiff perfect fluid" can always be interpreted as a massless scalar field, then (1.2.22) results from $\omega = 0$ and (1.2.20)–(1.2.21) follow from $T^{\alpha\beta}{}_{;\beta} = 0$. The popularity of these two entities among equation solvers stems from the similarity of the Einstein equations with $T = T^s$ to vacuum equations; see Section 6.1.

The general scalar field can be interpreted as a perfect fluid, too, with $\epsilon = [\tfrac{1}{2}\phi_{,\rho}\phi^{,\rho} - V(\phi)]$, $p = [\tfrac{1}{2}\phi_{,\rho}\phi^{,\rho} + V(\phi)]$, and u^α given by (1.2.22). Equation (1.2.20) guarantees that the equations of motion of a perfect fluid are obeyed, but the equation of state is undetermined when $V \neq 0$.

When $V(\phi) = \tfrac{1}{2}m^2\phi^2$, the scalar field is said to be massive, and m is interpreted as the mass of the field carrier.

(VI) T^e is the energy-momentum tensor of electromagnetic field,

$$T^e{}_{\alpha\beta} = \frac{1}{4\pi}\left(F_\alpha{}^\mu F_{\mu\beta} + \frac{1}{4}g_{\alpha\beta}F_{\mu\nu}F^{\mu\nu}\right), \tag{1.2.23}$$

where the electromagnetic field tensor F obeys the Maxwell equations:

$$F^{\alpha\beta}{}_{;\beta} = \frac{4\pi}{c}j^\alpha, \tag{1.2.24}$$

$$F_{[\alpha\beta,\gamma]} = 0, \tag{1.2.25}$$

j^α being the current. The field is called null when $F_{\mu\nu}F^{\mu\nu}=0$ and $F_{[\alpha\beta}F_{\gamma\delta]}=0$. Then vector fields k and w exist with the properties:

$$k_\alpha k^\alpha = k_\alpha w^\alpha = 0, \quad w_\alpha w^\alpha = -1,$$
$$F_{\alpha\beta} = \lambda(w_\alpha k_\beta - w_\beta k_\alpha),$$

(1.2.26)

and the energy-momentum tensor of the field has the form of (1.2.18), with $\tau = \lambda^2/(4\pi)$ (see Kramer *et al.* 1980, section 5.2, for more details).

When solutions with a combined perfect fluid/electromagnetic field source are considered, it is usually assumed that:

$$j^\alpha = \rho u^\alpha,$$

(1.2.27)

where ρ is the electric charge density, that is, that the electromagnetic field is generated by charges on the matter particles. Solutions where the current is not collinear with matter flow are rarities, see the paper by Bali, Singh and Tyagi (1987) and our Section 7.2.

In connection with solutions obeying (1.2.27), the problem of symmetry inheritance is often raised, that is, how isometries of the metric tensor influence the invariance transformations of F. This problem was solved by Wainwright and Yaremovicz (1976); earlier contributions to it were published by Woolley (1973), Ray and Thompson (1975) and Michalski and Wainwright (1975). An isometry of g, that is, the existence of a Killing vector field k for which $\pounds_k g_{\alpha\beta} = 0$, does impose limitations on $\pounds_k F_{\alpha\beta}$, but in general does not imply $\pounds_k F_{\alpha\beta} = 0$, and examples of solutions are known in which $\pounds_k g_{\alpha\beta} = 0 \neq \pounds_k F_{\alpha\beta}$ for certain fields k. One such example is given by Wainwright and Yaremovicz (1976a,b), and another by Li and Liang (1985), but none of them are within the scope of this book. However, groups of isometries of g of dimension larger than one impose stronger limitations on the invariances of F. In particular, if g is invariant under the group of rotations, $O(3)$, then so is F.

1.3 The FLRW and Kantowski–Sachs limits of solutions

1.3.1 Limits of spacetimes

The geometrical interpretation of limiting transitions in spacetimes and some of the related problems were discussed by Geroch (1969), and examples were considered by Bampi and Cianci (1980) (see also the remarks by Mignani, 1978). In this book, we shall only be concerned with the following problem of local geometry. Suppose the family of metrics $g(\lambda)$, each defined on a manifold $M(\lambda)$, is labelled by the set of arbitrary constants and/or functions λ. Let the metric h of the manifold N be given. Can the free parameters and/or functions in λ be specialized so that the limiting metric $\bar{g} \stackrel{\text{def}}{=} \lim_{\lambda \to \lambda_0} g(\lambda)$ is equivalent to h? (Equivalent means that either $\bar{g} \equiv h$ or else h results from $\bar{g}$ by a coordinate transformation.) An answer to this question in most cases follows from the following rather trivial observation: a sequence with nonzero elements can possibly have zero as its limit, but a sequence whose elements are all zero cannot have nonzero limit.[3]

It may happen that in a certain limit, the family of vector fields $v(\lambda)$ defined on $M(\lambda)$ goes over into a vector field on $\bar{M}=\lim_{\lambda\to\lambda_0}M(\lambda)$ that is identically zero. This means that the integral curves of $v(\lambda)$ are mapped into single points of $\bar{M}$, that is, that the limiting transition $\lambda\to\lambda_0$ is a projection that reduces the dimension of $M(\lambda)$ and makes the metric $g(\lambda)$ singular. Such limits do indeed occur, and we shall term them singular. In the following, we shall consider only nonsingular limits for which $\dim M(\lambda)\underset{\lambda}{\equiv}\dim \bar{M}=\dim N$. For such limits, the following hold true:

1. In a limiting transition, a spacetime can acquire a larger group of isometries, but can never lose any isometry it had at the beginning. The elements of the sequence to consider are the sets of components of the Lie derivative of the metric along a given vector field. Note that the structure of the isometry group need not be preserved in a limit; for the limiting metric more commutators between the symmetry generators may be zero.
2. In a limit, the Petrov or Segre type can become more special, but cannot become more general (the elements of the sequence to consider are the values of a given component of the Weyl tensor or of the Ricci tensor).
3. A perfect fluid solution with nonzero shear can become shearfree in a limit; the reverse is impossible.

By arguments of this type one can either prove that h is not a limit of $g(\lambda)$ or else convince oneself that the hypothesis $h=\lim_{\lambda\to\lambda_0}g(\lambda)$ has a sufficient degree of probability to be worth verifying directly (this is often difficult).

As noted by Geroch (1969), limits are in general coordinate-dependent. Therefore, we shall place emphasis on invariant definitions of limits: a scalar or tensor quantity (e.g. the shear scalar or the stress tensor) acquiring a geometrically distinguished form (e.g. being zero or its eigenvalues being equal respectively). Unfortunately, for some limits their invariant interpretation is unknown.[4] The point we wish to make is this: if a solution (N, h) can be represented as a limit of $(M(\lambda), g(\lambda))$ in any coordinate system, then (N, h) is not a separate solution, but, rather, a member of the $(M(\lambda), g(\lambda))$ family.

1.3.2 The invariant definitions of the FLRW and Kantowski–Sachs spacetimes

Every FLRW spacetime is homogeneous and isotropic, that is, it has a six-dimensional isometry group with three-dimensional spacelike orbits. The groups are direct

---

[3] As indicated, we shall deal with families of metrics labelled by continuous parameters and/or functions rather than with sequences, but the statement holds true for such families.

[4] Paiva, Rebouças and MacCallum (1993) demonstrated on the example of the Schwarzschild solution how very different the limits of the same spacetime can be when taken in different coordinates. The authors developed an invariant approach to limits based on the Karlhede algorithm for comparing metrics.

products of the form $B_3 \otimes O(3)$, where B_3 is one of the Bianchi groups. Only the Bianchi types I, V, $\mathrm{VII_0}$, VII_h and IX are compatible with spherical symmetry (Grishchuk 1967). One of the standard representations of the FLRW metrics is:

$$ds^2 = dt^2 - (1 + \tfrac{1}{4}kr^2)^{-2}R^2(t)[dr^2 + r^2(d\vartheta^2 + \sin^2\vartheta d\varphi^2)]$$
$$\equiv dt^2 - (1 + \tfrac{1}{4}kr^2)^{-2}R^2(t)(dx^2 + dy^2 + dz^2), \qquad (1.3.1)$$

where $r^2 = x^2 + y^2 + z^2$, but several others are used; see Section 1.3.4. In the above, $R(t)$ (the scale factor) is an arbitrary function to be determined from the Einstein equations, and k (the curvature index) is an arbitrary constant. The metric (1.3.1) automatically defines a pure perfect fluid energy-momentum tensor (see Section 1.2 point (I)), and the only limitations on $R(t)$ may come from an equation of state. When $k=0$, the spatial sections $t=$const of (1.3.1) are flat. This case is called the flat FLRW model; its group B_3 can be of Bianchi type I or $\mathrm{VII_0}$. When $k<0$, the spatial sections $t=$const have a constant negative curvature. This case is called the open FLRW model; its B_3 can be of Bianchi type V or VII_h. When $k>0$, the spatial sections $t=$const have a constant positive curvature. This is the closed FLRW model, its B_3 is necessarily of Bianchi type IX.

The following set of properties is a necessary and sufficient condition for a space-time to be FLRW:

1. The metric obeys the Einstein equations with a perfect fluid source.
2. The velocity field of the perfect fluid source has zero rotation, shear and acceleration.□

(Some authors have considered the FLRW metrics with nonperfect fluid and/or shearing/rotating/accelerating contributions to the energy-momentum tensor, but then those contributions always cancel each other. We shall not take such cases into account, see point 2 of the list in Section 1.1).

The necessity of these properties follows by calculation. The proof of their sufficiency follows from the equations of evolution of $\omega_{\alpha\beta}$, $\sigma_{\alpha\beta}$ and $\dot{u}^\alpha$ (Ellis 1971). Those equations imply that a perfect fluid solution with $\sigma = \omega = 0 = \dot{u}^\alpha$ must be conformally flat. All conformally flat perfect fluid solutions were found by Stephani (1967a; see Kramer *et al.* 1980, theorem 32.15). They have $\sigma = \omega = 0$, but in general $\dot{u}^\alpha \neq 0$. Specializing them to the case $\dot{u}^\alpha = 0$ we obtain the FLRW models (Krasiński 1981).

In this book we shall only be concerned with expanding and nonempty FLRW models for which $\theta \neq 0 \neq \epsilon + p$, and these inequalities will always be tacitly assumed. The FLRW model with $\theta = 0$ is the Einstein static Universe; the ones with $\epsilon + p = 0$ are the de Sitter vacuum Universes (the Minkowski spacetime is included among them as the limit $\Lambda = 0$); neither is appropriate for describing the observed Universe.

Note that condition 1, that is, the perfect fluid source, is essential. Examples will be given of solutions for which $\omega = \sigma = 0 = \dot{u}^\alpha$, but which are not FLRW because the source is not a perfect fluid (Singh and Bhamra 1990, Mitskievič and Senin 1981, Dodson 1972, Bergmann 1981). The consequences of the existence of an irrotational, geodesic and shearfree timelike congruence in spacetime were thoroughly investigated by Coley and McManus (1994). In particular, these authors recovered the above characterization of the FLRW spacetimes and identified some of the nonperfect fluid solutions in this class. For completeness let us note that, in a separate paper (McManus and Coley 1994), the same authors investigated the implications of the existence of such a congruence in the case of an anisotropic fluid source.

There is another invariant definition of the FLRW spacetimes which makes no use of the field equations. The following set of properties is a necessary and sufficient condition for a spacetime to be FLRW:

3. The spacetime admits a foliation into spacelike hypersurfaces of constant curvature.
4. The congruence of lines orthogonal to the leaves of the foliation are shearfree geodesics.
5. The expansion scalar of the geodesic congruence has its gradient tangent to the geodesics.□

A Kantowski–Sachs (K–S) spacetime has a four-dimensional symmetry group acting multiply transitively on three-dimensional orbits, and has no three-dimensional subgroup that would be simply transitive on the orbits (if it had, it would be one of the Bianchi spacetimes). The standard representation of the K–S spacetimes is:

$$ds^2 = dt^2 - R^2(t)dr^2 - S^2(t)(d\vartheta^2 + \sin^2\vartheta\, d\varphi^2), \tag{1.3.2}$$

where $R(t)$ and $S(t)$ are to be determined from the field equations. In the original paper by Kantowski and Sachs (1966), the solutions of the Einstein equations for the metric (1.3.2) were considered with a dust source, but they were generalized (in fact earlier!) by Kompaneets and Chernov (1964) to the general perfect fluid source. Other generalizations will be discussed in Section 2.11. In this book, we shall be using the name K-S for every spacetime that has the symmetries of (1.3.2). The full symmetry group is $\mathbb{R}^1 \otimes O(3)$, and it has no simply transitive subgroup of three dimensions because $O(3)$ has no two-dimensional subgroups at all. The metric (1.3.2) automatically defines an energy-momentum tensor of a fluid with anisotropic pressure, and the coordinates of (1.3.2) are comoving, that is, $u^\alpha = \delta^\alpha{}_0$. The rotation and acceleration are zero, but, if the source is to be a perfect fluid, then the shear is necessarily nonzero: in the limit of zero shear, R/S=const, the Einstein equations are incompatible with the pressure-isotropy condition $T^\alpha{}_\rho(\delta^\rho{}_\beta - u^\rho u_\beta) = -p(\delta^\alpha{}_\beta - u^\alpha u_\beta)$.

The metric (1.3.2) is spherically symmetric. There exist its counterparts with plane and hyperbolic symmetry in which $\sin\vartheta$ is replaced by ϑ and $\sinh\vartheta$ respectively. However, they are of Bianchi types I and III, respectively (that is, they do have simply transitive three-dimensional subgroups), and so will be basically excluded from this book (see point 1 of the list in Section 1.1). In fact, the hyperbolically symmetric counterpart of (1.3.2) was considered by Kantowski and Sachs (1966) and erroneously claimed to be another example of a group with no three-dimensional simply transitive subgroups, but awareness of this error is now common.

Note that the subspaces t=const of (1.3.2) do not contain their centres of symmetry, that is, have the local geometry of $\mathbb{R}^1 \otimes S_2$ (they are sometimes called hypercylinders).

While testing more general solutions for a FLRW limit (see the next subsection), the K-S metric sometimes emerges as another homogeneous limit, and this is the main reason for including it in this book.

1.3.3 Criteria for a FLRW limit

A given metric has a FLRW limit if one of the sets of necessary and sufficient conditions specified in the preceding section can be imposed on it, leading to a nontrivial and nonsingular result. These conditions are not always easy to apply. It is more practicable to apply several necessary conditions in succession until the solution investigated is either reduced to a FLRW limit or proved not to have it. The necessary conditions are:

1. The source must be a perfect fluid. If the energy-momentum tensor of the solution investigated has more components, for example heat-flow, then the additional quantities must be set to zero. Note that a pure scalar field source and dust are special cases of perfect fluid and are compatible with the FLRW geometry. Pure null fluid and pure electromagnetic field are not, and so such solutions have no FLRW limit. Also, solutions with a "tachyon fluid" source, for which the energy-momentum tensor has the form of (1.2.5), but where the vector field u is spacelike, have no FLRW limit.

2. The acceleration must be zero. In comoving and synchronous coordinates in which, if they exist, $u^\alpha \propto \delta^\alpha_{\ 0}$ and $g_{0i}=0$, $i=1, 2, 3$, this condition has the particularly simple form:

$$g_{00,i}=0, \quad i=1, 2, 3. \tag{1.3.3}$$

3. The rotation must be zero. The condition $\omega=0$ is equivalent to the existence of coordinates that are simultaneously comoving and synchronous.

4. The shear must be zero. In synchronous and comoving coordinates, the condition $\sigma=0$ is equivalent to:

$$u^0 g_{ij,0}=\tfrac{2}{3}\theta g_{ij}, \quad i,j=1, 2, 3, \tag{1.3.4}$$

and for every pair of nonzero components (g_{ij}, g_{kl}) this may be written as:

$$(\ln g_{ij}),_t=(\ln g_{kl}),_t=\tfrac{2}{3}\theta g_{00}^{\ 1/2}. \tag{1.3.5}$$

Should any $(\ln g_{kl}),_t$ turn out to be zero, then $\theta=0$ automatically in the $\sigma=0$ limit, and the metric has no FLRW limit. If the coordinates are not comoving, but the velocity field is tangent to two-surfaces S_2 with coordinates (t, x^1), and the spacetime is orthogonally transitive so that $g_{tA}=g_{1A}=0$, where $A=2, 3$, then $\sigma=0$ implies:

$$u^\rho(\ln g_{AB}),_\rho = u^\rho(\ln g_{CD}),_\rho = \tfrac{2}{3}\theta \qquad (1.3.6)$$

for every pair of nonzero (g_{AB}, g_{CD}); $A, B, C, D = 2, 3$. Again, if any $u^\rho(\ln g_{AB}),_\rho = 0$, then $\theta = 0$ in the $\sigma = 0$ limit, and no FLRW limit exists.

5. The gradient of pressure must be collinear with the velocity field:

$$u_{[\alpha} p,_{\beta]} = 0. \qquad (1.3.7)$$

For a solution with a pure perfect fluid source, (1.3.7) is equivalent to $\dot{u}^\alpha = 0$.

6. The gradients of matter-density and of the expansion scalar must be collinear with velocity.
7. The barotropic equation of state, $\epsilon,_{[\alpha} p,_{\beta]} = 0$, must hold.
8. The Weyl tensor must vanish. This condition is the easiest one to verify with an algebraic computer program at hand.
9. The hypersurfaces orthogonal to the velocity field must have constant curvature. This is easy to verify on a computer, too.

Examples have already been mentioned that showed in what ways a solution may fail to have a FLRW limit. The arbitrary parameters of a solution (constants or functions) often enter several physical quantities, and forcing a certain limit upon one of the quantities may result in simultaneously trivializing others. For example, the limit $\sigma = 0$ may automatically imply $\theta = 0$ (a static solution) or $\epsilon = 0$ (a vacuum solution) or $\epsilon + p = 0$ (then, with a pure perfect fluid source, the Bianchi identities imply $\epsilon = -p = \text{const}$, and the spacetime is a vacuum with cosmological constant). In such cases, no FLRW limit exists. It may happen that one of the conditions will produce a K–S limit instead; the properties 2, 3, 5, 6 and 7 are in fact necessary conditions for a K–S geometry.

It may also happen that a certain limit (for example $\sigma = 0$) implies values of constants or functions with which the metric components become singular. This may indicate a singular limit (see Section 1.3.1) and thus no FLRW limit, but not necessarily. In some cases it has been possible to rename the constants and functions in such a way that the limit (sometimes preceded by a coordinate transformation) was becoming nonsingular.

1.3.4 Different representations of the FLRW metrics

Several representations of the FLRW metrics are met in the literature, and some of these will show up in this book. The form of a FLRW metric depends on the solution from which it was obtained as a limit. The form (1.3.1) naturally results from the Stephani–Barnes family (see Sections 4.1 to 4.14). From the Szekeres–Szafron family (Sections 2.1 to 2.16) the FRLW limit preferably results in the form:

$$\mathrm{d}s^2 = \mathrm{d}t^2 - R^2(t)[\mathrm{d}r^2/(1 - kr^2) + r^2(\mathrm{d}\vartheta^2 + \sin^2\vartheta\,\mathrm{d}\varphi^2)], \qquad (1.3.8)$$

or in the form:

$$ds^2 = dt^2 - R^2(t)\left\{dr^2/(1-kr^2) + r^2(dx^2 + dy^2)/[1 + \tfrac{1}{4}(x^2 + y^2)]\right\}. \tag{1.3.9}$$

Equations (1.3.1), (1.3.8) and (1.3.9) conveniently cover all the signs of k in one formula. Another frequently used representation is:

$$ds^2 = dt^2 - R^2(t)[dr^2 + f^2(r)(d\vartheta^2 + \sin^2\vartheta\, d\varphi^2)], \tag{1.3.10}$$

where

$$f(r) = \begin{cases} \sin r & \text{for } k>0, \\ r & \text{for } k=0, \\ \sinh r & \text{for } k<0. \end{cases} \tag{1.3.11}$$

The three cases of (1.3.11) can be written in one formula as:

$$ds^2 = dt^2 - R^2(t)[dr^2 + k^{-1}\sin^2(k^{1/2}r)(d\vartheta^2 + \sin^2\vartheta\, d\varphi^2)]. \tag{1.3.12}$$

The range of r may be finite or infinite, depending on the sign of k and on the coordinates used, but this is easy to recognize in each case. Also, the coordinate transformations between the various representations are elementary and we shall not discuss them. Note that in the case $k>0$ the coordinates of (1.3.8) and (1.3.9) cover only half of the three-sphere $t=$const ($0 \le r < 1/k^{1/2}$) and are unsuitable for considering the geometry in the vicinity of the equator $r = k^{-1/2}$.

Less familiar is the form of the FLRW metric that results from solutions with two commuting Killing vector fields. The easiest way to find it is to demand that the metric be independent of y and z while the spaces $t=$const have constant curvature. Then

$$ds^2 = dt^2 - R^2(t)[dx^2 + f_{,x}^2 dy^2 + f^2(x)dz^2], \tag{1.3.13}$$

where:

$$f(x) = \begin{cases} \sin(k^{1/2}x) & \text{for } k>0, \\ x & \text{for } k=0, \\ \sinh[(-k)^{1/2}x] & \text{for } k<0. \end{cases} \tag{1.3.14}$$

Still another form of the FLRW metrics follows as a limit of plane symmetric solutions. Then:

$$ds^2 = dt^2 - R^2(t)[dx^2 + e^{2Cx}(dy^2 + dz^2)], \tag{1.3.15}$$

where C is a constant. When $C=0$, this is the flat FLRW metric; when $C \neq 0$, it is the open FLRW metric. The closed FLRW metric is incompatible with plane symmetry.

The FLRW models result in quite unfamiliar forms from the Goode and Wainwright (1982b) representation of the Szekeres models. One of the G–W forms is:

$$ds^2=dt^2-S^2[e^{2\nu}(dx^2+dy^2)+W^2f^2\nu_{,z}{}^2dz^2], \qquad (1.3.16)$$

where:

$$W^2=(\varepsilon-kf^2)^{-1},$$
$$e^\nu=f(z)[a(z)(x^2+y^2)+2b(z)x+2c(z)y+d(z)]^{-1}, \qquad (1.3.17)$$

ε and k are arbitrary constants, and $S(t),f(z),a(z),b(z),c(z),d(z)$ are arbitrary functions subject to

$$ad-b^2-c^2=\varepsilon/4. \qquad (1.3.18)$$

The slices $t=$const of the metric (1.3.16)–(1.3.18) are spaces of constant curvature equal to k/S^2. The t-coordinate lines are shearfree geodesics with the expansion scalar depending only on t. This is a characteristic property of the FLRW spacetimes (see points 3, 4, and 5 in Section 1.3.2). The constants ε and k can be scaled by coordinate transformations and reparametrizations of the functions so that each one of them is either 0 or $+1$ or -1. Such values of ε and k will be assumed further in the book.

The other G–W form is:

$$ds^2=dt^2-S^2[e^{2\nu}(dx^2+dy^2)+A^2dz^2], \qquad (1.3.19)$$

where:

$$e^\nu=[1+\tfrac{1}{4}k(x^2+y^2)]^{-1},$$
$$A=e^\nu\{a(z)[1-\tfrac{1}{4}k(x^2+y^2)]+b(z)x+c(z)y\}, \qquad (1.3.20)$$

$S(t),a(z),b(z)$ and $c(z)$ are arbitrary functions and k is an arbitrary constant. This metric has exactly the same properties as (1.3.16)–(1.3.18).

1.4 Spherically symmetric perfect fluid spacetimes

Since we will often discuss spherically symmetric solutions, here we will present a few properties of spacetimes that are spherically symmetric and whose metrics obey the Einstein equations with a perfect fluid source. The consideration below is usually credited to Misner and Sharp (1964), although the idea was originated by Lemaître[5] (1933a) (it was also presented by Podurets (1964); Ruban (1983) presented the most detailed interpretation of all the formulae, and Misner (1965) generalized it to include null radiation).

Let the metric be:

$$ds^2=e^{\gamma(t,r)}dt^2-e^{\alpha(t,r)}dr^2-R^2(t,r)(d\vartheta^2+\sin^2\vartheta\,d\varphi^2), \qquad (1.4.1)$$

---

[5] In fact, Lemaître derived the generalization of (1.4.2)–(1.4.6) to the case when the pressure has different eigenvalues in the radial and in the azimuthal directions.

and suppose the coordinates used above are comoving so that $u^\alpha = e^{-\gamma/2}\delta^\alpha{}_0$. Then, in the course of integrating the Einstein equations for (1.4.1) with a perfect fluid source, the following turns out to be true:

$$(4\pi/c^2)\epsilon R^2 R,_r = \frac{\partial m}{\partial r}, \tag{1.4.2}$$

$$(4\pi/c^2)p R^2 R,_t = -\frac{\partial m}{\partial t}, \tag{1.4.3}$$

where

$$m \stackrel{\text{def}}{=} (c^2/2G)[R + e^{-\gamma}RR,_t^2 - e^{-\alpha}RR,_r^2 - \tfrac{1}{3}\Lambda R^3], \tag{1.4.4}$$

and Λ is the cosmological constant. The remaining field equations for (1.4.1) to be considered together with (1.4.2)–(1.4.4) are:

$$\gamma,_r = -2p,_r/(\epsilon+p), \tag{1.4.5}$$

$$\alpha,_t = -2\epsilon,_t/(\epsilon+p) - 4R,_t/R. \tag{1.4.6}$$

From (1.4.2), the function $m(t,r)$ is easily recognized as the mass equivalent to the total energy contained within the comoving shell of radial coordinate r at time t. For a body of finite radius bordering on vacuum, $p=0$ at the surface and (1.4.3) implies that the total mass of such a body is conserved during evolution. For dust, when $p \equiv 0$ throughout the volume, the mass is conserved within every shell $r=$const.

An interesting thing happens when $R,_r = 0$. The metric (1.4.1) is then an inhomogeneous perturbation of the Kantowski–Sachs (1966) metric: it is spherically symmetric, but does not contain the centres of symmetry in the hypersurfaces $t=$const. In Russian literature, such solutions are called "T-models" (Novikov 1963, Ruban 1968, 1969 and 1983). For them, the "standard" Schwarzschild coordinates (in which $R=r$) do not exist. Then, as can be seen from (1.4.2) and (1.4.3), $\partial m/\partial r=0$, and $m(t)$ is determined by its initial value, the same on all shells. If, in addition, $p=0$, then $m=$const. Ruban (1968, 1969 and 1983) observed that this can be interpreted as follows: in a T-model of dust, any amount of mass added to the source has its effect on the gravitational field exactly cancelled by its gravitational mass defect, so that the active gravitational mass remains the same for all shells and all time. Indeed, when $p=0$, then $\gamma=\gamma(t)$ from (1.4.5), and so $\gamma=0$ can be achieved by a transformation of t. When $R,_r=0$ in addition, then (1.4.4) may be rewritten as:

$$R,_t^2 = -1 + 2Gm/(c^2R) + \tfrac{1}{3}\Lambda R^2, \tag{1.4.7}$$

and this is the equation of motion of any shell $r=$const, in which $m=$const plays the role of the active gravitational mass contained within the shell, and is the same for all shells.

Note that with $R,_r=0$, eqs. (1.4.2)–(1.4.6) are not sufficient to determine the solution: eq. (1.4.2) does not determine ϵ then, and (1.4.6) only provides a connection between α and ϵ. This is so because eq. (1.4.2) is obtained from the (0 0) field equation by multiplying it by $R^2R,_r$, and so becomes an identity when $R,_r=0$. In this case, in order to obtain a solution, one has to go back to the Einstein equations in their original form. The solution will be presented and discussed in Section 2.6.

The vacuum *T*-model is:

$$ds^2 = \left(\frac{2Gm}{c^2 t} - 1\right)^{-1} dt^2 - \left(\frac{2Gm}{c^2 t} - 1\right) \ dr^2 - t^2(d\vartheta^2 + \sin^2\vartheta d\varphi^2). \qquad (1.4.8)$$

This is within the K–S class and covers that part of the Schwarzschild manifold which is inside the event horizon. It does not result from (1.4.1)–(1.4.6) as the limit $\epsilon = p = 0$ because some of the equations in the set (1.4.1)–(1.4.6) were obtained under the assumption $(\epsilon + p) \neq 0$; the vacuum limit then has to be considered separately.

2

The Szekeres–Szafron family of solutions

2.1 The formulae and general properties

These solutions were obtained by assuming that there exist coordinates in which the metric form is:

$$ds^2 = dt^2 - e^{2\alpha}dz^2 - e^{2\beta}(dx^2 + dy^2), \qquad (2.1.1)$$

where α and β are functions of (t, x, y, z) to be determined from the Einstein equations. The source is taken to be the perfect fluid (but generalizations are known, see Sections 2.10, 2.14 and 2.15), and the coordinates of (2.1.1) are assumed to be comoving so that

$$u^\mu = \delta^\mu{}_0. \qquad (2.1.2)$$

This implies at once that $\dot{u}^\mu = 0$ and $p = p(t)$; see Section 1.3.3.

Historically, the earliest solution belonging to this family (the spherically symmetric dust with cosmological constant) was found by Lemaître (1933a), and it still plays a prominent role as an example of applications of exact solutions to physics and cosmology (see the whole of Chapter 3 of this book). The boldest step was taken by Szekeres (1975a), who solved the Einstein equations for (2.1.1) with dust source and with $\Lambda = 0$, without any further assumptions. Szekeres' result was generalized by Szafron (1977) to arbitrary $p \neq 0$.

Several different parametrizations are in use for these solutions, and some notations are conflicting. We shall follow Szafron's exposition in this section, and transcribe the results of most other papers into it. The whole derivation simply consists in careful integration of the Einstein equations for (2.1.1) and is presented by Szafron (1977) with sufficient clarity.[6] Only the results are reproduced here.

The cases $\beta' = 0$ and $\beta' \neq 0$ have to be considered separately ($\beta' \overset{\text{def}}{=} \partial\beta/\partial z$) because the limit $\beta' \to 0$ of the case $\beta' \neq 0$ is singular (see later). The relation between these two cases is closely analogous to the relation between ordinary spherically symmetric spacetimes and the "T-models" from Section 1.4. In fact, the T-models are all contained in the $\beta' = 0$ subfamily.

---

[6] The paper by Das (1981) simply repeats the whole of Szafron's reasoning, even though the author cites Szafron's paper.

2.1.1 The $\beta'=0$ subfamily

The solution here is given by:

$$e^\beta = \Phi(t)/[1+\tfrac{1}{4}k(x^2+y^2)], \tag{2.1.3}$$

$$e^\alpha = \lambda(t,z) + \Phi\Sigma, \tag{2.1.4}$$

$$\Sigma = [\tfrac{1}{2}U(z)(x^2+y^2) + V_1(z)x + V_2(z)y + 2W(z)]/[1+\tfrac{1}{4}k(x^2+y^2)], \tag{2.1.5}$$

where k is an arbitrary constant, $U(z)$, $V_1(z)$, $V_2(z)$ and $W(z)$ are arbitrary functions, the function $\Phi(t)$ is determined by the equation:

$$2\Phi_{,tt}/\Phi + \Phi_{,t}^2/\Phi^2 + \kappa p + k/\Phi^2 = 0, \tag{2.1.6}$$

and $\lambda(t,z)$ is determined by the equation:

$$\lambda_{,tt}\Phi + \lambda_{,t}\Phi_{,t} + \lambda\Phi_{,tt} + \lambda\Phi\kappa p = U(z) + kW(z). \tag{2.1.7}$$

Equations (2.1.6) and (2.1.7) can be solved once the pressure $p(t)$ is specified, and examples of solutions will be given in the following sections. Note that (2.1.6) is identical to one of the equations governing the FLRW models. Note also that $W=0$ can be assumed without loss of generality because $W=0$ results from the reparametrization $U=\tilde{u}+kW$, $\lambda=\tilde{\lambda}-2\Phi W$; $\tilde{u}$ then replaces U and $\tilde{\lambda}$ replaces λ in the equations. The matter-density is:

$$\kappa\epsilon = 2(\lambda\Phi_{,tt}/\Phi - \lambda_{,tt})e^{-\alpha} + 3\Phi_{,t}^2/\Phi^2 + 3k/\Phi^2. \tag{2.1.8}$$

In general, this family of spacetimes has no symmetry (Bonnor, Sulaiman and Tomimura 1977). However, when $U=kW$ and $V_1=V_2=0$, it acquires a three-dimensional symmetry group acting on two-dimensional orbits; the symmetry is spherical, plane or hyperbolic when $k>0$, $k=0$ or $k<0$ respectively. In general, a certain quasi-symmetry is present: the surfaces given by $\{t=\text{const}, z=\text{const}\}$ have constant curvature proportional to k. The lack of symmetry in the spaces $t=\text{const}$ is due to the spheres being placed nonconcentrically (when $k>0$) and to the planes being non-parallel (when $k=0$). This subfamily appeared in the paper by Stephani (1987, solution 5 in our Section 2.9) within a broader context.

This subfamily contains all the FLRW models. They result in the form (1.3.19)–(1.3.20) when $\lambda=0$ and $U=-kW$. Further simplifications may be introduced by coordinate transformations within the FLRW limit. One of these is $U=W=V_2=0$, $V_1=1$, then:

$$ds^2 = dt^2 - \Phi^2(t)[1+\tfrac{1}{4}k(x^2+y^2)]^{-2}(x^2dz^2 + dx^2 + dy^2), \tag{2.1.9}$$

and the transformation:

$$\Phi=R, \quad x=r\sin\vartheta, \quad y=r\cos\vartheta, \quad z=\varphi \tag{2.1.10}$$

reduces (2.1.9) to (1.3.1). The interesting property of (2.1.9) is that the orbits of the $O(3)$ group in this limit have nothing to do with the two-surfaces of constant curvature present in the general Szafron spacetime: the former are the spheres on which $x^2+y^2=$constant, the latter are surfaces of constant z. From this point of view, the FLRW limit results unnaturally in this subfamily: the additional symmetries of the FLRW models appear from nowhere.

The $\beta'=0$ subfamily also contains the Kompaneets–Chernov (1964) limit that results when $U=V_1=V_2=W=0=\lambda_{,z}$. Note that this limit follows "naturally": with $k=1$, the surfaces $\{t=$const, $z=$const$\}$ are then concentric spheres so that the spaces $t=$const and the whole spacetime inherit their symmetry; with $k=0$ these surfaces are then parallel planes. The "proper" Kompaneets–Chernov spacetimes are those with $k=+1$. However, in connection with other solutions of the $\beta'=0$ subfamily, we shall also consider the hyperbolic counterparts of these spacetimes (with $k=-1$), even though they are of Bianchi type III.

2.1.2 The $\beta'\neq0$ subfamily

In this case, the Einstein equations imply for the metric (2.1.1):

$$e^\beta=\Phi(t,z)e^{\nu(x,y,z)}, \tag{2.1.11}$$

$$e^\alpha=h(z)e^{-\nu}(e^\beta)_{,z}, \tag{2.1.12}$$

$$e^{-\nu}=A(z)(x^2+y^2)+2B_1(z)x+2B_2(z)y+C(z), \tag{2.1.13}$$

where $\Phi(t,z)$ is defined by the equation:

$$2\Phi_{,tt}/\Phi+\Phi_{,t}^2/\Phi^2+\kappa p(t)+k(z)/\Phi^2=0, \tag{2.1.14}$$

identical to (2.1.6), except that here Φ and k are allowed to depend on z. The functions $p(t)$, $A(z)$, $B_1(z)$, $B_2(z)$, $C(z)$ and $h(z)$ are arbitrary, while $k(z)$ is determined by:

$$AC-B_1^2-B_2^2=\tfrac{1}{4}[h^{-2}(z)+k(z)]. \tag{2.1.15}$$

As stated before, the limit $\beta_{,z}\to 0$ of this case is singular. Equation (2.1.14) can be integrated once $p(t)$ is specified. It can be formally integrated to the equation:

$$\Phi_{,t}^2=2M(z)/\Phi-k(z)-\tfrac{1}{3}\kappa\Phi^{-1}\int p\left(\frac{\partial}{\partial t}\,\Phi^3\right)dt, \tag{2.1.16}$$

and then the matter-density is given by:

$$\kappa\epsilon=e^\nu[(e^\beta)_{,z}]^{-1}\left\{2M_{,z}/\Phi^2+6M\nu_{,z}/\Phi^2-\tfrac{1}{3}\kappa\Phi^{-2}\int p\left(\frac{\partial^2}{\partial t\partial z}\,\Phi^3\right)dt\right.$$
$$\left.-\kappa\nu_{,z}\Phi^{-2}\int p\left(\frac{\partial}{\partial t}\,\Phi^3\right)dt\right\}. \tag{2.1.17}$$

Our presentation differs slightly from Szafron's. The coordinate z can be transformed by $z=f(z')$, and this freedom can be used to give $h(z)$ any form we wish.

Szafron chose z so that $h=1$, but those coordinates are inconvenient for considering the FLRW limit. Therefore we shall keep z unspecified.

This subfamily, like the preceding one, has in general no symmetry, and acquires a G_3 with two-dimensional orbits when A, B_1, B_2 and C are all constant. The sign of $(AC-B_1{}^2-B_2{}^2)$ has the same consequences as the sign of k had before, but here, with A, B_1, B_2 and C being functions of z, not all the surfaces $z=$const within a single space $t=$const must have the same symmetry, and their geometries are independent of the sign of k, so the $\beta' \neq 0$ subfamily has a greater flexibility built in.

The FLRW limit in the form (1.3.16)–(1.3.18) follows when

$$\Phi(t,z)=f(z)R(t), \quad k=kf^2, \quad k_0=\text{const.} \tag{2.1.18}$$

If $B_1=B_2=0$, $C=4A=1$ and $f=z$ in addition, then the FLRW limit is "natural"; its $O(3)$ orbits are the spheres from the Szafron $\beta' \neq 0$ spacetime, made concentric in the limit.

2.2 Common properties of the two subfamilies

Two common properties have already been mentioned: the lack of any symmetry in general and the existence of the surfaces of constant curvature $\{t=\text{const}, z=\text{const}\}$. In fact, the lack of symmetry was proved by Bonnor, Sulaiman and Tomimura (1977) for the Szekeres solutions, but since they are the limit $p=0$ of the Szafron spacetimes, it follows immediately that the latter have in general no symmetry either. Other properties in common are as follows.

The Weyl tensor of the Szafron spacetimes has its magnetic part with respect to the velocity field of the source equal to zero (Szafron and Collins 1979, Barnes and Rowlingson 1989) and is in general of Petrov type D (Szafron 1977). It degenerates to zero in the FLRW limit only.

Rotation and acceleration of the fluid source are zero. The expansion and the shear are nonzero, two of the eigenvalues of the shear tensor are equal. Two eigenvalues of the three-dimensional Ricci tensor, $^{(3)}R_{ab}$, of the slices $t=$const are also equal, and its eigenframe coincides with that of shear (Collins and Szafron 1979a). The eigenspaces corresponding to the degenerate eigenvalues of shear and of $^{(3)}R_{ab}$ are the surfaces of constant curvature mentioned above.

The slices $t=$const of these spacetimes are conformally flat (Berger, Eardley and Olson 1977). This indicates that the spacetimes are nonradiative in the sense of York (1972). There are further hints of their being nonradiative, but these have only been verified for the Szekeres limit $p=0$ (see Section 2.4.2).

In Wainwright's (1979 and 1981) classification scheme, both the subfamilies are intrinsically of class V, that is, their hypersurfaces orthogonal to the fluid flow are conformally flat. Extrinsically, they are in general of class C_1: the acceleration is zero, the shear is nonzero, and the spatial gradient of the expansion scalar, $h^p{}_\alpha \theta_{,p} \neq 0$, is in general not an eigenvector of shear. However, in the limit of a G_3/S_2 symmetry, the intrinsic class degenerates to C_2, that is, $h^p{}_\alpha \theta_{,p}$ becomes an eigenvector of shear.

Note that the curvature of the $\{x,y\}$-surfaces is a global constant only in the $\beta'=0$ subfamily, and there it is equal to the curvature index of the $t=$const slices in the resulting FLRW limit. In the $\beta'\neq0$ subfamily, the curvature of the $\{x,y\}$-surfaces is determined by $\Delta=(AC-B_1^2-B_2^2)$ and is independent of the curvature index of the FLRW limit, which is determined by $k(z)$. Both $k(z)$ and Δ are only constant within each $\{x,y\}$-surface and can vary within a $t=$const slice. The variation of k over a $t=$const slice has an interesting consequence: if observers in different spatial locations of the Szafron $\beta'\neq0$ Universe are trying to approximate it by FLRW models, then each of them may choose a different FLRW model. Even the sign of k is not a global property of a general Universe. Its global constancy is a peculiarity of the FLRW class. This observation will be developed further in Chapter 3, where astrophysical properties of the L–T model will be discussed.

In the Szafron spacetimes in their full generality, with arbitrary $p(t)$, the dynamic equations are in fact not yet solved. The missing link is the equation of state. The favourite equation of state for physicists and astronomers used to be the barotropic one, $f(\epsilon,p)=0$. It produces nontrivial results when $p=$const, in particular $p=0$, and this case will be discussed in Sections 2.4 and 2.5. When $\epsilon=\epsilon(p)$ and $p=p(t)$, the Szafron spacetimes trivialize: those with $\beta'\neq0$ become FLRW, and those with $\beta'=0$ acquire either a FLRW or a K–S geometry, or the plane and hyperbolic counterparts of the latter (Spero and Szafron 1978).[7]

2.3 The invariant definitions of the family

In Section 2.1, we began with the historically earliest definition, through the metric (2.1.1). Later, Szafron and Collins (1979) worked out the invariant definition that is presented below. The following conditions must be obeyed simultaneously:

1. The metric is a solution of the Einstein equations with a perfect fluid source (the assumption $\Lambda=0$ made in Szafron and Collins' paper is unnecessary, it can be always fulfilled by redefining the matter-density and pressure).
2. The flow-lines of the fluid source are geodesic and nonrotating (the last property implies that the flow-lines are orthogonal to a family S_t of spacelike hypersurfaces).
3. Each hypersurface of the family S_t is conformally flat.
4. Two of the eigenvalues of the Ricci tensor $^{(3)}R_{ab}$ of the hypersurfaces S_t are equal.
5. Two of the eigenvalues of the shear tensor $\sigma_{\alpha\beta}$ are equal.$\square$

For spherically symmetric spacetimes (and, more generally, for all spacetimes that have a G_3/S_2 symmetry group), properties 4, 5 and $\omega=0$ are automatically fulfilled, and property 3 results from $\dot{u}^\alpha=0$ via the field equations. Hence, the subcases of the Szekeres–Szafron spacetimes with a G_3/S_2 symmetry are defined only by property 1 and $\dot{u}^\alpha=0$.

---

[7] See "Notes added in proof", Note 2.

If conditions 1–5 are fulfilled, then there exist coordinates in which the metric has the form (2.1.1). These coordinates are defined up to the following transformations:

$$t=t'+C, \quad z=f(z'), \quad u=g(u')+\overline{g(u')}, \tag{2.3.1}$$

where C is an arbitrary constant, f is an arbitrary real function, g is an arbitrary complex function, and:

$$u=x+iy, \quad u'=x'+iy'. \tag{2.3.2}$$

Another invariant definition of the whole family was given by Wainwright (1977) for the Szekeres limit $p=0$, and then extended by Szafron (1977) to the general case. The Szafron spacetimes with nonzero Weyl tensor comprise all perfect fluid solutions for (2.1.1) which satisfy the following properties:

1. The velocity field of the fluid is geodesic and irrotational.
2. The Weyl tensor is of type D, and the velocity vector of the fluid at every point of the spacetime lies in the two-plane spanned by the principal null directions.
3. Any vector orthogonal to both the repeated principal null directions is an eigenvector of shear.
4. The two-surfaces generated by the principal null directions admit orthogonal two-surfaces.□

It follows from here that the repeated principal null directions are orthogonal to the surfaces of constant curvature, $\{t=\text{const}, z=\text{const}\}$.

The principal null directions are not in general geodesic. They become geodesic if and only if the spacetimes acquire local rotational symmetry, that is, a G_3/S_2 symmetry group.

The proper Szekeres solutions result from the two definitions when the additional requirement $p=0$ is made.

Still one more invariant definition was provided by Barnes and Rowlingson (1989): the Szafron (1977) spacetimes result from the Einstein equations if it is assumed that:

1. the source is a geodesic and rotation-free perfect fluid;
2. the Weyl tensor is purely electric and of Petrov type D;
3. the shear tensor has two equal eigenvalues and its degenerate eigensurface coincides with that of the Weyl tensor.

2.4 Properties of the Szekeres solutions

2.4.1 The $\beta'=0$ subfamily

When $\kappa p=-\Lambda=\text{const}$, eq. (2.1.6) is integrated to:

$$\Phi,{_t}^2=-k+2M/\Phi+\tfrac{1}{3}\Lambda\Phi^2, \tag{2.4.1}$$

and, using (2.1.6), eq. (2.1.7) can be integrated to:

$$\lambda_{,t}\Phi\Phi_{,t}+\lambda M/\Phi-\tfrac{1}{3}\Lambda\lambda\Phi^2=(U+kW)\Phi+X(z), \qquad (2.4.2)$$

where M is an arbitrary constant and $X(z)$ is an arbitrary function (the integral (2.4.2) is valid only when $\Phi_{,t}\neq0$, but this will be tacitly assumed to always be the case, otherwise the solution has no FLRW limit).

The solution of (2.4.2) can be formally represented as:

$$\lambda=\left(-k+2M/\Phi+\tfrac{1}{3}\Lambda\Phi^2\right)^{1/2}\left\{\int\frac{(U+kW)\Phi+X(z)}{\Phi(-k+2M/\Phi+\tfrac{1}{3}\Lambda\Phi^2)^{3/2}}\,\mathrm{d}\Phi+Y(z)\right\}, \qquad (2.4.3)$$

where $Y(z)$ is another arbitrary function.

If we want to interpret the constant Λ as the cosmological constant, then the energy-density has to be redefined by:

$$\epsilon=\tilde{\epsilon}-\Lambda, \qquad (2.4.4)$$

where $\tilde{\epsilon}$ is the new energy-density. Then, from (2.1.8), (2.4.1) and (2.4.2) we obtain:

$$\kappa\tilde{\epsilon}=(2X+6M\Sigma)/(\Phi^2\mathrm{e}^\alpha). \qquad (2.4.5)$$

With $\Lambda\neq0$, the solutions of (2.4.1) involve elliptic functions. Each solution will contain an additional arbitrary constant which defines the initial moment of evolution; it can be assumed zero with no loss of generality. The evolution may or may not begin with a Big Bang singularity (a sufficiently large positive Λ would prevent it), but if it does, then, just as in the FLRW models, the singularity is necessarily simultaneous in the comoving time t for all observers.

An additional singularity of infinite density occurs where (and if) $\mathrm{e}^\alpha=0$. This is an analogue of the shell-crossing singularity which is discussed in more detail for the $\beta'\neq0$ subfamily and for the L–T model, see also Section 2.5.

Equation (2.4.1) is identical to the Friedmann equation, and so are its solutions. Note that M becomes the mass integral in the FLRW limit, so in fact it should be assumed positive for "physical reasons". However, a solution exists when $M<0$ and $\{k<0$ or $\Lambda>0\}$, and it should not be discarded without deeper consideration. The case with $M<0$ and $\Lambda=0$ necessarily has negative matter-density in some part of the space (Bonnor and Tomimura 1976).

The solution of (2.4.2) will contain one more arbitrary function $Y(z)$, but one of the functions $\{U, V_1, V_2, W, X, Y\}$ can be specified by a choice of the coordinate z. Hence, the general solution depends on two arbitrary constants (k and M) and five arbitrary functions of z.

The solution of (2.4.1)–(2.4.2) for the case $\Lambda\neq0=k$ was given by Barrow and Stein-Schabes (1984); it is expressed through elementary functions. Note that when $M\neq0$, one can redefine $\lambda=\tilde{\lambda}+X\Phi/(3M)$, $W=\tilde{W}-X/(3M)$ and $U=\tilde{U}-kX/(3M)$, and the result is the same as if $X=0$. Thus $X=0$ can be assumed without loss of generality whenever $M\neq0$, and the Barrow–Stein-Schabes result is given in such a parametrization. (We shall keep X for the sake of comparison with other solutions later.) That solution tends asymptotically to the de Sitter spacetime.

Both equations (2.4.1) and (2.4.2) have elementary solutions when $\Lambda = 0$; they define the Szekeres (1975a) models with $\beta' = 0$. The solutions of (2.4.1) with $\Lambda = 0$ are identical to the well-known Friedmann solutions that can be found in any textbook on relativity and cosmology. Given Φ and $\Lambda = 0$, eq. (2.4.2) can be integrated to find $\lambda(t,z)$. The results are presented in the papers by Szekeres (1975a, with misprints) and Bonnor and Tomimura (1976, misprints corrected, but notation not uniform from case to case). We shall avoid displaying the explicit solutions because then the various possible signs of k and M have to be considered separately (but see Section 2.5). Bonnor and Tomimura (1976) discussed the evolution of the solution in each case in detail. In short, the discussion revealed that, depending on the case, the metric or the matter-density may or may not approach homogeneity as the initial singularity is approached, and as the model evolves to infinity or to a final singularity. This independent asymptotic behaviour of expansion and density reflects the freedom to prescribe the initial density and the initial velocity of expansion independently in the Einstein equations. It shows that the Szekeres models are more generic than FLRW (not only with respect to symmetry); this feature is found in most of the inhomogeneous models. The different behaviour of the different subcases shows the presence of growing and decaying inhomogeneities, just as in the perturbative calculations. Still, these models "represent very special cases of the general initial value problem".

Barrow and Silk (1981) discussed the case $k = +1$ in somewhat more detail. They showed that the spheres $\{t = \text{const}, z = \text{const}\}$ expand and recollapse just as in a FLRW model, but along the z-direction the space first collapses and then expands to infinity as the final singularity is reached. The maximum of expansion of the spheres and the minimum of collapse in the z-direction do not in general occur simultaneously. Several formulae are given, but no other qualitative conclusions are drawn.

Tomimura (1981) showed that in the solution with $k = 0$ there exist null geodesics along which the dependence of redshift on the luminosity distance r_A is the same as in the FLRW models up to terms of order $r_A{}^2$. This should be a warning against careless reliance on the formula in question when testing the FLRW models, but the paper does not indicate how typical or how exceptional such geodesics are.

2.4.2 *The $\beta' \neq 0$ subfamily*

When $\kappa p = -\Lambda = \text{const}$, eq. (2.1.14) has an integral analogous to (2.4.1)

$$\Phi,{}_t{}^2 = -k(z) + 2M(z)/\Phi + \tfrac{1}{3}\Lambda \Phi^2, \tag{2.4.6}$$

the only difference being the dependence of k, M and Φ on z. As before, the density is modified to $\tilde{\epsilon} = \epsilon - \Lambda$, and:

$$\kappa\tilde{\epsilon}=(2Me^{3\nu})_{,z}/[e^{2\beta}(e^{\beta})_{,z}].\qquad(2.4.7)$$

Again, with $\Lambda\neq 0$ the solutions of (2.4.6) involve elliptic functions. A general formal integral of (2.4.6) was presented by Barrow and Stein–Schabes (1984). Equation (2.4.6) was derived and briefly discussed by Covarrubias (1983 and 1984). In particular, the last author showed how the solutions of de Sitter, L–T, Eardley–Liang–Sachs (see Section 2.12), the Einstein static Universe and the Schwarzschild solution with its plane and hyperbolic counterparts result from (2.4.6). Any solution of (2.4.6) will contain one more arbitrary function of z, usually denoted $t_0(z)$. The instant $t=t_0(z)$ defines the initial moment of evolution; when $\Lambda=0$ it is necessarily a singularity corresponding to $\Phi=0$, and it goes over into the Big Bang singularity in the FLRW limit. Therefore, $t_0(z)$ is sometimes called the "bang time". However, when $t_{0,z}\neq 0$ (that is, in general) the instant of singularity is position-dependent, that is, the Big Bang in these models is not simultaneous in the comoving and synchronous time t. With this t, the Big Bang is a process extended in time rather than a single event in spacetime. The consequences of this kind of singularity have been discussed in many papers for the L–T (spherically symmetric) limit of (2.4.6), and we shall postpone that discussion until Chapter 3.

As before, an additional singularity may occur where $(e^{\beta})_{,z}=0$ (if this equation has solutions). In the L–T limit, this is the now familiar shell-crossing singularity where different dust particles collide and stick together afterwards. Note, however, that it is qualitatively different in the nonsymmetric case. In the spherically symmetric L–T model, if this singularity is present, whole spherical shells stick together at it. Here, as can be seen from (2.1.11) and (2.1.13), the equation $(e^{\beta})_{,z}=0$ may have a solution for some range of $\{x,y\}$, and may possibly not have one for some other range, that is, the singular sheets do not necessarily have the S^2 topology (they may be plane disks, for example).

Equation (2.4.6) is formally analogous to the Friedmann equation, but, with k and M depending on z, each surface $z=$const evolves independently of the others. This flexibility results in a rich variety of possible physical effects that have been discussed for the L–T limit; they will be described in Chapter 3.

Altogether, the models defined by (2.1.11)–(2.1.13) and (2.4.6) contain eight functions of z (this includes $t_0(z)$), but only six of them are arbitrary; one can be specified by a choice of z (still arbitrary up to now), the other is determined by (2.1.15).

As before, the elementary solution of (2.4.6) that results when $k(z)=0$ was presented by Barrow and Stein–Schabes (1984), but no properties were discussed (their formula for ϵ has a misprint).

The solutions of (2.4.6) with $\Lambda=0$ can be expressed in terms of elementary functions. These are strictly analogous to the Friedmann solutions, and define the Szekeres (1975a) models with $\beta'\neq 0$. We shall avoid displaying them for the same reason as before, but they will be presented in the Goode–Wainwright representation (Section 2.5) and in the L–T limit (Section 2.12).

The following gives the list of physical properties deduced for this family.

Szekeres (1975b) investigated the subcase with $AC - B_1{}^2 - B_2{}^2 > 0$, that is, the situation when the $\{t = \text{const}, z = \text{const}\}$ surfaces are spheres. He first showed that if one defines the surface mass-density within such a sphere by $\sigma_s = \epsilon e^\alpha$ (so that $dM = \int_{S_2(z)} \sigma_s e^{2\beta}\, dx\, dy$ is the mass between the spheres of radii z and $(z + dz)$), then σ_s has the form of a mass dipole and the dipole axis changes from one sphere to another. This observation was later refined by de Souza (1985; see below).

Then, defining the mass within the sphere $z = r$ at the time $t = t_0$ by:

$$m(t_0, r) = \int_\epsilon (-g)^{1/2} d_3 x, \Big|_{t = t_0} \tag{2.4.8}$$

Szekeres showed that:

$$m(t_0, r) = 4\pi \int_0^r 2M_{,z} h(z)\, dz = m(r), \tag{2.4.9}$$

that is, the mass in a comoving volume does not depend on time and is given by the same formula as in the spherically symmetric case. The density may die off with increasing r so fast that $\lim_{r \to \infty} m(r) < \infty$. Szekeres' conclusion that one cannot set $\epsilon = 0$ for $r > r_0$ except in the spherically symmetric case was contradicted soon after by Bonnor (1976, see below).

The rest of Szekeres' paper is devoted to the discussion of singularities. Since most of that discussion was done by Goode and Wainwright (1982b, see Section 2.5) in greater detail and at a higher level of generality, we shall quote only the most illuminating points. Szekeres considered collapse to a singularity and observed that it may occur at $\Phi = 0$ (today it would be called the Big Crunch) or at $(e^\beta)_{,z} = 0$ (which is the shell-crossing singularity in the current terminology). He further observed that the first one is inevitable along every flow-line and that the second one, if it exists, may possibly occur before the Big Crunch. He drew qualitative conclusions for two special cases only: the zero energy case (i.e. $k = 0$, then the z-coordinate may be chosen so that $h = 1$) and the time-symmetric case, $\Phi_{,t}(0, r) = 0$. In these two cases, the shell-crossing singularity will occur somewhere on the sphere $z = r$ earlier than the Big Crunch if and only if:

$$\bar{\rho}_0(r) > \langle \bar{\rho}_0(r) \rangle, \tag{2.4.10}$$

where

$$\bar{\rho}_0(r) \overset{\text{def}}{=} \frac{dm}{dr} \Big/ \frac{dV(t_0, r)}{dr}, \tag{2.4.11}$$

$$V(t, r) \overset{\text{def}}{=} \int_0^r dz \int_{S_2(z)} dx\, dy\, e^{\alpha + 2\beta}, \tag{2.4.12}$$

$$\langle \bar{\rho}_0(r) \rangle \overset{\text{def}}{=} \langle \bar{\rho}(t_0, r) \rangle, \tag{2.4.13}$$

$$\langle \bar{\rho}(t_0, r) \rangle \overset{\text{def}}{=} \left[\int_0^r \bar{\rho}(t, r') h^{-1}(r')\, dV(t, r') \right] \Big/ \int_0^r h^{-1}(r')\, dV(t, r'). \tag{2.4.14}$$

The quantities defined above have the following interpretation: $\bar{\rho}_0(r)$ is the average of matter-density at $t = t_0$ over the sphere $z = r$, V is the volume inside $z = r$, and $\langle \bar{\rho}_0(r) \rangle$

is the average of $\bar{\rho}_0(r)$ over the interior of $z=r$, weighted by $h^{-1}(z)$. Condition (2.4.10) means that the mean density over the sphere $z=$const as a function of z rises between $z=0$ and $z=r$ steeply enough that at $z=r$ it is greater than its (h^{-1})-average over the interior of $z=r$. This is the condition of "strong increase".

In the same two cases, Szekeres formulated the condition for the occurrence of trapped surfaces. A sphere $\{t=t_s,\ z=r\}$ will be trapped for all $t_s>t_H$, where t_H is determined by:

$$\Phi(t_H,r)=2M(r). \tag{2.4.15}$$

(Compare this to the Barnes (1970) condition for the formation of a black hole in the L–T model, Section 3.4 of this book. The existence of t_H obeying (2.4.15) is guaranteed, since for collapse Φ tends to zero after a finite time along every world-line.) This shows that in the two cases considered, the Big Crunch will always be preceded by the formation of a trapped surface, that is, it will be hidden inside an event horizon (this is not always so; see examples based on the L–T model in Section 3.6). However, a shell-crossing singularity may precede the trapped surface and thus be naked. It may be locally naked (visible only out to a finite distance) or globally naked (visible out to any distance). Szekeres' paper contains a quantitative criterion for the occurrence of a locally naked singularity. Its physical interpretation is this: the initial density distribution must be sufficiently asymmetric (the dipole moment defined by $\sigma_s(t_0,r)$ must be sufficiently large). However, a naked (locally or globally) singularity may also occur in the spherically symmetric subcase if the "strong increase" condition is further strengthened (Yodzis, Seifert and Müller zum Hagen 1973 gave an example). This situation is illustrated in Szekeres' paper by a diagram.

The paper is concluded by the statement that a naked singularity, after it forms, will not persist, that is, it will only be visible for a time comparable to the collapse time of a material body, and will then be swallowed up by the event horizon. This is true for a nearby observer, but at infinity, the process of observing this finite existence time may take an infinite amount of the observer's time; see the Penrose diagram given by Eardley and Smarr (1979) for the L–T model with a naked singularity. This is analogous to the well-known effect of observing the collapse to a black hole state from infinity in the Schwarzschild spacetime.

In a later paper (Szekeres 1980), the same author presented the main results of this study in a broader context. That paper will be mentioned again in Section 3.6.

Bonnor (1976a,b) showed that if the tube interior to $z=$const is cut out of the Szekeres spacetime with $AC-B_1^2-B_2^2>0$, then the resulting metric can be matched to the Schwarzschild solution, even though the interior metric still has no symmetry. The Schwarzschild spacetime contains no gravitational radiation, so in such a configuration the Szekeres model does not radiate. This proves that there exist configurations of collapsing dust that have no symmetry and do not produce gravitational waves.

Exactly the same subcase of the Szekeres solutions was shown to be nonradiative by Covarrubias (1980). In that paper, conditions were found under which the solution with $\beta' \neq 0$, $AC - B_1{}^2 - B_2{}^2 > 0$ is a small perturbation of the Minkowski metric, and then the Szekeres solution obeying those conditions was linearized with respect to the Minkowski background. It follows that the third time-derivatives of the quadrupole moment vanish for a portion of the Szekeres dust contained within a comoving surface of any shape. This implies no radiation, according to the classic Einstein formula.

Again for the same subcase, Gleiser (1984) showed how a reparametrization followed by a limiting transition can reduce the Szekeres metric to a Robinson–Trautman form, and its dust source to a null fluid of the Vaidya type. The physical interpretation of this limit is: the dust particles are rendered massless and moving with the speed of light.

De Souza (1985) refined Szekeres' (1975b) observation that the density distribution over each single sphere {t=const, z=const} in the case $\beta' \neq 0$, $AC - B_1{}^2 - B_2{}^2 > 0$ has a dipole form. De Souza showed that although the dipole axis indeed changes from sphere to sphere, the separation of ϵ into the monopole and the dipole can be done globally, and the surface where the dipole contribution is zero is comoving.

When $AC - B_1{}^2 - B_2{}^2 > 0$, we can redefine the arbitrary functions so that:

$$AC - B_1{}^2 - B_2{}^2 = \tfrac{1}{4}, \tag{2.4.16}$$

which means $h^{-2} = 1 - k$. Then, the further transformations:

$$x + iy = e^{i\psi'}\cot(\vartheta'/2), \quad z = r, \tag{2.4.17}$$

$$x' = r\sin\vartheta' \cos\psi', \quad y' = r\sin\vartheta' \sin\psi', \quad z' = r\cos\vartheta' \tag{2.4.18}$$

bring the expression for matter-density (2.4.7) to the form:

$$\kappa\epsilon = \frac{[2M(r)/\bar{P}^3]_{,r}}{[\Phi(t,r)/\bar{P}]^2(\Phi/\bar{P})_{,r}}, \tag{2.4.19}$$

where:

$$r\bar{P} \overset{\text{def}}{=} (A - C)z' + 2B_1 x' + 2B_2 y' + r(A + C). \tag{2.4.20}$$

The expression (2.4.19) is now separated into a spherically symmetric part, depending only on t and r, and $\Delta\epsilon$, chosen so that the hypersurface where $\Delta\epsilon = 0$ includes the point $x' = y' = z' = 0$, that is, $r = 0$. Such a separation is unique and yields:

$$\epsilon = \epsilon_s(t,r) + \Delta\epsilon(t,x',y',z'), \tag{2.4.21}$$

where:

$$\kappa\epsilon_s = \frac{2M_{,r}(A + C) - 6M(A + C)_{,r}}{\Phi^2[\Phi_{,r}(A + C) - \Phi(A + C)_{,r}]}, \tag{2.4.22}$$

$$\kappa\Delta\epsilon = \frac{(A + C)_{,r}\bar{P} - (A + C)\bar{P}_{,r}}{\Phi_{,r}\bar{P} - \Phi\bar{P}_{,r}} \cdot \frac{6M\Phi_{,r} - 2M_{,r}\Phi}{\Phi^2[\Phi_{,r}(A + C) - \Phi(A + C)_{,r}]}$$

$$= \frac{(A+C)^2 [\bar{P}/(A+C)]_{,r}}{\Phi^2 (\bar{P}/\Phi)_{,r}} \cdot \frac{(2M/\Phi^3)_{,r}}{[(A+C)/\Phi]_{,r}}. \qquad (2.4.23)$$

The definitions given above make sense except when $(\bar{P}/\Phi)_{,r} \equiv 0$ or $[(A+C)/\Phi]_{,r} \equiv 0$. In both these cases, the Szekeres model degenerates into a FLRW dust. Hence, (2.4.21) – (2.4.23) make sense whenever the Szekeres model is inhomogeneous. Now, $\Delta\epsilon = 0$ has two solutions: $[\bar{P}/(A+C)]_{,r} = 0$ and $(2M/\Phi^3)_{,r} = 0$. The second one defines a hypersurface that depends on t, that is, it is not comoving except when $(2M/\Phi^3)_{,r} \equiv 0$, but then the matter-density r becomes spatially homogeneous. The first hypersurface, defined by:

$$[\bar{P}/(A+C)]_{,r} = 0, \qquad (2.4.24)$$

has its equation independent of t, that is, it is a world-sheet of a comoving surface. Moreover, $\Delta\epsilon$ changes sign when the hypersurface (2.4.24) is crossed, although it is not antisymmetric in $\{x',y',z'\}$. Hence, $\Delta\epsilon$ is a dipole-like contribution to matter-density. Although the separation (2.4.21) is global, the orientation of the dipole axis is indeed different on every sphere $\{t=\text{const}, r=\text{const}\}$.

De Souza then showed that there exists such an r-dependent rotation of the coordinates $(x',y',z') \mapsto (x,y,z)$ that brings eq. (2.4.23) to a form in which $\Delta\epsilon = 0$ on $x=0$ and $\Delta\epsilon\,(x,-y,z) = \Delta\epsilon(x,y,z)$. The dipoles thus have a mirror-like symmetry with respect to the surfaces $y=0$.

Królak *et al.* (1994) showed that the Big Bang singularity (the one at $\Phi=0$) in the $k(z)=0$ subcase of the Szekeres solution is a naked strong curvature singularity. Therefore, the Szekeres solutions are another counterexample to the oldest and simplest formulation of the cosmic censorship hypothesis. In fact, Królak *et al.*'s result shows that the Szekeres solutions are not sufficiently generic from the point of view of the cosmic censorship paradigm. The shell-crossing singularity (the one at $\Phi_{,z}=0$), although naked as well, is not strong. Several other papers have been published in which the spherically symmetric limit of the $\beta' \neq 0$ Szekeres solutions (i.e. the Lemaître–Tolman model) has been discussed as a testing ground for cosmic censorship; see our Section 3.6.

2.5 Common characteristics and comparisons of the two subfamilies of Szekeres solutions

Goode and Wainwright (1982b) introduced a description of the Szekeres solutions in which many properties of the two subfamilies can be considered at one go. We shall first present their result, and then show how it can be obtained from the representation used in the preceding sections.[8]

The metric is:

---

[8] The G–W notation became a standard which is, unfortunately, partly in conflict with the notation widely used for the L–T models (see a comment in Section 2.12). The G–W notation will be used only in Section 2.5.

$$ds^2 = dt^2 - S^2[e^{2\nu}(dx^2 + dy^2) + H^2 W^2 dz^2], \tag{2.5.1}$$

where $S(t,z)$ is defined by:

$$S_{,t}^2 = -k + 2\mathcal{M}/S, \tag{2.5.2}$$

$k=0,\pm1$, and $\mathcal{M}(z)$ is an arbitrary function,

$$H = A(x,y,z) - \beta_+ f_+ - \beta_- f_-, \tag{2.5.3}$$

A, e^ν and W will be defined below (they differ for each subfamily), $\beta_+(z)$ and $\beta_-(z)$ are functions of z defined below, and $f_+(t,z)$ and $f_-(t,z)$ are the two linearly independent solutions of the equation:

$$F_{,tt} + 2(S_{,t}/S)F_{,t} - (3\mathcal{M}/S^3)F = 0. \tag{2.5.4}$$

The solutions of (2.5.2) for which $S_{,t}>0$ can be represented parametrically as:

$$S = \mathcal{M}g'(\eta), \tag{2.5.5}$$
$$t - T(z) = \mathcal{M}g(\eta),$$

where $T(z)$ is an arbitrary function (the bang time) and $g(\eta)$ is:

$$g(\eta) = \begin{cases} \eta - \sin\eta & \text{when } k=+1, \\ \sinh\eta - \eta & \text{when } k=-1, \\ \frac{1}{6}\eta^3 & \text{when } k=0. \end{cases} \tag{2.5.6}$$

With $k=0$, S can be rescaled so that $\mathcal{M}=$const (see later), and this choice of S will be assumed. The corresponding solutions of (2.5.4) are:

$$f_+ = \begin{cases} \dfrac{6\mathcal{M}}{S}\left[1 - \dfrac{\eta}{2}\cot\dfrac{\eta}{2}\right] - 1 & \text{when } k=+1, \\[2mm] \dfrac{6\mathcal{M}}{S}\left[1 - \dfrac{\eta}{2}\coth\dfrac{\eta}{2}\right] + 1 & \text{when } k=-1, \\[2mm] \dfrac{1}{10}\eta^2 & \text{when } k=0. \end{cases} \tag{2.5.7}$$

$$f_- = \begin{cases} \dfrac{6\mathcal{M}}{S}\cot\dfrac{\eta}{2} & \text{when } k=+1, \\[2mm] \dfrac{6\mathcal{M}}{S}\coth\dfrac{\eta}{2} & \text{when } k=-1, \\[2mm] 24/\eta^3 & \text{when } k=0. \end{cases} \tag{2.5.8}$$

The coefficients in $f_\pm$ were chosen for later convenience. The solutions with $S_{,t}<0$ are time reverses of (2.5.7)–(2.5.8), and $\mathcal{M}>0$ is assumed for correspondence of the results with the FLRW models. With these assumptions, f_+ are the solutions that increase with time and f_- are the solutions that decrease with time.

The two subfamilies are now defined as follows:

$$\beta' \neq 0:$$

$T_{,z}^2 + \mathcal{M}_{,z}^2 \neq 0$ so that $S_{,z} \neq 0$, and then:

$$e^{\nu} = f(z)[a(z)(x^2+y^2)+2b(z)x+2c(z)y+d(z)]^{-1}, \tag{2.5.9}$$

where $f(z)$ is arbitrary and a, b, c, d are functions of z subject to:

$$ad - b^2 - c^2 = \tfrac{1}{4}\varepsilon, \quad \varepsilon = 0 \pm 1, \tag{2.5.10}$$

$$W^2 = (\varepsilon - kf^2)^{-1}, \tag{2.5.11}$$

$$\beta_+ = -kf\mathcal{M}_{,z}/(3\mathcal{M}), \quad \beta_- = fT_{,z}/(6\mathcal{M}), \tag{2.5.12}$$

$$A = f\nu_{,z} - k\beta_+, \tag{2.5.13}$$

and it is understood that $\mathcal{M}_{,z} = 0$ when $k=0$ (see later).

$$\beta' = 0:$$

$\mathcal{M}_{,z} = T_{,z} = 0$ so that $S_{,z} = 0$ and:

$$e^{\nu} = [1 + \tfrac{1}{4}k(x^2+y^2)]^{-1}, \quad W=1, \tag{2.5.14}$$

$$A = \begin{cases} e^{\nu}\{a(z)[1 - \tfrac{1}{4}k(x^2+y^2)] + b(z)x + c(z)y\} - k\beta_+ & \text{for } k = \pm 1, \\ a(z) + b(z)x + c(z)y - \tfrac{1}{2}\beta_+(x^2+y^2) & \text{for } k=0, \end{cases} \tag{2.5.15}$$

a, b, c, β_+ and β_- being arbitrary functions of z.

The remarkable thing about the Goode–Wainwright (G–W) parametrization is that eqs. (2.5.1)–(2.5.8) all hold for both subfamilies, the $\beta' = 0$ subfamily differing from the other one only by $\mathcal{M}$ and T being both constant. In both subfamilies the subcase $\beta_+ = \beta_- = 0$ are the FLRW models represented in the form (1.3.16)–(1.3.20).

In the $\beta' \neq 0$ subfamily, the Goode–Wainwright representation arises as follows. Define $\phi(z)$ by:

$$k(z) = K\phi^2(z), \tag{2.5.16}$$

where $k(z)$ is the function from (2.1.14)–(2.1.15), and $K = 0, \pm 1$. When $k \neq 0$, $\phi = |k|^{1/2}$, when $k=0$, eq. (2.5.16) does not define $\phi(z)$, and in this case we take $\phi(z)$ to be a new arbitrary function. Then define $S(t,z)$ by:

$$\Phi = \phi S, \tag{2.5.17}$$

so that S obeys:

$$2S_{,tt}/S + S_{,t}^2/S^2 + K/S^2 = 0, \tag{2.5.18}$$

resulting from (2.1.14) with (2.5.16), (2.5.17) and $p=0$. Equation (2.5.2) is the integral of (2.5.18). Equation (2.1.15) now becomes:

$$AC - B_1{}^2 - B_2{}^2 = \tfrac{1}{4}[h^{-2}(z) + K\phi^2]. \tag{2.5.19}$$

Let us define $G(z)$ by:

$$h^{-2}(z) + K\phi^2 = \varepsilon G^2, \tag{2.5.20}$$

where $\varepsilon = 0, \pm 1$ so that $G = |h^{-2} + K\phi^2|^{1/2}$ when $h^{-2} + K\phi^2 \neq 0$ and $0 \neq G(z)$ is an arbitrary function otherwise. Then let us define a, b, c, d, f and W by:

$$(A, B_1, B_2, C, \phi) = (a, b, c, d, f)\, G, \tag{2.5.21}$$
$$h = W/G;$$

the functions a, b, c, d will then obey (2.5.10). In the new variables, from (2.1.11), (2.1.13) and (2.5.17), we obtain:

$$e^\beta = Se^{\bar{\nu}}, \tag{2.5.22}$$

where $\bar{\nu}$ is the ν given by (2.5.9) (not equal to the one from (2.1.13)), and from (2.1.11)–(2.1.13), (2.5.17) and (2.5.20)–(2.5.21) we obtain:

$$e^\alpha = WS(fS_{,z}/S + f\nu_{,z}), \tag{2.5.23}$$

and so:

$$H = fS_{,z}/S + f\nu_{,z}. \tag{2.5.24}$$

When $k = 0$, the function $\phi(z)$ (and with it $f(z)$) is not yet defined. In this case, let us define ϕ by:

$$\phi = M^{1/3}, \tag{2.5.25}$$

where M is the function from (2.4.6). Then, S as defined by (2.5.17) will obey (2.5.2) with $k = 0$ and $\mathcal{M} = 1$, all other equations (2.5.1)–(2.5.13) remaining unchanged. Hence, when $k = 0$, one can assume $\mathcal{M} = $ const with no loss of generality and this will be assumed from now on in the whole of Section 2.5.

Now it can be verified, case-by-case from (2.5.6)–(2.5.8), (2.5.12), (2.5.13) and (2.5.25) that:

$$F \stackrel{\text{def}}{=} \beta_+ f_+ + \beta_- f_- = A - H = -fS_{,z}/S - k\beta_+ \tag{2.5.26}$$

obeys (2.5.4). Note that with $k = 0$, necessarily $\beta_+ = 0$. In the case under consideration, the formula for matter-density (2.4.7) becomes:

$$\kappa\epsilon = 6\mathcal{M}S^{-3}(1 + F/H) \equiv 6\mathcal{M}A/(S^3 H). \tag{2.5.27}$$

In the $\beta' = 0$ subfamily, the constant k can be rescaled to $+1$ or -1 when $k \neq 0$ by simple rescalings of x, y, ϕ, U and W. We will therefore assume that $k = 0, \pm 1$. The G–W representation is then introduced in a different way when $k = \pm 1$ and when $k = 0$.

When $k = \pm 1$ we can redefine U, W and λ by:

$$U = \tfrac{1}{2}(u - kw),$$

$$W = \tfrac{1}{2}(ku + w), \tag{2.5.28}$$

$$\lambda = \tilde{\lambda} - ku\Phi,$$

and then $\tilde{\lambda}$ obeys (2.1.7) with $p=0$ and zero on the right-hand side. The result is as though $U + kW = 0$, and this can be assumed with no loss of generality. The G–W representation then arises by:

$$\Phi = S, \quad e^\beta = Se^\nu,$$

$$W = a/2, \quad U = -ka/2, \quad V_1 = b, \quad V_2 = c, \tag{2.5.29}$$

$$e^\alpha = (A + k\beta_+)S + \lambda \overset{\text{def}}{=} HS,$$

$$\beta_+ = -kX(z)/(3M),$$

where e^ν and A are given by (2.5.14)–(2.5.15) and $X(z)$ is the function from (2.4.2). Hence:

$$F = A - H = -\lambda/S - k\beta_+, \tag{2.5.30}$$

and it can be verified, case-by-case again, with the help of (2.4.3), that the F defined above obeys (2.5.4) and $F = \beta_+ f_+ + \beta_- f_-$, where f_+ and f_- are given by (2.5.7)–(2.5.8) and $\beta_-(z)$ is an arbitrary function.

When $k=0$, the procedure is different. We first observe that when $\Lambda = k = 0$, the function $X(z)$ in (2.4.2) can be assumed zero with no loss of generality because $X=0$ results after the reparametrization $\lambda = \tilde{\lambda} + X\Phi/3$, $W = w - X/6$ (then $e^\alpha = \tilde{\lambda} + \Phi\Sigma$, as before). Then, we take $\Phi = S = e^\beta$ and $e^\alpha = \lambda + S\Sigma$, where Σ is given by (2.1.5) and S is given by (2.5.5)–(2.5.6). Next, we find λ from (2.4.3) in the case $k = \Lambda = 0$. Finally, we define:

$$F = -\lambda/S = \beta_+ f_+ + \beta_- f_- + A, \tag{2.5.31}$$

where $\beta_+ = -U$, $b = V_1$, $c = V_2$, $a = 2W$ and $\beta_-(z)$ is arbitrary. The functions f_+ and f_- are as in (2.5.7)–(2.5.8), while A is as in (2.5.15). Using the information given in this paragraph, one can verify that such F obeys (2.5.4). The matter-density is given by (2.5.27) for the $\beta' = 0$ subfamily as well.

As Goode and Wainwright observe, eq. (2.5.4) is linear and has the same form as the equation derived for the density perturbation in the linearized perturbation scheme around the FLRW dust background, if the perturbed solution is also dust. This holds both in relativity and in Newtonian theory; see eqs. (15.9.23) and (15.10.57) from Weinberg (1972) and eqs. (12.10) and (12.31) from Raychaudhuri (1979). However, this is only a formal similarity and whether there is any physics behind it remains to be seen. The following differences should be noted:

1. In the corresponding equation of the linearized perturbation scheme, the coefficients $(2S_{,t}/S)$ and $3M/S^3 = \kappa\rho/2$ are taken from the background FLRW model,

while here they come from the perturbed model. Assuming F small, this means, not surprisingly, that the equation of the linearized perturbation is the linear approximation to (2.5.4).

2. In the linearized perturbation scheme, the solution of (2.5.4) is $\delta=(\rho_p-\rho_b)/\rho_b$, where ρ_p is the perturbed density and ρ_b is the FLRW background density. In the Szekeres models, assuming F small and linearizing about the background, one obtains $F=A(\rho_p-\rho_b)/\rho_b$, that is, $F\cong\delta$ only if $A\cong1$, which is a limitation imposed on the model. Otherwise, F has no clear physical meaning, although it does define a certain deformation on the FLRW background.

3. In the linearized perturbation scheme, every solution of (2.5.4) generates a perturbed model. In the Szekeres models of the $\beta'\neq0$ subfamily the coefficients β_+ and β_- are provided by other field equations (see (2.5.12)), and when $k=0$ the growing mode is necessarily absent ($\beta_+\equiv0$). In the $\beta'=0$ subfamily with $k\neq0$, the coefficients are truly arbitrary, but with $k=0$ the β_+ is again provided from elsewhere. This means that the Szekeres models are generated only by some solutions of (2.5.4), but not by all of them.[9]

Still, the G–W representation is enlightening in several ways, as shown by the authors themselves. From eq. (2.5.12) one can see that the growing mode of perturbation is generated by a departure from the FLRW relation between mass and energy ($\mathcal{M}_{,z}\neq0$), and the decaying mode is generated by the nonsimultaneity of the Big Bang ($T_{,z}\neq0$).

The expansion and shear of the dust source are in both cases:

$$\theta=3S_{,t}/S-F/H, \tag{2.5.32}$$

$$2\sigma_{11}=2\sigma_{22}=-\sigma_{33}=\tfrac{2}{3}F/H. \tag{2.5.33}$$

From the above and from (2.5.7)–(2.5.8) one can now show that (2.5.4) is the Raychaudhuri equation:

$$\theta_{,\nu}u^{\nu}+\tfrac{1}{3}\theta^2+2\sigma^2+\tfrac{1}{2}\epsilon=0, \tag{2.5.34}$$

which, surprisingly, becomes linear for the Szekeres models in the G–W representation.

Equation (2.5.33) shows that $(\sigma=0)\Leftrightarrow(F=0=\beta_+=\beta_-)$, which is further evidence that in the limit $\beta_+=\beta_-=0$ the Szekeres models become FLRW.

In general, there can be two independent scalar polynomial curvature singularities in the Szekeres models. They occur where $S=0$ and $H=0$. The first one is the Big Bang and it occurs inevitably. It is a pointlike singularity when $\beta_-=0$ or a cigar-type singularity when $\beta_-\neq0$. In the $\beta'\neq0$ subfamily, $\beta_-=0$ implies a simultaneous Big Bang, in the $\beta'=0$ subfamily, the Big Bang is always simultaneous. The second singularity, $H=0$, may or may not occur, depending on the initial conditions. It is the shell-crossing singularity, mentioned in Section 2.4.2 (for more

---

[9] The three critical comments come from the author (A.K.) of this book.

about shell-crossing singularities – see Sections 3.6 and 3.8). When $\beta_-(z_0)>0$, the shell-crossing singularity along the flow-lines with $z=z_0$ will occur later than the Big Bang, and so is astrophysically relevant. Whenever it occurs, it is of the pancake type.

For a simultaneous Big Bang, the metric in the neighbourhood of the Big Bang can be written as:

$$ds^2 = dt^2 - t^{4/3}[g_{ij}^{(0)}dx^i dx^j + O(t^{2/3})],\qquad(2.5.35)$$

where $g_{ij}^{(0)}$ does not depend on t (use was made of T=const to reset the time of singularity to t=0). This kind of singularity is called Friedmann-like (Goode and Wainwright 1982a), and, according to Eardley, Liang and Sachs (1972) the $g_{ij}^{(0)}$ is the metric of the singular hypersurface. In approaching that singularity, the following hold:

$$\sigma\xrightarrow[t\to 0^+]{}+\infty,\qquad \sigma/\theta\xrightarrow[t\to 0^+]{}0,\qquad 3\epsilon/\theta^2\xrightarrow[t\to 0^+]{}1,$$

$$C_{\alpha\beta\gamma\delta}\xrightarrow[t\to 0^+]{}+\infty,\qquad C_{\alpha\beta\gamma\delta}C^{\alpha\beta\gamma\delta}/\epsilon^2\xrightarrow[t\to 0^+]{}0.\qquad(2.5.36)$$

The singularities described so far may be called "initial" in the sense that matter emerging from them is expanding (with $\theta>0$). Singularities may also occur at recollapse, and Goode and Wainwright call them "final". The criteria for a final singularity are as follows:

When k=0, a final singularity occurs if and only if $\beta_+>0$, and it is of the pancake type (shell-crossing, H=0). This may happen only in the β'=0 subfamily.

When k=−1, a final singularity occurs if and only if $H_0<0$, and it is of the pancake type again (H_0 is the value of H when $\beta_+=\beta_-=0$).

When k=+1, the final singularity of the Big Crunch type (S=0) occurs inevitably. It is pointlike if and only if $\pi\beta_+-\beta_-=0$, and it is of the cigar type otherwise. If and only if $\pi\beta_+(z_0)-\beta_-(z_0)$ will the world-lines with $z=z_0$ first encounter the shell-crossing singularity H=0 which is of the pancake type.

A pointlike final singularity is Friedmann-like, in the sense of (2.5.35). A world-line of the dust source may begin and end in a pointlike singularity if and only if $\beta_+=\beta_-=0$, but then the region containing the line must be FLRW.

When there is no final singularity, asymptotic behaviour as $t\to+\infty$ may be considered. In this case, the following is true:

$$\left(\sigma/\theta\xrightarrow[t\to\infty]{}0\right)\Leftrightarrow(\text{either }\{k=-1, H_0>0\}\text{ or }\{k=0=\beta_+\}),\qquad(2.5.37)$$

$$\left(C_{\alpha\beta\gamma\delta}C^{\alpha\beta\gamma\delta}/\epsilon^2\xrightarrow[t\to\infty]{}0\right)\Leftrightarrow(\beta_+=0\text{ and }\{k=-1\text{ or }k=0\}).\qquad(2.5.38)$$

In the first case, the solution is asymptotically isotropic (the metric is asymptotically FLRW). In the second case, the matter-density becomes asymptotically homogeneous.

Singularities at which the Ricci scalar of the surfaces $t=$const and their trace-free Ricci tensor $S^{(3)}_{\alpha\beta}$ have the properties $R^{(3)}/\theta^2 \to 0$ and $S^{(3)}_{\alpha\beta}/\theta^2 \to 0$ are called velocity-dominated, after Eardley, Liang and Sachs (1972; because the spatial curvature then becomes negligible compared with expansion). Goode and Wainwright show that all the singularities that may occur in the Szekeres models, whether initial or final, Big Bang or shell-crossing, are velocity-dominated. However, curvature dominates over shear since $\sigma^2/R^{(3)} \xrightarrow[t \to \text{singularity}]{} 0$, and $S^{(3)}_{\alpha\beta}$ has the same rate of growth as $R^{(3)}$.

The metric of the singularity, $g^{(0)}_{ij}$, defines the following curvature:

$$R^{(0)} = 6\, (k + \tfrac{2}{3}\beta_+/A), \qquad\qquad (2.5.39)$$

$$2S^{(0)}_{11} = 2S^{(0)}_{22} = -S^{(0)}_{33} = -\tfrac{2}{3}\beta_+/A,$$

and so is in general not a metric of constant curvature. As can be seen from the above, the initial singularity already contains a seed for the growing mode of perturbation, unless it is pointlike ($\beta_+ = 0$).

The G–W representation was very nearly rediscovered by Kasai (1992), who discussed the Einstein equations for irrotational dust under the assumption that the deviation of the solutions away from a FLRW background obeys the same equation as does the linearized perturbation. The author generalized the consideration to nonzero cosmological constant in a subsequent paper (Kasai 1993).

Bonnor and Pugh (1987) identified those of the Szekeres solutions that obey the postulate of uniform thermal histories (PUTH). The postulate states that matter has the same thermal history in every position in the Universe. Technically, a function $f(x)$ on a spacetime M (treated as a model of the Universe) is called uniform if, when restricted to a world-line of the matter source, $\{x^\alpha(s)\}$, it is the same function $f(\{x^\alpha(s)\})$ of the proper time s on every world-line (assuming that the initial moment of time, $s=0$, is chosen on the same hypersurface of constant density $\epsilon = \epsilon_0$ for every world-line). The PUTH requires that all hydrodynamical scalars (energy-density, pressure, entropy per baryon, etc.) are uniform. It was hoped that PUTH would imply spatial homogeneity in the Bianchi sense and thus provide a physical definition of homogeneity. However, some of the L–T models (see Bonnor and Ellis 1986) turned out to be counterexamples, see Section 3.8. This discovery was so unexpected to its authors that further counterexamples were in demand. The Bonnor and Pugh paper provided more of them among the Szekeres solutions. It turned out that in the $\beta'=0$ subfamily, PUTH does imply spatial homogeneity (either Bianchi or K–S type). In the $\beta' \neq 0$ subfamily, PUTH implies a G_3/S_2 symmetry (i.e. the spacetime has to be a L–T model or its plane- or hyperbolically symmetric counterpart) and in addition it requires:

$$k\mathcal{M},_z = 0, \qquad T(z) = \mu\ln z + \omega, \qquad\qquad (2.5.40)$$

where μ and ω are arbitrary constants (with $\mu=0$, this subcase becomes a FLRW model). Bonnor and Pugh's paper contains a detailed list of those Szekeres solutions that obey PUTH and are inhomogeneous $((\delta^{\alpha}_{\beta}-u^{\alpha}u_{\beta})\epsilon,_{\alpha}\neq0)$. The result demonstrates that they are rare. A mathematical criterion for identifying them, as yet unknown, is called for.

Bonnor (1986) used the Szekeres models to test an idea of Penrose's (1979), later corrected by Wainwright (1984). The idea was that the quantity $P\overset{\text{def}}{=}C^{\alpha\beta\gamma\delta}C_{\alpha\beta\gamma\delta}/R^{\mu\nu}R_{\mu\nu}$ provides an arrow of time in cosmology, that is, that $\partial P/\partial t\geq0$ during the evolution and $P=0$ at the Big Bang. In the Szekeres models, $P=\frac{4}{3}A^{-2}(\beta_{+}f_{+}+\beta_{-}f_{-})$ (in the notation of Goode and Wainwright, 1982b). Then, $P=0$ at the Big Bang only if $\beta_{-}=0$, that is, if the decaying mode of inhomogeneity is absent (in the $\beta'\neq0$ subfamily this means that the Big Bang is simultaneous; in the $\beta'=0$ subfamily, the Big Bang is always simultaneous). In that case, $\partial P/\partial t\geq0$ follows. If $\partial P/\partial t=0$, then the model degenerates into FLRW. The model with $k=0\neq\beta'$ is exceptional because in it $\beta_{+}=0$ necessarily and so $\beta_{-}=0$ reduces the model to FLRW. In this case, the arrow of time is not defined.

The vector

$$S^{\alpha}=(C_{\beta\gamma\delta\epsilon}C^{\beta\gamma\delta\epsilon})^{2}(R_{\mu\nu}R^{\mu\nu})^{-3/2}u^{\alpha}/(8\pi) \tag{2.5.41}$$

could be interpreted as an entropy vector in the L–T limit because in that case it fulfils $S^{\alpha}_{;\alpha}\geq0$ everywhere. For the Szekeres models:

$$S^{\alpha}=\epsilon P^{2}\delta^{\alpha}_{0}, \tag{2.5.42}$$

where ϵ is the matter-density. Then $S^{\alpha}_{;\alpha}=\epsilon\partial(P^{2})/\partial t$, and so $S^{\alpha}_{;\alpha}\geq0$ because $\partial P/\partial t\geq0$.

Bonnor discussed the asymptotic behaviour of the quantities P, σ/θ and σ^{2}/ϵ. They all vanish at the Big Bang, and so give no indication of initial inhomogeneity. An indication of inhomogeneity is given by $Z\overset{\text{def}}{=}h^{\alpha\beta}\epsilon,_{\alpha}\epsilon,_{\beta}/\epsilon^{2}$; Z is finite, nonzero and position-dependent at the Big Bang.

Physically, the arrow of time is provided as follows. In the ever-expanding models, the transformation $t\longrightarrow-t$ changes expansion into contraction, and these motions are distinguishable observationally. In the recollapsing models of the $\beta'\neq0$ subfamily, $\beta_{-}=0$ makes the Big Bang simultaneous, but the final singularity is not simultaneous. Thus the Universe does not contain white holes at the beginning and may develop black holes before recollapsing. This is again distinguishable from the time-reversed situation. In the $\beta'=0$ subfamily, the initial and final singularities are always simultaneous, and no physical arrow of time is provided.

In an earlier paper (Bonnor 1985c) the author applied the same idea and the same reasoning to the recollapsing L–T model with a simultaneous Big Bang. The results of the earlier paper follow from those described above.

Lawitzky (1980) showed that the Szekeres models with $\beta'\neq0$ have a closely analogous Newtonian counterpart.

Carminati (1990) investigated in detail the Einstein–Maxwell equations for a perfect fluid in a non-null electromagnetic field under the assumptions that: 1. the Weyl tensor is of type D and the velocity field of the fluid lies in the two-space spanned by the principal null directions of the Weyl tensor; 2. the principal null directions of the complex electromagnetic tensor $F^+_{\alpha\beta}=F_{\alpha\beta}+i(-g)^{-1/2}\epsilon_{\alpha\beta}{}^{\gamma\delta}F_{\gamma\delta}$ coincide with those of the Weyl tensor; 3. the magnetic part of the Weyl tensor relative to the fluid velocity is zero. Properties 1 and 3 are characteristic for the Szekeres and Szafron spacetimes (see Section 2.3), so Carminati's class should include electromagnetic generalizations of them. The ultimate aim of Carminati's paper was apparently to derive explicit solutions. However, the solutions actually presented are stationary.

Bruni, Matarrese and Pantano (1995) described the evolution of the Szekeres models by investigating the phase space of solutions of the autonomous set of ordinary differential equations for expansion, shear and the electric part of the Weyl tensor. These are Ellis' (1971) evolution equations, and they are equivalent to the Einstein equations. This discussion was an illustrative subcase of the dynamics of irrotational dust spacetimes with $H_{\alpha\beta}=0$ called silent Universes ($H_{\alpha\beta}$ is the magnetic part of the Weyl tensor). The main qualitative result of Bruni *et al.* was that a generic model collapses to a spindle singularity rather than to a pancake singularity.

2.6 The spherically symmetric and plane symmetric Szafron spacetimes with $\beta'=0$

From this point on, this book should be read with reference to the diagrams. Until further notice is given, the diagram to consult is Figure 2.1.

The "natural" spherically symmetric limit results from the $\beta'=0$ Szafron spacetimes when $k=+1$, the spheres $\{t=\text{const}, z=\text{const}\}$ are made concentric so that they become orbits of the $O(3)$ group, and the group becomes an isometry group of the spacetime. We recall that, from this point of view, the FLRW limit of these spacetimes results "unnaturally", see Section 2.1. Whether there exist more general "unnatural" spherically symmetric limits in this subfamily was not investigated. In this section we shall consider only the natural limits.

From (2.1.3)–(2.1.5) it is easily found that, with $k=+1$, the spheres $\{t=\text{const}, z=\text{const}\}$ become orbits of an isometry group when:

$$V_1=V_2=0, \quad U=W, \tag{2.6.1}$$

and then the metric simplifies to:

$$ds^2=dt^2-e^{2\alpha}dr^2-\Phi^2(t)(d\vartheta^2+\sin^2\vartheta\,d\varphi^2), \tag{2.6.2}$$

where:

$$x+iy=2e^{i\varphi}\tan(\vartheta/2), \quad z=r, \tag{2.6.3}$$

$$e^\alpha=\lambda(t,r), \tag{2.6.4}$$

and $\{\Phi,\lambda\}$ are determined by (2.1.6) with $k=+1$ and (2.1.7) with $U=W=0$. The assumption $U=W=0$ is no loss of generality because, with $V_1=V_2=0$ and $k=+1$, it can be fulfilled by redefining $\lambda=\tilde{\lambda}-2U\Phi$; the λ in (2.6.4) is in fact the redefined $\tilde{\lambda}$.

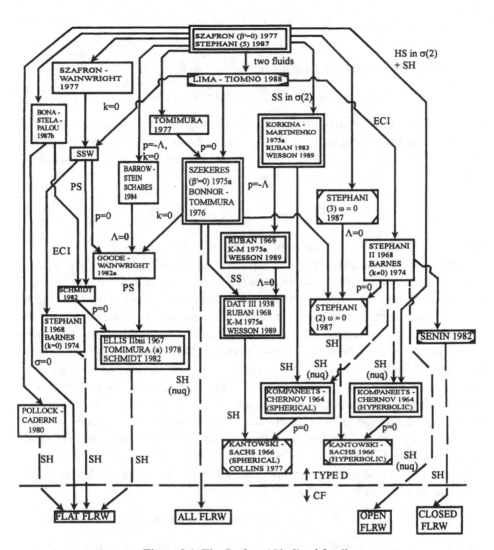

Figure 2.1. The Szafron ($\beta'=0$) subfamily.

This limit, in almost full generality, was first considered by Korkina and Martinenko (1975a). The limitation they assumed was $\Phi_{,t} \neq 0$, but when $\Phi_{,t}=0$, the metric (2.6.2) is not able to reproduce the Kantowski–Sachs (1966) solutions and this case is not considered in this book (the Einstein equations (2.1.6)–(2.1.7) then imply $\epsilon=p=$const). With $\Phi_{,t} \neq 0$, one can choose the time coordinate τ so that:

$$ds^2 = e^{\nu(\tau)}d\tau^2 - e^{\mu(\tau,r)}dr^2 - \tau^2(d\vartheta^2 + \sin^2\vartheta \, d\varphi^2), \qquad (2.6.5)$$

then $\nu(\tau)$ remains an arbitrary function, μ is determined by:

$$e^{-\nu}\nu_{,\tau}(-\mu_{,\tau}+2/\tau) + e^{-\nu}(2\mu_{,\tau\tau}+\mu_{,\tau}^2+2\mu_{,\tau}/\tau-4/\tau^2) - 4/\tau^2 = 0, \qquad (2.6.6)$$

and the matter-density and pressure are:

$$\kappa\epsilon = e^{-\nu}(\mu_{,\tau}/\tau + 1/\tau^2) + 1/\tau^2, \qquad (2.6.7)$$
$$\kappa p = e^{-\nu}(\nu_{,\tau}/\tau - 1/\tau^2) - 1/\tau^2.$$

The solution of (2.6.6) can be formally represented as follows:

$$e^{\mu} = e^{\mu_0}[g(r)\int e^{-\mu_0}\tau^{-1}e^{\nu/2}d\tau + f(r)]^2, \qquad (2.6.8)$$

where $f(r)$ and $g(r)$ are arbitrary functions and $\mu_0(\tau)$ is the solution of (2.6.6) obtained with the assumption $\mu_{,r} = 0$. Note that (2.6.8) is just a different way of writing (2.4.3) with $U = W = 0$ and $k = +1$. The Korkina–Martinenko parametrization results from (2.6.2)–(2.6.4) as follows:

$$\Phi = \tau, \quad \lambda = e^{\mu/2}, \quad \frac{\partial}{\partial t} = \Phi_{,t}\frac{\partial}{\partial\tau}, \quad \Phi_{,t} = e^{-\nu/2}; \qquad (2.6.9)$$

then the equivalence of the other formulae may be verified.

Korkina and Martinenko's paper is important for historical reasons: it was published simultaneously with that of Szekeres (1975a) and earlier than that of Szafron (1977), and thus its result was a new solution at the time. The paper contains more results, but most of the others were not new or not really noteworthy. The case $\epsilon = kp$, $k = const$ in fact necessarily implies $\epsilon = p$, but the authors overlooked this. As given, the solution is within the Kompaneets–Chernov (1964) class, and we shall come back to it in Section 2.11. The cases $\epsilon = const$ are not relevant for cosmology. The case $p = const$ was a rediscovery of the Ruban (1969) solution (with misprints, see below). The paper is remarkable for its language (Ukrainian!), which may have prevented the recognition of its importance. In the limit of homogeneous matter-density, $\mu_{,\tau\tau} = 0$, the K–M spacetimes reproduce the spherically symmetric class of Kompaneets and Chernov (1964). This happens if and only if $f_{,r}g - fg_{,r} = 0$.

Ruban (1983) reobtained the result (2.6.5)–(2.6.8), but under weaker assumptions. He showed that if the metric has the form:

$$ds^2 = e^{\gamma(t,r)}dt^2 - e^{\alpha(t,r)}dr^2 - R^2(t)(d\vartheta^2 + \sin^2\vartheta\,d\varphi^2), \qquad (2.6.10)$$

the source in the Einstein equations is a perfect fluid and the coordinates in (2.6.10) are comoving ($u^{\alpha} = e^{-\gamma/2}\delta^{\alpha}_0$), then $\gamma_{,r} = 0$ follows from the Einstein equations (see eqs. (1.4.2)–(1.4.5)); Korkina and Martinenko assumed $\gamma_{,r} = 0$ from the beginning.

The solutions (2.6.10) are called "T-models" (after Novikov 1963) because they do not admit the "standard" Schwarzschild coordinates; the latter are admissible only in the "R-regions". Ruban showed that these solutions can be interpreted as such configurations of matter in which the gravitational mass defect of any matter added exactly cancels its contribution to the active mass. If such a "T-model" is matched to the Schwarzschild spacetime, then the active mass is specified by initial conditions and does not depend on the amount of matter in the spacetime (see Section 1.4).

Still the same kind of spacetimes were again rediscovered by Wesson (1989), who gave an example of an explicit solution with $0 > p \cong -t^{-2}$ which does not allow the limit $p = 0$, and so does not contain the proper K–S models.

The subcase $\kappa p = -\Lambda = $ const of the above, that is, spherically symmetric dust with the cosmological constant, was first found by Ruban (1969). Then, Φ in (2.6.2) is determined by (2.4.1) with $k = +1$, and λ in (2.6.4) is determined by (2.4.3) with $k = +1$ and $U = W = 0$. With these simplifications, eq. (2.4.5) simplifies to:

$$\kappa\epsilon = 2X/(\Phi^2\lambda), \tag{2.6.11}$$

and Ruban (1969) derived precisely these equations, with:

$$(r_0, \chi, r, \lambda, \mathcal{M}')_{\text{Ruban}} = (2M, z, \Phi, Y, X)_{\text{ours}}. \tag{2.6.12}$$

Ruban's paper contains the expression for $\Phi(t)$ in terms of the Weierstrass elliptic function. When $\Lambda = 1/(9M^2)$, the right-hand side of eq. (2.4.1) (with $k = +1$) has a double root, then $\Phi(t)$ can be expressed in terms of elementary functions and the paper presents this case, too.

The Ruban (1969) solution (without the explicit expressions for $\Phi(t)$) was later rederived by Korkina and Martinenko (1975a) and by Wesson (1989). The vacuum limit, $X = 0$, of this solution is the Kottler solution, that is, the generalization of the Schwarzschild metric to arbitrary Λ. The coordinates of (2.6.2) then cover only that part of the Kottler manifold which is inside the Schwarzschild horizon. For more about the vacuum limit – see below. With $X/Y = $ const, this solution goes over into the generalization of the K–S solutions for arbitrary Λ.

The limit $\Lambda = 0$ of the Ruban (1969) solution was first found by Datt (1938). Datt's paper actually introduces three spherically symmetric dust solutions; the other two will be mentioned in Sections 2.12 and 2.13. Here we refer to Datt's third solution, the only one that was still new in 1938 (!). Since it can be defined by explicit formulae in terms of elementary functions, we shall present it in full, transcribed to our notation of (2.6.1)–(2.6.4):

$$ds^2 = dt^2 - \lambda^2(t,r)dr^2 - \Phi^2(t)(d\vartheta^2 + \sin^2\vartheta d\varphi^2), \tag{2.6.13}$$

where $\Phi(t)$ is the inverse function to:

$$t - t_0 = 2M \arcsin [\Phi/(2M)]^{1/2} - \Phi^{1/2}(2M - \Phi)^{1/2}, \tag{2.6.14}$$

t_0 is an arbitrary constant,

$$\lambda = 2X(r)(1 - Z \cot Z) + Y(r) \cot Z, \tag{2.6.15}$$

$X(r)$ and $Y(r)$ are the arbitrary functions from (2.4.3), and:

$$Z \overset{\text{def}}{=} \arcsin [\Phi/(2M)]^{1/2}. \tag{2.6.16}$$

Datt's formula for matter-density is equivalent to (2.6.11).

Being a dust solution with $\Lambda = 0$, the Datt III metric is the "natural" spherically symmetric limit of the Szekeres $\beta' = 0$ solutions. With $X/Y = $ const it goes over into the Kantowski–Sachs (1966) solution.

The limit $\epsilon = 0$, that is, $X = 0$, of the Datt III solution is a vacuum solution. Its metric can be written in the form:

$$ds^2 = (2M/\Phi - 1)^{-1}d\Phi^2 - Y^2(r)(2M/\Phi - 1)dr^2 - \Phi^2(d\vartheta^2 + \sin^2\vartheta\, d\varphi^2), \quad (2.6.17)$$

and this is easily recognized as the Schwarzschild metric inside the horizon with Φ as the time coordinate; $Y(r)$ can be transformed away by $r' = \int Y(r)\, dr$. This is the vacuum "T-model" mentioned at the end of Section 1.4. The coordinates of (2.6.13)–(2.6.16) are, in the Schwarzschild limit, comoving with a test cloud of dust. Such coordinates were first used by Lemaître (1933a) to demonstrate that the "Schwarzschild singularity" was spurious.

The spatially homogeneous limit of the Datt III solution ($\epsilon_{,r} = 0$) is the Kantowski–Sachs (1966) solution (the proper one, with spherical symmetry).

Datt's paper is of major historical importance, but was published much too early to be properly appreciated. At that time, the FLRW models were still a novelty, barely digestible to the astronomical community, and potential readers were not yet ready for their generalizations. The K-S geometry, implicitly contained in (2.6.13), was not yet recognized. The solution (2.6.13)–(2.6.16) was later rediscovered by Ruban (1968), Korkina and Martinenko (1975a) and Wesson (1989). Ruban (1983) showed that the Datt solution can be matched to the Schwarzschild solution, and then interpreted as follows: it is a ball of dust that explodes out of a singularity at $t = t_0$ (then $\Phi = 0$), expands to $\Phi = 2M$ at $t = t_0 + M\pi$, and then again collapses to $\Phi = 0$ at $t = t_0 + 2M\pi$. In the moment of maximal expansion, the ball of dust fills the whole Schwarzschild sphere.

The Datt solution matched to the Schwarzschild solution is a model of an ideal "gravitational machine" that can transform the entire rest-mass of accreted matter into radiation energy. In such a model, an arbitrarily large amount of matter can be bound under the Schwarzschild horizon, its energy equivalent being completely radiated away so that the active gravitational mass of the configuration is finite and unchanged. That mass has the pure field nature of Wheeler's geometrodynamic "massless mass" (Ruban 1968).

The Datt model and its generalizations have no analogues in the Newtonian theory and do not show up in linear approximations to Einstein's theory (Ruban 1968 and 1969).

Spherically symmetric solutions usually have their counterparts with plane and hyperbolic symmetry. The solutions by Datt, Ruban (1969) and Korkina–Martinenko (1975a) presented so far are united with their other G_3/S_2 counterparts within the Szekeres–Szafron solutions with $\beta' = 0$. The hyperbolically symmetric solutions of this class were not published separately. The plane symmetric dust solution appeared in the paper by Ellis (1967).

Ellis considered, among other things, dust solutions with local rotational symmetry. His case I are stationary solutions, case IIai is the Einstein static Universe, case III are Bianchi-type solutions, none of them are considered in this book. Case II includes in addition two important families that are relevant for this book. Case IIaii will be mentioned in Section 2.12, case IIb are the subcases of the Szafron ($\beta' = 0$) spacetimes with $\kappa p = -\Lambda = $ const and a natural G_3/S_2 symmetry. However, explicit formulae are displayed only for the case $p=\Lambda=0=k$, that is, dust with plane symmetry. They result from the Szekeres–Szafron family as follows: from (2.1.1)–(2.1.7) it can be seen that the surfaces $\{t=\text{const}, z=\text{const}\}$ are planes when $k=0$, and their internal symmetries become isometries of the whole spacetime when, further, $U=V_1=V_2=0$; then by a reparametrization of λ one can achieve $W=0$ without loss of generality, and the solution is found from (2.1.1)–(2.1.5) and (2.4.1)–(2.4.3) with $k=0=\Lambda$ to be:

$$ds^2=dt^2-\lambda^2(t,z')dz'^2-t^{4/3}(dx^2+dy^2), \qquad (2.6.18)$$
$$\lambda=[t+Y(z')/X(z')]/t^{1/3},$$

where use was made of the freedom to scale Φ so that $M=2/9$, and of the freedom to transform z by $z'=\frac{2}{3}\int X(z)\mathrm{d}z$. In doing so, we assumed $X\neq0\neq M$, but $X=0$ is the vacuum limit (being a subcase of the Kasner solution) and $M=0$ is the static limit; neither of them will be considered here. Equation (2.6.18) defines the Ellis (1967) case IIbiii (a misprint is corrected here); the matter-density is:

$$\kappa\epsilon=2t^{-1}(t+Y/X)^{-1}. \qquad (2.6.19)$$

In the limit $Y=0$ (i.e. $\sigma=0$), the solution (2.6.18)–(2.6.19) becomes the flat FLRW dust.

The same solution was later rediscovered by Tomimura (1978, her case (a)) and Schmidt (1982, his dust subcase $\alpha=0$). Tomimura found that the function Y/X measures the initial density contrast $\bar{\sigma}=\epsilon_{,z}/\epsilon=-(Y/X)_{,z}/(t+Y/X)$. The contrast becomes infinite at the Big Bang, but tends to zero as $t\to\infty$, and so the model approaches the flat FLRW dust in the asymptotic future. This behaviour agrees exactly with that found by Bonnor and Tomimura (1976) for the general Szekeres $\beta'=0$ model with $k=0$.

Zakharov (1987) discussed properties of this solution in the linear approximation and found that the growing mode of perturbation is absent here (as indeed follows from (2.5.15) in the case of plane symmetry).

Lake (1992) showed that this solution can be matched to the plane symmetric subcase of the Kasner solution. Lake calculated the redshift along light rays connecting a comoving emitter with a comoving observer, in particular when the light ray originates and terminates within dust, but passes a slot of vacuum on the way. The redshift is anisotropic, and the plane symmetry of the configuration would not be evident to the observer.

The solution (2.6.18)–(2.6.19) was mentioned by Collins (1979) and by Collins and Szafron (1979a) under the name of "the Taub plane-symmetric model", but with no reference.

2.7 The Szafron spacetimes of embedding class 1

Suppose, a manifold M_n of dimension n can be locally embedded in a flat manifold E_N of dimension N (i.e. M_n can be represented as a subspace of E_N). For the smallest

N with this property, the number $(N-n)$ is called the class of M_n. A spacetime, that is, a manifold underlying a given solution of the Einstein equations, is thus of class 1 when it can be embedded in a flat five-dimensional space. There is no reason to expect any simple connection between spacetimes of class 1 and the Szekeres–Szafron solutions. However, it turned out that a large subset of the spacetimes of class 1 whose metrics obey the Einstein equations with a perfect fluid source is within the Szafron $\beta'=0$ subfamily. They were found by Stephani (1968) and Barnes (1974).

Stephani was looking for the perfect fluid solutions that are of class 1, of Petrov type D and have zero acceleration of the perfect fluid source. He obtained two classes of solutions. In class I, after transcribing to our usual notation, the result was:

$$ds^2 = dt^2 - e^{2\alpha}dz^2 - t(dx^2 + dy^2), \tag{2.7.1}$$

where:

$$e^\alpha = G_1(z)t^{3/2} + G_5(z)$$
$$+ t^{1/2}[\tfrac{3}{4}G_1(z)(x^2+y^2) + G_2(z)x + G_3(z)y + G_4(z)], \tag{2.7.2}$$

$G_1,..., G_5$ are arbitrary functions,

$$\kappa p = 1/(4t^2), \tag{2.7.3}$$

$$\kappa\epsilon = 3\kappa p - (4G_1t^{3/2} + G_5)/(2t^2e^\alpha). \tag{2.7.4}$$

This is easily recognized as the subcase of (2.1.1)–(2.1.7) determined by:

$$(k, \Phi, U, V_1, V_2, W, \lambda)$$
$$= (0, t^{1/2}, 3G_1/2, G_2, G_3, G_4/2, G_1t^{3/2} + G_5). \tag{2.7.5}$$

Matter-density in this solution will become spatially homogeneous ($\epsilon_{,x} = \epsilon_{,y} = \epsilon_{,z} = 0$) when either $G_1 = G_5 = 0$ or $\{G_1 = G_2 = G_3 = 0, G_5 = CG_4, C=\text{const}\}$. In the first case the flat FLRW limit is obtained (with the equation of state $\epsilon = 3p$). The limit is of the form (1.3.19)–(1.3.20) with $k=0$. In the second case, the solution becomes a Bianchi I metric, and becomes the same flat FLRW model as before when, in addition, $C=0$. The relation of this solution to the others considered so far is best summarized in the accompanying tables.

The Stephani (1968) solution I is not contained as a limit in the solution of	Because
Szekeres ($\beta'=0$) (1975a) Ellis IIbiii (1967) Datt III (1938)	All three have zero pressure
Ruban (1969)	The Ruban solution has constant pressure
Korkina–Martinenko (1975a)	The K–M solution is spherically symmetric while the Stephani solution is not

The Stephani (1968) solution I does not contain as a limit the solution of	Because
Szekeres ($\beta'=0$) (1975a) Ellis IIbiii (1967) Datt III (1938)	The Stephani solution has no limit $p=0$
Ruban (1969)	The Stephani solution has no limit $p=$const
Korkina–Martinenko (1975a)	The pressure in the K–M solution is an arbitrary function of time

A similar consideration has to be made whenever a new solution is added to the collection already included in this book. I omit it because it is always based on fairly obvious arguments; the tables are meant to be an example of the method. On the basis of such considerations, it is implied that if two references in the diagrams throughout are not joined by an arrow, then the solutions they contain are *not* related by any limiting transition.

Stephani's (1968) class II is the following solution[10] (transcribed to our notation, but the transcription this time involves a coordinate transformation):

$$ds^2=dt^2-e^{2\alpha}dz^2-F^2(t)(dx^2+dy^2)/[1-\tfrac{1}{4}\varepsilon(x^2+y^2)]^2, \qquad (2.7.6)$$

where:

$$F^2 \overset{\text{def}}{=} \varepsilon(t^2-eb^2), \qquad (2.7.7)$$

$\varepsilon=\pm1$ and $e=\pm1$, b is an arbitrary constant, and:

$$e^\alpha=G_1(z)\int F^{-1}dt+G_2(z)+F(t)\Big\{ G_3(z)x+G_4(z)y$$
$$+G_5(z)[1+\tfrac{1}{4}\varepsilon(x^2+y^2)]\Big\}/[1-\tfrac{1}{4}\varepsilon(x^2+y^2)], \qquad (2.7.8)$$

$G_1,..., G_5$ being arbitrary functions;

$$\kappa p=eb^2/F^4, \qquad (2.7.9)$$
$$\kappa\epsilon=3\kappa p+2[G_1(z)(\varepsilon tF-eb^2\int F^{-1}dt)-eb^2G_2(z)]/(F^4e^\alpha).$$

This is the subcase of (2.1.1)–(2.1.8) defined by:

$$(\Phi, k, U, V_1, V_2, W)=(F,-\varepsilon, G_5\varepsilon/2, G_3, G_4, G_5/2), \qquad (2.7.10)$$
$$\lambda=G_1\int F^{-1}dt+G_2.$$

The solution becomes spatially homogeneous in the following three cases:

---

[10] Equations (2.7.6)–(2.7.9) are based on Kramer *et al.*'s (1980) eqs. (32.52) because Stephani's original paper contains errors (private communication from H. Stephani). Kramer *et al.*'s book had another misprint, corrected in (2.7.9).

I. When $G_1=b=0$, but then it just degenerates to a vacuum solution.
II. When $G_1=G_2=0$; then it degenerates to the open and closed FLRW models with a specific equation of state, in the form (1.3.19)–(1.3.20).
III. When $G_3=G_4=G_5=0$, $G_2=CG_1$, C=const. Then it has the Kantowski–Sachs (1966) geometry when $\varepsilon=-1$, or its hyperbolically symmetric counterpart when $\varepsilon=+1$, and is within the class considered by Kompaneets and Chernov (1964). With $b=0$ in addition, pressure vanishes. This can happen only with $\varepsilon=+1$. This limit of the Stephani (1968) solution 2 is a degenerate case of the (hyperbolic) Kantowski–Sachs (1966) case 2, corresponding to $Y_{,t}=1$, and was overlooked in the K–S paper.

Note that Stephani's (1968) class II solution is a counterexample to Szekeres' (1966b) theorem, which says that the only dust solutions of class 1 are FLRW. The limit $b=0$ of the Stephani class II solution is dust, but it is inhomogeneous (and in fact has no symmetry in general).

Barnes (1974) considered perfect fluid solutions of embedding class 1 without further limiting assumptions. He found that they can be of three types:

(a) Conformally flat. This is the Stephani (1967a) solution which belongs to the Stephani–Barnes family; it will be considered in Section 4.9.
(b) Geodesic – see below.
(c) Having a G_3/S_2 symmetry. No explicit nonstatic solution was found in this case.

The geodesic solutions (case b) are those by Stephani (1968). The Barnes solution with $K=0$ results by a coordinate transformation from the Stephani solution I (this makes it possible to correct a misprint in Barnes' paper); the other solutions are coordinate transforms of various subcases of the Stephani solution II.

2.8 Other subcases of the Szafron $\beta'=0$ spacetime

So far, only those subcases of the Szafron $\beta'=0$ spacetimes have been considered here that can be defined by invariant properties of the spacetime. Several other subcases have been published whose invariant definitions are unknown, even though some of them imply physically meaningful conclusions.

Bona, Stela and Palou (1987b) found and investigated those subcases of the Szafron family in which the spaces orthogonal to the fluid flow are flat. This is in fact an invariant definition, but one that refers to a class of subspaces rather than to the whole spacetime (it is an example of the application of the "intrinsic symmetries" approach of Collins 1979). It turns out that in the $\beta'=0$ subfamily, the Bona–Stela–Palou subclass is defined by $k=0=U(z)$ in (2.1.1)–(2.1.8); then the functions $\Phi(t,z)$ and $\lambda(t,z)$ can be represented as:

$$\Phi(t)=f_{,t}^{-1/3},$$

$$\lambda(t,z)=f_{,t}^{-1/3}[X_1(z)f(t)+X_2(z)], \tag{2.8.1}$$

where $f(t)$, $X_1(z)$ and $X_2(z)$ are arbitrary functions. The pressure is then:

$$\kappa p(t) = \tfrac{2}{3} f_{,ttt}/f_{,t} - (f_{,tt}/f_{,t})^2. \tag{2.8.2}$$

The spacetimes (2.8.1)–(2.8.2) reproduce the flat FLRW model in its full generality in the limit $X_1(z)=0$. The corresponding solution for the $\beta′ \neq 0$ subfamily will be presented in Section 2.13.

Szafron and Wainwright (1977) considered the following subcase of the Szafron $\beta′ = 0$ class:

$$k=0=W(z), \quad U=(1-2q)C(z)/2,$$

$$\Phi(t)=[Q(t)]^{2/3}, \quad Q(t)=C_1 t^{1-q} + C_2 t^q,$$

$$\lambda(t,z)=(\lambda_1 t^{1-q} + \lambda_2 t^q)/[Q(t)]^{1/3}, \tag{2.8.3}$$

$$\lambda_1 = B(z) + C(z)\int Q^{-1/3} t^q \mathrm{d}t,$$

$$\lambda_2 = A(z) - C(z)\int Q^{-1/3} t^{1-q} \mathrm{d}t,$$

where q, C_1, C_2 are arbitrary constants and $A(z)$, $B(z)$, $C(z)$ are arbitrary functions. The pressure is:

$$\kappa p = \tfrac{4}{3} q(1-q)/t^2. \tag{2.8.4}$$

This was the first investigation of a subcase of the Szafron $\beta′=0$ class in which the pressure was nonzero but the evolution was determinate. Equation (2.8.4) is a substitute for an equation of state that closes the set of equations (2.1.6)–(2.1.7) and allows us to calculate a definite form of $\Phi(t)$.[11] Szafron and Wainwright investigated several properties of this solution; this was just a preliminary investigation of a representative of a new class of solutions, meant to gain more familiarity with the whole class. The most important of the properties follow from those established for the whole Szafron $\beta′=0$ family, discussed in preceding sections of this book.

The subcase $C_1=0$, $C_2=1$ of the Szafron–Wainwright (1977) solution frequently appears as a subcase or parent of other solutions in Figures 2.1 and 2.2, where it is denoted by SSW.

The case $C=V_1=V_2=C_1=0$, $C_2=1$ of (2.8.3), discussed separately by Szafron and Wainwright, was later reobtained by Schmidt (1982); it is plane symmetric. In the further limit $q=1$ (i.e. $p=0$), the Schmidt subcase reduces to Ellis' (1967) case IIbiii. Stephani's (1968) solution I results from the Szafron–Wainwright solution as the limit $C_1=0$, $C_2=1$, $q=3/4$.

Tomimura (1977) considered the subcase of the Szafron $\beta′=0$ class defined by:

$$p(t)=\bar{C}/\Phi^2(t), \tag{2.8.5}$$

---

[11] In fact, (2.8.3) does not exhaust all solutions of (2.1.6) with pt^2=const and $k=0$. The remaining solutions contain terms of the form $t^{1/2}\ln t$ (when $q=1/2$) and $[t^a \cos (b \ln t)]$ and $[t^a \sin (b \ln t)]$. Those including cos and sin might be interesting: they describe a recollapsing Universe in which different cycles of Big Bang and recollapse are not isometric to each other.

where $\bar{C}$=const. This is another substitute for an equation of state that closes the
set (2.1.6)–(2.1.7) and allows one to obtain definite solutions. Tomimura's solutions
are a generalization of the Szekeres $\beta'=0$ class, the latter results from (2.8.5) when
$\bar{C}=0$. Tomimura's paper contains the complete listings of explicit solutions of
(2.1.1)–(2.1.7) with (2.8.5).

Goode and Wainwright (1982a) discussed a certain subcase of the Szekeres solutions. This paper in fact repeats the discussion made in their other paper (Goode
and Wainwright 1982b) on a simple example in which the number of arbitrary functions is kept to a minimum, and was probably meant to be an easier-to-read version
of the (1982b) paper. The case it considers is the limit $\Lambda=0$ of the case $\beta'=0$ of the
Barrow and Stein-Schabes (1984) solutions, and at the same time the limit $k=0$ of
the Szekeres $\beta'=0$ class. It is also contained in the SSW subcase of the
Szafron–Wainwright (1977) solution as the limit $p=0$.

Lima and Tiomno (1988) invented another way of closing the set of equations
(2.1.6)–(2.1.7). They interpreted the source as a mixture of inhomogeneous dust and
a homogeneous perfect fluid with the equation of state $p(t)=(\gamma-1)\epsilon_{\text{hom}}(t)$, γ=const.
With the further assumption that $\Phi(t)$ is the same as in the FLRW model with
$p=(\gamma-1)\epsilon$, the set of equations (2.1.1)–(2.1.7) becomes determinate.

It must be stressed that the Szafron models were in fact tailor-made for this kind
of interpretation. As long as everyone believed that the cosmic background radiation was homogeneous and isotropic, the Lima–Tiomno interpretation with $\gamma=4/3$
could be a model of an inhomogeneous Universe permeated by homogeneous and
isotropic radiation. Today, however, when anisotropies in the background radiation
have been positively identified, the Lima–Tiomno model is too simple to be **the**
model. Still, the paper is an important contribution that covers several other subcases (see later). Lima and Tiomno considered the following cases:

1. The homogeneous component degenerates into the cosmological constant
   ($\gamma=0$). Here, the assumptions made earlier have rather drastic consequences: all
   such solutions are static. Hence, contrary to what might be expected, the
   Lima–Tiomno solutions will not reproduce the Barrow and Stein-Schabes (1984)
   solutions with $\beta'=0$, nor the Ruban (1969) solution.
2. The homogeneous component is radiation ($\gamma=4/3$) – see the comment below on
   Pollock and Caderni (1980).
3. The homogeneous component is another dust. A simple reinterpretation of the
   source allows one to recover the whole Szekeres $\beta'=0$ class here. The Tomimura
   (1977) solutions are not contained in Lima and Tiomno's class because the $\bar{C}$ in
   (2.8.5) results from it with a definite negative value.

The solutions for $\Phi(t)$ are determined by $t(\Phi)$, which is represented by hypergeometric functions; the formulae for $e^{\beta(\Phi)}$ involve hypergeometric functions, too.
Lima and Tiomno's paper contains a listing of explicit results and a qualitative discussion of evolution. At the Big Bang singularity, the source behaves like a FLRW
model with $\epsilon=p$. At the asymptotic future, the models evolve towards FLRW except

for one subcase of the $k=0$ model which develops negative matter-density in the inhomogeneous component. The energy-momentum tensors of the two components are not conserved separately, but the evolution is adiabatic: they can only exchange entropy. Lima and Tiomno do not consider shell-crossing singularities.

The Lima–Tiomno approach has a certain philosophical consequence that is uncomfortable. Assumptions made about the homogeneous component of the source fix the time-evolution of $\Phi(t)$, and then the time-evolution of the whole model is determined. When the homogeneous component is radiation, this implies that background radiation drives the evolution of the Universe even today.

In a subsequent paper (Lima and Tiomno 1989), the authors investigated the thermodynamical properties of the parabolic (i.e. $k=0$) subcase of their (1988) solutions. They showed that even if the model approaches FLRW asymptotically, its temperature will evolve differently than in the FLRW models unless a position-dependent equation of state is properly chosen.

Pollock and Caderni (1980) were historically the first to interpret the Szafron models as a mixture of homogeneous radiation and inhomogeneous dust. However, they made additional *ad hoc* assumptions about the solutions that forced their result to be a special subcase of the $\beta'=0$ subfamily only. The assumptions were:

$$\Phi(t,r)=t^{\nu}\psi(r), \quad k=0, \quad p(t)=A/\Phi^4, \tag{2.8.6}$$

where ν and A are constants and ψ is a function to be determined from the Einstein equations. The solution finally obtained by Pollock and Caderni is not a subcase of those considered by Lima and Tiomno (1988), even though its homogeneous component obeys the equation of state $p(t)=\epsilon(t)/3$. This is because Pollock and Caderni did not assume that $\Phi(t)$ must be exactly the same as in the FLRW model of radiation. In fact, the Lima–Tiomno solution with $k=0$ and $\gamma=4/3$ is the degenerate subcase $b=0$ of the Pollock and Caderni solution; this subcase cannot be obtained by a limiting transition in the solution, but has to be derived separately from the field equations. The spatially homogeneous limit of the Pollock–Caderni solution is a flat FLRW model with a definite equation of state.

Similarly to Lima and Tiomno, Pollock and Caderni found that their solution has the equation of state $\epsilon=p$ close to the singularity. The paper contains a more detailed discussion, but it concentrates on analytic rather than qualitative properties, and readers are herewith advised to consult the original paper for details.

The SSW limit of the Szafron–Wainwright (1977) solution is the subcase $k=0$ of the Lima–Tiomno (1988) solution. The Stephani (1968) solutions II are the subcases $\gamma=4/3$, $k=\pm1$ of the Lima–Tiomno solution.

Senin's (1982) solution deserves special attention even though it has no invariant definition. Senin found a solution whose local geometry (not just topology!) is that of a three-dimensional torus. In the original notation, the solution is:

$$ds^2=a^2(t)\left\{dt^2-d\rho^2-\sin^2\rho\,d\xi^2-[\cos\rho+\tfrac{1}{2}b(t)/a(t)]^2d\varphi^2\right\}, \tag{2.8.7}$$

where:

$$a=a_0\sin t, \quad b=C_1t+C_2, \tag{2.8.8}$$

a_0, C_1 and C_2 are arbitrary constants. This is a perfect fluid solution with:

$$u^\alpha=a^{-1}\delta^\alpha_0, \quad \kappa p=a_0^2/a^4,$$

$$\kappa\epsilon=a_0^{-2}(a\cos\rho+b/2)^{-1}[C_1\cos t/\sin^3t+3a_0\cos\rho/\sin^3t \tag{2.8.9}$$

$$+(C_1t+C_2)/(2\sin^4t)],$$

and it becomes conformally flat when $C_1=C_2=b=0$. Then it reduces to the closed FLRW model in the form (1.3.13)–(1.3.14) with the equation of state $\epsilon=3p$. Each surface $\{t=\text{const}, \xi=\text{const}\}$ has the local geometry of a torus with large radius $b/2$ and small radius a; the two radii evolve according to different laws. Each surface $\{t=\text{const}, \varphi=\text{const}\}$ is locally a sphere of radius a, but globally it is a pair of spheres that touch each other at $\rho=\pi$ and whose points $\rho=0$ and $\rho=2\pi$ are identified. The model has a finite duration, as does the closed FLRW model. At $t=0$, it has a Big Bang singularity; the space $t=\text{const}$ begins its expansion from a ring of radius $C_2/2$. The large radius of the torus keeps growing from $C_2/2$ to $(\pi C_1+C_2)/2$; the small radius grows from zero at $t=0$ to the maximum a_0 at $t=\pi/2$, then collapses again to zero at the final singularity $t=\pi$. The final singularity is again a ring, of radius $(\pi C_1+C_2)/2$. With $C_2=0$, the Big Bang is pointlike. The energy-density will be positive throughout the evolution with $C_2=0$ if $C_1/(2a_0)$ is sufficiently large, but the dominant energy condition will be violated in the recollapse phase.

In order to relate the Senin solution to the others, it is convenient to transform the coordinates as follows:

$$t=\arccos(-T/a_0), \quad \rho=2\arctan(r/2). \tag{2.8.10}$$

Then the metric becomes:

$$ds^2=dT^2-a^2(T)(1+r^2/4)^{-2}(dr^2+r^2d\xi^2)$$

$$-[a(T)(1-r^2/4)/(1+r^2/4)+\tfrac{1}{2}b(T)]^2d\varphi^2, \tag{2.8.11}$$

$$a^2(T)=a_0^2T^2, \quad b(T)=C_1\arccos(-T/a_0)+C_2,$$

and is the following limit of Stephani's (1968) solution II:

$$\varepsilon=-1=-e, \quad G_1=1/2, \quad G_2=G_3=G_4=0, \quad G_5=1,$$

$$x=r\cos\xi, \quad y=r\sin\xi, \quad z=\varphi. \tag{2.8.12}$$

Senin's (1982) solution seems to be the final refinement of the attempts by Mitskievič and Senin (1981) to obtain a spacetime with toroidal space. The other attempts resulted in nonperfect fluid solutions; see Sections 2.13 and 7.2.

2.9 Stephani's (1987) solutions

Stephani (1987) found a collection of solutions with no symmetries. Among them is a remarkable solution with nonzero expansion and rotation. It does not belong to the Szekeres–Szafron family, but is closely related to it. As shown in Figure 2.1, its limit $\omega=0$ belongs to the Szekeres–Szafron subfamily with $\beta'=0$. Figure 2.2 is an appendix to Figure 2.1 which shows the relation of the Stephani solution and of several nonperfect fluid solutions to the main entries of Figure 2.1.

The Stephani (1987) solutions result from the following assumptions:
1. The spacetime is foliated into *timelike* hypersurfaces of zero extrinsic curvature so that:

$$ds^2=-N^2(x^1,x^n)(dx^1)^2-g_{ab}(x^n)dx^adx^b, \tag{2.9.1}$$

where $a, b, n=2, 3, 4$, the signature of g_{ab} is $(++-)$, and x^4 is the time coordinate.
2. The three-spaces with the metric g_{ab} are conformally flat.
3. The velocity field u_a is an eigenvector of the Ricci tensor of each hypersurface $x^1=$const.
4. The Ricci tensor of the hypersurfaces $x^1=$const can be decomposed as follows:

$$R^{(3)}_{ab}=Su_au_b+\tfrac{1}{2}(-S+R^{(3)})(g_{ab}-u_au_b). \tag{2.9.2}$$

Already at this point it can be established that the magnetic part of the Weyl tensor of such solutions will be zero, and that they will be of Petrov type D or O.

Solution 1 (Stephani's eq. (4.5)) has no FLRW limit (in the limit of spatially homogeneous matter-density, $\epsilon,_{\mu}(\delta^\mu{}_\alpha-u^\mu u_\alpha)=0$, it degenerates into the Minkowski spacetime).

Solution 2 (Stephani's eq. (4.8)) has a rotating and expanding dust as a source. It is, in the original notation except for the signature:

$$ds^2=-N^2(dx^1)^2-\eta_{ab}dx^adx^b, \quad \eta_{ab}=\text{diag}\,(1, 1, -1),$$
$$N\overset{\text{def}}{=}\tfrac{1}{2}M\ln T+g_ax^a+h, \quad T^2\overset{\text{def}}{=}-\eta_{ab}(x^a-f^a)(x^b-f^b), \tag{2.9.3}$$

where M, f^a, g_a and h are arbitrary functions of x^1; the velocity field is:

$$u_a=T,_a, \quad u_1=0; \tag{2.9.4}$$

and the matter-density is:

$$\kappa\epsilon=M/(NT^2). \tag{2.9.5}$$

This solution has in general no symmetry, although the three-spaces $x^1=$const are flat and so have a six-dimensional symmetry group. This is an example of a spacetime with intrinsic symmetries in the sense of Collins (1979).

The rotation will vanish when $T,_{a1}=0$. This implies that all $f^a=$const, and then they can be transformed away by $x^a=x^{a'}+f^a$ (this will only result in redefining h). Then, the coordinate transformation:

$$x^1=z, \quad x^2=Txe^\nu, \quad x^3=Tye^\nu, \quad x^4=T(2e^\nu-1),$$
$$e^{-\nu}\overset{\text{def}}{=}1-\tfrac{1}{4}(x^2+y^2), \tag{2.9.6}$$

will transform the metric to the form:

$$ds^2=dT^2-(Te^\nu)^2(dx^2+dy^2)-N^2dz^2, \tag{2.9.7}$$

$$N=\tfrac{1}{2}M \ln T+h+\{g_2 x+g_3 y+g_4[1+\tfrac{1}{4}(x^2+y^2)]\}e^\nu T. \tag{2.9.8}$$

This is the limit $p=0$ of the Stephani (1968) solution II that results when the following specializations are made:

$$(\varepsilon,\, b,\, G_1,\, G_2,\, G_3,\, G_4,\, G_5,\, t)=(1,\, 0,\, M/2,\, h,\, g_2,\, g_3,\, g_4,\, T). \tag{2.9.9}$$

The coordinates of (2.9.7)–(2.9.8) are comoving. The matter-density, still given by (2.9.5), will become spatially homogeneous when $g_2=g_3=g_4=0$, $h=$const, $M=$const. Then, (2.9.7)–(2.9.8) becomes the same degenerate subcase of the Kantowski–Sachs (1966) hyperbolic solution that was mentioned after eq. (2.7.10).

Solution 3 (eq. (4.23) in Stephani's paper) is said to be a perfect fluid solution with constant pressure. In fact, it can be interpreted as dust with cosmological constant, and then the condition $p>0$ need not be required. In the formulae given below, Stephani's p was replaced by Λ. Again in the original notation except for the signature, the solution is:

$$ds^2=-N^2(dx^1)^2-(3/\Lambda)\eta_{ab}dx^a dx^b/(1+r^2/4)^2,$$

$$N=\frac{3M}{2\Lambda}\left[1+\frac{1-r^2/4}{1+r^2/4}\ln(r/2)\right]+[g_a x^a+h(1-r^2/4)]/(1+r^2/4),$$

$$r^2=-\eta_{mn}(x^m-f^m)(x^n-f^n), \tag{2.9.10}$$

$$\kappa\epsilon+\Lambda=(M/Nr^2)(1+r^2/4)^2,$$

$$u_a=(3/\Lambda)^{1/2}r_{,a}/(1+r^2/4),\quad u_1=0,$$

η_{ab}, M, f^a, g_a and h are as in (2.9.3). A probable misprint in the factor $(1+r^2/4)^{-2}$ in the metric has been corrected;[12] without the correction u_a would not obey $u_a u^\alpha=1$. In order to include the case $\Lambda<0$ and the limit $\Lambda\to 0$, the following reparametrization should be made:

$$(x^a, f^a, r)=(\Lambda/3)^{1/2}(x'^a, f'^a, T),$$

$$M=(\Lambda/3)^{3/2}M',\quad x^1=(\Lambda/3)^{-1/2}x^{1'}, \tag{2.9.11}$$

$$h=(\Lambda/3)^{1/2}(h'-\tfrac{1}{2}M' \ln|(\Lambda/3)^{1/2}/2|-M'/2),$$

and $\ln(r/2)$ should be replaced by $\ln|(\Lambda/3)^{1/2}r'/2|$. Then, with primes dropped, the limit $\Lambda=0$ of (2.9.10) is (2.9.3). The limit $\omega=0$ of (2.9.10) is calculated similarly to that of (2.9.3) except that, in addition to (2.9.11), the following transformation must be made:

$$T=2(\Lambda/3)^{-1/2}\tan[(\Lambda/3)^{1/2}t/2]. \tag{2.9.12}$$

Then, $\omega=0$ leads to the Szafron $\beta'=0$ solution with:

---

[12] See "Notes added in proof", Note 3.

$$\Phi = (\Lambda/3)^{-1/2}\sin[(\Lambda/3)^{1/2}t],$$

$$k=-1, \quad 2W=U/2=g_4, \quad V_1=g_2, \quad V_2=g_3, \tag{2.9.13}$$

$$\lambda = \cos[(\Lambda/3)^{1/2}t](\tfrac{1}{2}M \ln T + h - M/2) + M/2,$$

and the further limit $\Lambda \rightarrow 0$ reproduces (2.9.7)–(2.9.8). The spatially homogeneous limit of (2.9.13) is within the Kompaneets–Chernov (1964) class.

Solution 4 (eq. (16.10) in Stephani's paper) degenerates into vacuum with cosmological constant in the spatially homogeneous limit, and so is not within the scope of this book. Solution 5 (section 7 in Stephani's paper) is equivalent to the whole Szafron $\beta'=0$ class.

Solution 2 is remarkable because it seems to be the only example existing in the literature of a perfect fluid solution with expansion and rotation both nonzero and with the property that its limit $\omega=0$ is still nonstatic, and even inhomogeneous. Several solutions with $\omega\theta\neq0$ have been published, but most of them are spatially homogeneous from the beginning, and all the others have the property that $\theta=0$ in the limit $\omega=0$. No solution is known to the present author that would reproduce the FLRW models in the limit $\omega=0$, and through the absence of such a result, the Stephani (1987) solution 2 seems to be the best achievement so far in the quest for rotating cosmological models.[13] Work on its interpretation still remains to be done.

2.10 Other generalizations of various subcases of the β'=0 subfamily

The specialization diagrams for the solutions from this section are shown in Figure 2.2. It includes only those entries from Figure 2.1 that are directly reproduced by the nonperfect fluid solutions.

2.10.1 Solutions with heat-flow

Goode (1986) generalized the whole Szafron $\beta'=0$ subfamily for nonzero heat-flow. His class of solutions has no invariant definition so far, but nearly all the other solutions given in Figure 2.1 can be obtained from it by taking various limits.

Goode assumed the same form of the metric as in the Szafron solutions, and in addition:

---

[13] Another interesting result is that of Demiański and Grishchuk (1972). They discussed a Bianchi VII$_0$ model with expansion and rotation that degenerates into a Bianchi I model when $\omega=0$. The latter admits a flat FLRW limit. However, the model is defined through a set of ordinary differential equations, and only asymptotic solutions are explicitly derived and discussed in the paper.

One more example of a solution with $\omega\theta\neq0$ was found by Korotkii and Krechet (1988). The solution is of Bianchi type (which type is not specified in the paper), and the source is a mixture of anisotropic rotating fluid, null radiation and two scalar fields. When $\omega=0$, the pressure anisotropy disappears and the scalar fields degenerate into the cosmological constant; the resulting spacetime is a flat FLRW model. When any of the two scalar fields is set to zero, then the second scalar field, pressure anisotropy and the rotation automatically vanish, too. In the limit of isotropic pressure, rotation vanishes automatically and the two scalar fields merge into one.

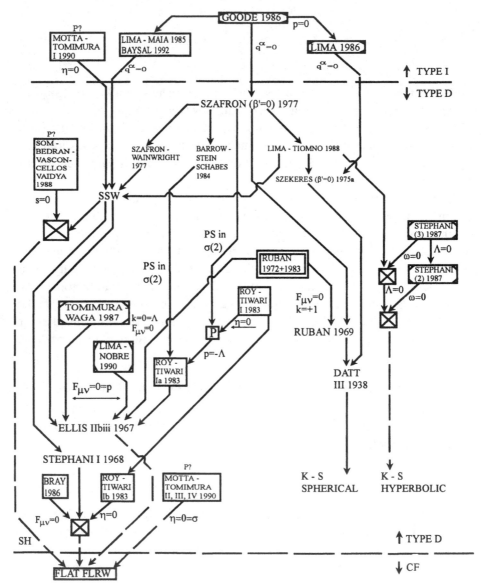

SYMBOLS USED:
η - viscosity,   ω - rotation,   s - "spin"
REFERENCES WITHOUT FRAMES ARE PERFECT FLUID SOLUTIONS FROM FIGURE 2.1
DEFINITIONS OF LIMITING TRANSITIONS IN BETWEEN THEM - SEE FIGURE 2.1

Figure 2.2. Generalizations to Figure 2.1 (heat-flow, electromagnetic field,
viscosity, rotation).

1. $\beta_{,tx} = \beta_{,ty} = 0$,
2. The surfaces {t=const, z=const} have constant curvature.

With a perfect fluid source, both these properties follow from the Einstein equations, but this is not the case when the source includes heat-flow. It turns out that $\beta' \neq 0$ implies zero heat-flow under the above assumptions, hence only the $\beta'=0$ subfamily allows such a generalization.

The solution is (in the original notation except for the signature):

$$ds^2 = dt^2 - S^2(t)[e^{2\nu}(dx^2+dy^2) + H^2 dz^2], \qquad (2.10.1)$$

where:

$$e^{-\nu} = 1 + k(x^2+y^2)/4, \quad k = \pm 1, 0,$$

$$H = e^{\nu}[A(x,y,z) + B(x,y,z)Q(t)] + F(t,z),$$

$$A(x,y,z) = a(z)(x^2+y^2) + b(z)x + c(z)y, \qquad (2.10.2)$$

$$B(x,y,z) = h_1(z)(x^2+y^2) + h_2(z)x + h_3(z)y,$$

$$Q(t) = \int S^{-3}(t)dt,$$

a, b, c, h_1, h_2, h_3 are arbitrary functions of z, $S(t)$ is defined by:

$$2S_{,tt}/S + S_{,t}^2/S^2 + k/S^2 + \kappa p = 0 \qquad (2.10.3)$$

(Goode considered $\Lambda \neq 0$, but Λ can always be included into p and ϵ, so we shall omit it), and $F(t,z)$ is defined by:

$$F_{,tt} + 3S_{,t}F_{,t}/S - kF/S^2 = 2[h_1(z)Q(t) + a(z)]/S^2. \qquad (2.10.4)$$

The nonzero components of the heat-flow vector are:

$$q_1 = e^{2\nu}\left\{2h_1 x + h_2[1 - \tfrac{1}{4}k(x^2+y^2)] - \tfrac{1}{2}kh_3 xy\right\}/(S^3 H), \qquad (2.10.5)$$

$$q_2 = e^{2\nu}\left\{2h_1 y + h_3[1 + \tfrac{1}{4}k(x^2-y^2)] - \tfrac{1}{2}kh_2 xy\right\}/(S^3 H), \qquad (2.10.6)$$

the matter-density is:

$$\kappa\epsilon = 3[(S_{,t}/S)^2 + k/S^2] - 2(H_{,tt} + 2H_{,t}S_{,t}/S)/H, \qquad (2.10.7)$$

and the pressure is an arbitrary function of time.

The heat-flow vanishes and the Szafron $\beta'=0$ subfamily is recovered when $h_1 = h_2 = h_3 = 0$ (and consequently $B=0$). Properties of the solution were only discussed in the case $p=0$.[14] Just as in the limit $q^\alpha = 0$, the hypersurfaces t=const are conformally flat. The difference is that with $q^\alpha \neq 0$ the magnetic part of the Weyl tensor is nonzero. Its frame components in the natural orthonormal frame of (2.10.1) are:

---

[14] This subcase was published independently by Lima (1986). Both papers contain complete listings of explicit solutions. Goode follows the notation of Goode and Wainwright (1982b), Lima's notation is different. Goode's discussion is more detailed. In the limit $q^\alpha=0$, the Lima solutions reproduce the whole Szekeres $\beta'=0$ subfamily.

$$H_{31} = -q_{(2)}/2, \quad H_{32} = -q_{(1)}/2, \tag{2.10.8}$$

where $q_{(i)}$ are the frame components of the heat-flow vector, related to the coordinate components q_i by $q_{(i)} = S^{-1}e^{-\nu}q_i$, $i=1, 2$. The Petrov type of the solutions with $q^\alpha \neq 0$ is I. The shear is zero in the FLRW limit, and only then. The presence of heat-flow gives an algebraically independent contribution to the decaying mode of inhomogeneity. This new contribution, called the β-mode, becomes smaller than the ordinary α-mode (the decaying one) as the initial singularity is approached. As long as $q^\alpha \neq 0$, the α-mode cannot be eliminated, except along isolated world-lines. Goode's paper contains a detailed list of conditions under which the initial and final singularities may be pointlike, pancake or cigar type. With $q^\alpha \neq 0$, pointlike singularities may occur only along isolated world-lines, while with $q^\alpha = 0$ they could be occurring on open sets.

At an initial singularity, $q^\alpha q_\alpha/\theta^4$ tends to zero, but σ/θ tends to a nonzero limit when $q^\alpha \neq 0$ (except possibly at isolated world-lines). Hence, heat-flow makes the initial singularity more anisotropic. Asymptotically towards the future, both these quantities may tend to zero.

This much can be said without imposing the thermodynamic conditions (1.2.12)–(1.2.14) on q^α. After they are taken into account, the solution is restricted further. Goode shows that the temperature will not in general diverge at a shell-crossing singularity. Equations (1.2.12)–(1.2.14) are fulfilled, for example, when $b=c=0=h_2=h_3=\beta_+$, $h_1=$const, $a=$const, but then the metric acquires a two-dimensional Abelian symmetry group.

The subcase $k=0$, $p=c(R_{,t}/R)^2$ of the Goode (1986) solutions was derived by Lima and Maia (1985). The zero heat-flow limit of that solution is the SSW subcase of the Szafron–Wainwright (1977) solution.

Baysal's (1992) solution is the subcase $k=0$, $p=cR_{,tt}/R$ of the Goode (1986) solutions. It is in fact identical to the Lima–Maia (1985) solution because in both solutions $S(t)$ is a power function of t; the two cases differ only by renaming the exponent.

2.10.2 Solutions with electromagnetic field

Bray (1986) found two solutions. Both have no invariant definitions. The first of them in fact does not acquire a perfect fluid source in the limit $F_{\mu\nu}=0$ unless $k=0$, so we consider only the subcase $k=0$. Then, both solutions have the same limit $F_{\mu\nu}=0$ and this is the subcase $G_1=G_2=G_3=0$ of the Stephani (1968) solution I. A flat FLRW limit results when $G_5=0$ in addition. Both solutions are plane symmetric, and thus are of Petrov type D.

Tomimura and Waga (1987) considered generalizations of the Szafron $\beta'=0$ subfamily to the case when there is a free electromagnetic field in the spacetime, with the electric and magnetic fields both being tangent to the z-coordinate lines. They assumed that the matter in the spacetime is a mixture of inhomogeneous dust and homogeneous radiation with density $\kappa\epsilon_{rad}=3c^2e^{-\beta(t)}$, where c is a constant. Under these assumptions there are three classes of solutions. The first is just the flat FLRW model, the second class is static. In the third class, the spacetime is plane symmetric and of Petrov type D. The solutions for e^α in the third class are hypergeometric series in e^β (see Tomimura and Waga's paper). In the limit $F_{\mu\nu}=0=p$, they go over into the Ellis IIbiii (1967) solution. Tomimura and Waga find that the magnetic field makes the initial singularity more isotropic.

Lima and Nobre's (1990) solution is very similar to that found by Tomimura and Waga except that here the matter source is a perfect fluid with $p=3(\gamma-1)\beta_{,t}^2(t)$, and this results in a different dependence of β on t and of e^α on β. The limit $p=0=F_{\mu\nu}$ is again the Ellis IIbiii (1967) solution. Lima and Nobre find that for $\gamma<4/3$ the magnetic field will prevent the Big Bang singularity and the model will evolve toward FLRW, while for $\gamma>4/3$ the Big Bang is present and anisotropy survives in the asymptotic future.

Ruban (1972) considered the case when the Szafron $\beta'=0$ spacetime has an arbitrary G_3/S_2 symmetry, constant pressure (i.e. the cosmological constant) and free electric field tangent to the r-coordinate lines. Transcribed to our notation (2.6.1)–(2.6.2) the solution is:

$$ds^2=dt^2-\lambda^2(t,r)dr^2-\Phi^2(t)[d\vartheta^2+k^{-1}\sin^2(k^{1/2}\vartheta)\,d\varphi^2], \qquad (2.10.9)$$

where $k=\pm1, 0$; the spacetime is spherically symmetric for $k=+1$, plane symmetric for $k=0$ and hyperbolically symmetric for $k=-1$. Further, $\Phi(t)$ is defined by:

$$\Phi_{,t}^2=-k+2M/\Phi-E^2/\Phi^2+\tfrac{1}{3}\Lambda\Phi^2, \qquad (2.10.10)$$

where M and E are arbitrary constants and Λ is the cosmological constant. The function λ is given by a formula that is closely analogous to (2.4.3):

$$\lambda(t,r)=\Phi_{,t}[X(r)\textstyle\int\Phi^{-1}\Phi_{,t}^{-3}\,d\Phi+Y(r)], \qquad (2.10.11)$$

where $X(r)$ and $Y(r)$ are arbitrary functions, and the matter-density is:

$$\kappa\epsilon=2X\Phi^{-2}\lambda^{-1}. \qquad (2.10.12)$$

The constant E is the charge that generates the electric field. With $k=E=0=\Lambda$, this solution reproduces the Ellis (1967) IIbiii solution; with $E=0$ and $k=+1$ it reproduces the Ruban (1969) solution. Equation (2.10.10) is closely analogous to the corresponding one in the $\beta'\neq0$ subfamily (see Section 2.14).

This solution was briefly mentioned in a later paper (Ruban 1983) that was mostly devoted to perfect fluid solutions (see Section 2.6).

There is a problem in interpreting this solution geometrically. The function $\Phi(t)$ is the distance from the centre of symmetry, and the term $(-E^2/\Phi^2)$ in (2.10.10) is suggestive of an electric repulsive force. However, since Φ depends only on time, Φ is constant over every space $t=$const (and so is the electric field), and those spaces do not contain their centres of symmetry. In an ordinary space (an "R-region"), the electric term in (2.10.10) corresponds to the exterior electric field, and the source of the field is supposed to surround the centre of symmetry (or be a point charge in the centre). However, where is the source of the field in (2.10.10) when the space does not contain its centre?

By analogy with Shikin's (1972a) result we can observe that the Big Bang singularity ($\Phi=0$) will not occur when $\Lambda=0$. However, in the case $k=+1$ solutions will exist only if $E^2 \leq M^2$; the one with $E^2 = M^2$ is static.

2.10.3 Solutions with viscosity

Motta and Tomimura (1990) found four classes of solutions with viscosity that generalize certain subcases of the Szafron $\beta'=0$ subfamily. Their class I is similar to the Lima–Maia (1985) solution and in the limit $\xi \to 0$, $\eta \to 0$ reproduces the SSW subcase of the Szafron–Wainwright (1977) solution. The other three classes do have inhomogeneous limits $\xi \to 0$, $\eta \to 0$, but are not in any direct way related to other solutions from our Figures 2.1 and 2.2. In the limit $\sigma=0$ all four of Motta and Tomimura's classes reproduce the flat FLRW models with definite equations of state.

Tomimura and Motta (1990a) discussed asymptotic properties of several subcases of the solutions mentioned above, in the vicinity of the singularity and in the infinite future. It is a listing of several possibilities with no simple qualitative feature in common.

The solutions found in the next paper (Tomimura and Motta 1990b) differ from those by Motta and Tomimura (1990) only by the property that in the latter $p = c\Phi_{,t}^2/\Phi^2$, c=const, while in the former $p = c\Phi_{,tt}/\Phi$. The solutions that result have the same (power-law) dependence of Φ on t; only the exponent is parametrized in a different way. Hence, they are in fact the same solutions (compare Section 2.10.1 – the solutions of Lima–Maia, (1985), and those of Baysal (1992) are related in the same way).

Roy and Tiwari (1983) considered plane symmetric metrics of the Szafron type (both subfamilies) with viscous fluid source. Their case I, in the limit $\xi=\eta=0$, reproduces the plane symmetric subcase of the $\beta'=0$ subfamily with arbitrary $p(t)$. Their subcase Ia results when $\xi=\eta=0$ and $\kappa p = -\Lambda$. It is the plane symmetric limit of the Barrow–Stein-Schabes (1984) solution with $\beta'=0$, and reproduces the Ellis (1967) solution IIbiii when $\Lambda=0$. Roy and Tiwari's subcase Ib is a handpicked special case, and its limit $\xi=\eta=0$ happens to be the same as the limit $F_{\mu\nu}=0$ of the Bray (1986) solution.

2.10.4 Other solutions

Som, Bedran and Vasconcellos-Vaidya (1988) found a solution with a source that they called a spinning fluid. In addition to spin, the fluid has heat-flow. Vanishing spin automatically implies vanishing heat-flow, and then their solution reduces to the subcase $C(z)=0$ of the SSW subcase of the Szafron–Wainwright (1977) solution. The (flat only) FLRW limit results with a definite equation of state. This solution has been listed here for completeness, but it is not clear what it actually represents. The quantity that the authors called "spin" is defined as a tensor product of two spacelike vectors of an orthonormal tetrad which is limited only by $e_{(0)}{}^\alpha = u^\alpha$; otherwise it is arbitrary.

2.11 Solutions with Kantowski–Sachs geometry and their other generalizations

The solutions discussed here are shown in Figure 2.3. Since they are of secondary interest for this book (most of them are spatially homogeneous), they will be mentioned only briefly. In this section, we shall ignore the plane and hyperbolically symmetric counterparts of the K–S solutions.

2.11.1 The Kompaneets–Chernov (1964) spacetimes and their subcases

The Kompaneets–Chernov spacetimes have the K–S geometry and a general perfect fluid source. At this level of generality, there is just one equation to be obeyed by the two functions from (1.3.2), that is, the evolution is not determinate until an equation of state is imposed. These spacetimes are the spatially homogeneous limit of the Korkina–Martinenko (1975a) class. They were first considered in the context of the Einstein equations by Kompaneets and Chernov (1964), but their peculiar geometry was first recognized by Kantowski and Sachs (1966).

Several subcases, corresponding to various equations of state, have been considered in separate papers.[15] The linear barotropic case, $p = (\gamma - 1)\epsilon$, $\gamma = $const, was discussed by Goethals (1975) and by Ruban (1983), and then investigated by Goethals (1975) and by Weber (1984) by the method of qualitative analysis of autonomous sets of differential equations. The further subcase $\gamma = 2$, that is, the "stiff fluid", was discussed by Vajk and Eltgroth (1970, their section VII B.2), Korkina and Martynenko (1976), Ruban (1983), Lorenz (1983, his section 4.2), van den Bergh and Wils (1985, their case I) and by Xanthopoulos and Zannias (1992). The case when the source is a mixture of radiation (with the equation of state $\epsilon = 3p$) and dust was discussed qualitatively (in the same sense as before) by Weber (1986). The further subcase $\epsilon = 3p$, that is, pure radiation, was discussed by Kompaneets and Chernov (1964), Vajk and Eltgroth (1970, their section VII B.1), McVittie and Wiltshire (1975, their case b) and by Ruban (1983). The case $\kappa p = -\Lambda$, that is, dust with cosmological constant, was discussed by Ruban (1969), Lorenz (1982), and (again qualitatively) by Weber (1985). Finally, the proper K–S models with $p = 0$ were discussed by Kompaneets and Chernov (1964), Kantowski and Sachs (1966), Vajk and Eltgroth (1970, their section VII B.1) and by Collins (1977).

---

[15] Most of the solutions mentioned in this paragraph were derived by R. Kantowski in his PhD thesis in 1966. Specifically, Kantowski obtained the solutions with (I) $p = 0 = \Lambda$, (II) $p = 0 \neq \Lambda$, (III) $\epsilon = 3p$, (IV) $\epsilon = p$. However, Kantowski's results were apparently never published in any journal, so in this case the other authors may be excused for their rediscoveries. I am grateful to R. Kantowski for a copy of his thesis.

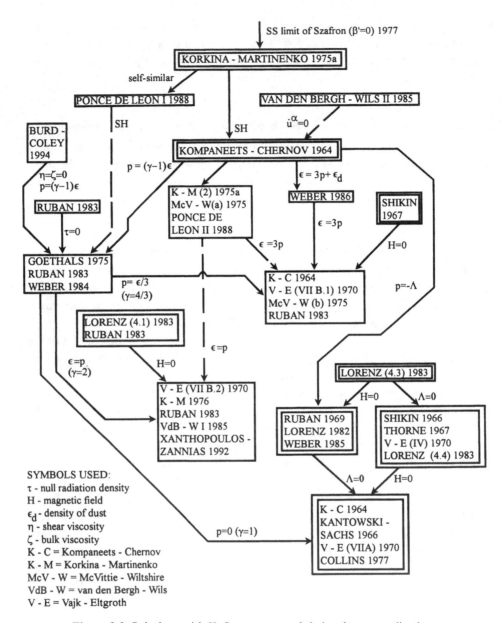

Figure 2.3. Solutions with K–S geometry and their other generalizations.

Over each of the subcases mentioned above, and over the whole class, several generalizations have been built that are described below.

2.11.2 The van den Bergh and Wils (1985) solution

Van den Bergh and Wils (1985) found the following perfect fluid solution:

$$ds^2 = e^{2\nu}dt^2 - 4\,e^{-6\nu}\nu_{,x}dr^2 - e^{-4\nu}(d\vartheta^2 + \sin^2\vartheta\,d\varphi^2), \qquad (2.11.1)$$

where ν is a function of:

$$x = t + r^2/r_0, \quad r_0 = \text{const}, \qquad (2.11.2)$$

and is defined by the equation:

$$2(e^{-\nu})_{,xx} = xe^{7\nu}/r_0 + e^{5\nu}. \qquad (2.11.3)$$

The solution has nonzero acceleration, and therefore it does not belong to Figure 2.1. The coordinates of (2.11.1) are comoving, so the acceleration will vanish when $\nu_{,r} = 0$, that is, $r_0 \to \infty$. The formulae for pressure and density are given in van den Bergh and Wils' paper.

Equation (2.11.3) implies a definite dependence of ν on x, and thus a definite equation of state; it is rather complicated and not of any form that is familiar in the literature. In the limit $r_0 \to \infty$, eq. (2.11.3) is simply incompatible with the linear barotropic equation of state $p = (\gamma - 1)\epsilon$. Hence, the limit $r_0 \to \infty$ of (2.11.1)–(2.11.3) is a subcase of the Kompaneets–Chernov (1964) class, different from all the other K–S-type solutions in Figure 2.3.

Equations (2.11.1)–(2.11.3) define what van den Bergh and Wils called class II. The other solutions in the paper are as follows. Class I is the Kompaneets–Chernov spacetime, and the example of a solution given explicitly has the equation of state $\epsilon = p$. The separable solutions and the class III solutions do not admit the limit $\dot{u}^{\alpha} = 0$, and so have neither FLRW nor K–S limits. The class IV solutions are described, but are not given explicitly, so they will not be taken into account.

2.11.3 Generalizations of the linear barotropic case, $p = (\gamma - 1)\epsilon$

Vajk and Eltgroth (1970, their section VA) investigated the Einstein–Maxwell equations for the linear barotropic perfect fluid with magnetic field, but produced no explicit solutions.

Ruban (1983) found and discussed a solution with the K–S geometry in which the source was a mixture of the linear barotropic perfect fluid and null radiation.

Ponce de Leon (1988, his case I) found an inhomogeneous self-similar solution that belongs to the Korkina–Martinenko (1975a) class. In the spatially homogeneous limit it goes over into a subcase of the linear barotropic Kompaneets–Chernov model.

Burd and Coley (1994) discussed qualitatively solutions of K–S geometry with bulk and shear viscosity. The pressure and both viscosities were assumed proportional to a power of matter-density. This case includes the perfect fluid with the linear barotropic equation of state as a subcase.

2.11.4 Generalizations of the "stiff fluid" case

The solutions with K–S geometry and a stiff fluid source were generalized in two different ways. Lorenz (1983) found a generalization to nonzero magnetic field (his section 4.1), and the same generalization was found by Ruban (1983). Korkina and Martinenko (1975a, their case 2) found a subcase of the Kompaneets–Chernov class that is not defined in any invariant way, but contains the $\epsilon = p$ and $\epsilon = 3p$ limits (both are severely restricted: with $\epsilon = p$, one of the scale factors is constant; with $\epsilon = 3p$ both pressure and matter-density are everywhere negative). The authors stated that this solution obeys the equation of state $p = (\gamma - 1)\epsilon$, but it does so only with further restrictions, and then is only a special case of such a solution. Properties of this solution were discussed in another paper by the same authors (Korkina and Martynenko 1975b).

The same solution was found independently by McVittie and Wiltshire (1975, their case a), and again by Ponce de Leon (1988, his case II). In fact, the expressions for the constants in the Ponce de Leon paper are incorrect, but this could just be a typographical error (the McVittie–Wiltshire version was verified by Ortocartan and shown to be correct).

2.11.5 Generalizations of the case $\epsilon = 3p$

Shikin (1967) found a generalization of this case to include free electromagnetic field.

2.11.6 Generalizations of the proper K–S model, $p = 0$

Lorenz (1983, his section 4.3) found a generalization of the proper K–S model to nonzero cosmological constant and free electromagnetic field. The limit $\Lambda = 0$ of that solution was found by Shikin (1966), Thorne (1967), Vajk and Eltgroth (1970, their section IV) and Lorenz (1983, his section 4.4). The limit of zero electromagnetic field of the Lorenz solution 4.3, that is, the K–S model with a cosmological constant, was already mentioned at the end of Section 2.11.1.

2.11.7 Other papers

Ray and Zimmerman (1977) investigated the problem, what sources are allowed for the K–S metric by the relativistic kinetic theory. It turned out that without any extra assumptions about the distribution function the source will in general be an imperfect fluid with zero heat-flow but nonzero anisotropic stress.

2.12 Solutions of the $\beta' \neq 0$ subfamily with a G_3/S_2 symmetry

This section refers to Figure 2.4. The most general set of solutions with a three-dimensional symmetry group acting on two-dimensional orbits was found by Ellis (1967, his case IIaii). They are dust solutions with cosmological constant. Ellis found all the solutions with a dust source and local rotational symmetry, that is, with a one-dimensional isotropy group at every point of the spacetime. There

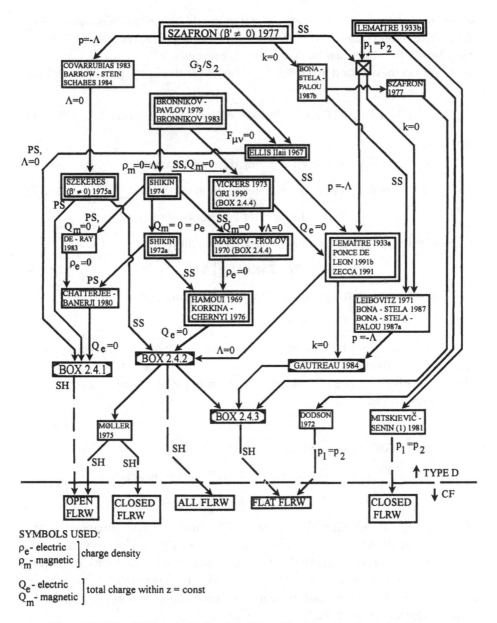

Figure 2.4. The Szafron ($\beta'\neq0$) subfamily and its generalizations with electro-magnetic field.

are several cases that emerge separately. Ellis' case I are stationary solutions, case III are Bianchi solutions. Case II exhausts all the situations in which $\kappa p = -\Lambda$ and the {t=const, z=const} surfaces in the Szafron spacetimes are symmetry orbits. Case IIai is the Einstein static Universe. Case IIb belongs to the $\beta'=0$ subfamily (only the IIbiii solution is given explicitly and displayed in Figure 2.1). Case IIaii will be described in this section. Note that the G_3/S_2 symmetry together with the assumptions $p=0 \neq \theta$ and $(\delta^\rho_{\ \alpha} - u^\rho u_\alpha)\ \theta,_\rho \neq 0$ uniquely define the Ellis classes IIaii and IIb. The other conditions of Szafron and Collins (1979, see Section 2.3) are automatically fulfilled when this symmetry is present.

Rewritten in the notation that will be used further in this book, the Ellis solutions are:

$$ds^2 = dt^2 - [\varepsilon + 2E(r)]^{-1} R,_r^2(t,r) dr^2 - R^2(t,r)[d\vartheta^2 + f^2(\vartheta) d\varphi^2], \qquad (2.12.1)$$

where:

$$\varepsilon = +1, \quad f(\vartheta) = \sin \vartheta \quad \text{for spherical symmetry,}$$

$$\varepsilon = 0, \quad f(\vartheta) = \vartheta \quad \text{for plane symmetry,} \qquad (2.12.2)$$

$$\varepsilon = -1, \quad f(\vartheta) = \sinh \vartheta \quad \text{for hyperbolic symmetry,}$$

$E(r)$ is an arbitrary function, and $R(t,r)$ is determined by:

$$R,_t^2 = 2E + 2M(r)/R + \tfrac{1}{3}\Lambda R^2, \qquad (2.12.3)$$

$M(r)$ being another arbitrary function.[16] The matter-density is:

$$\kappa\varepsilon = 2M,_r /(R^2 R,_r). \qquad (2.12.4)$$

Note that when $\varepsilon = +1$, that is, in the case of spherical symmetry, eqs. (2.12.1)–(2.12.4) follow from (1.4.1)–(1.4.6) with $p=0$ and $Gm/c^2 = M$.

The FLRW limit requires that $R(t,r)$ separates into $f(r)S(t)$; then the r coordinate can be chosen so that:

$$R(t,r) = r\, S(t), \qquad (2.12.5)$$

and then the FLRW limit follows when:

$$-2E/r^2 = k = \text{const} - \text{the FLRW curvature index,}$$
$$\qquad (2.12.6)$$
$$M(r)/r^3 = M_0 = \text{const} - \text{the FLRW mass integral,}$$

Note that only $\varepsilon = +1$ allows all three FLRW limits; the requirement of the Lorentzian signature forces the FLRW limit to be open ($k<0$) when $\varepsilon = 0$ or $\varepsilon = -1$.

From eqs. (2.1.1) and (2.1.11)–(2.1.17) we see that the Ellis solutions result from the Szafron metrics when $(e^\nu),_{,z} = 0$ and $\kappa p = -\Lambda$, the stereographic coordinates $\{x, y\}$ must then be transformed to the quasi-spherical coordinates $\{\vartheta, \varphi\}$, and the following identifications made:

$$\{z,\ \Phi, k,\ 4\,(AC - B_1^2 - B_2^2)\} = \{r, R,\ -2E,\ \varepsilon\}. \qquad (2.12.7)$$

This last identification is possible because $(AC - B_1^2 - B_2^2)$, when it is constant, can be scaled by scaling $\{x, z\}$, but its sign remains unchanged.

---

[16] Note that this $M(r)$ is *not* identical with $\mathcal{M}(z)$ of the Goode–Wainwright (1982b) representation (see Section 2.5). In the G–W approach, $\mathcal{M} = \text{const}$ in the FLRW limit, while here the FLRW limit is given by (2.12.6). This clash of notation could be prevented only by parting company with one or another important body of literature.

The plane symmetric subcase ($\varepsilon=0$) of the Ellis (1967) solutions, with $\Lambda=0$, was derived in the papers listed in Box 2.4.1. That subcase is at the same time the plane symmetric limit of the Szekeres $\beta' \neq 0$ solutions.

EARDLEY - LIANG - SACHS 1972
TOMITA 1975
HORSKY - LORENC - NOVOTNY 1977
TOMIMURA (b) 1978
NOVOTNY - HORSKY 1979

Box 2.4.1. Rediscoveries of the plane symmetric counterpart of the L–T model.

Eardley, Liang and Sachs (1972) gave a complete list of explicit solutions of (2.12.3) with $\Lambda=0=\varepsilon$ and discussed the properties of the Big Bang singularity. The analytic form of the solutions is in fact the same as in the spherically symmetric case (see below). In agreement with the remark following eq. (2.12.6), all the solutions are ever-expanding (hyperbolic); the different cases that arise correspond to the different signs of M. Eardley et $al.$ found that with $t_0(r)=$const, that is, for a simultaneous Big Bang, there is only the Big Bang singularity, while for $t_{0,r} \neq 0$ another singularity appears (which is the shell-crossing singularity). More exactly, when $t_0(r)=$const the two singularities coincide. Eardley et $al.$'s paper contains other results, among them a proposal to define the intrinsic geometry of a singular hypersurface. These will be described in Section 3.6. The authors' statement that the case $R_{,r}=0$ leads to the FLRW model is incorrect; this case is the Ellis IIbiii (1967) solution (see Figures 2.1 and 2.2).

Tomita (1975) found the solution (2.12.1)–(2.12.4) with $\Lambda=0=\varepsilon$ while discussing dust solutions in an anti-Newtonian approximation.

Horsky, Lorenc and Novotny (1977) rederived the solution (with $M<0$) and matched it to the Taub (1951) vacuum solution.

Tomimura (1978, her case b) rederived the solution once again and showed that the density contrast, defined as $\Delta\epsilon=\epsilon_{,z}/\epsilon$, tends to a fixed nonzero value as $t\rightarrow\infty$. This means "that at large times the dust has escaped from its own gravitational field and there is no longer anything causing the density distribution to change". Tomimura observed that the solution of (2.12.3) with $M<0=\Lambda$ has matter-density negative everywhere at all times, and so is not a reasonable cosmological model.

Finally, Novotny and Horsky (1979) elaborated further on the paper by Horsky, Lorenc and Novotny (1977) by finding the explicit solution for $M>0$ and showing that it can be matched to the Kasner vacuum solution. They pointed out that the case $R_{,r}=0$ requires a separate treatment, but incorrectly concluded that it leads to the Kasner solution only. In fact, it leads to the solution (2.6.18)–(2.6.19), which reduces to the plane symmetric subcase of the Kasner solution in the vacuum limit $X=0$.

Zakharov (1987) investigated the plane symmetric dust solutions of both sub-families (see Section 2.6 for the $\beta'=0$ subfamily), by developing them into linearized perturbations around the FLRW background in order to compare them with Lifshitz's (1946) perturbative results. Zakharov found that the solutions of the $\beta'\neq0$ subfamily contain both the growing and decaying modes of perturbation. Then, using the perturbative formulae, the author showed that a density perturbation $(\delta\epsilon/\epsilon)_r\cong10^{-4}$ formed at the time of recombination could have collapsed into a planar density singularity by the present day. This last conclusion obviously extends beyond the range of applicability of the linearized perturbation scheme.

The spherically symmetric limit of the Ellis IIaii solutions was first derived and discussed by Lemaître (1933a). This solution came to be known as the "Tolman" or "Tolman–Bondi" model, perhaps due to a lack of inclination of the majority of the scientific community to look up non-English-language literature. Tolman (1934) first made the model popular, and stated clearly that the solution he discussed was Lemaître's. That paper by Lemaître is sometimes cited as a description of the FLRW models, but this is a drastic underappreciation. The paper contains, among other things, the definition of mass for a spherically symmetric system (see our Section 1.4), a proof that the Schwarzschild horizon is not a singularity (by transforming the Schwarzschild solution to coordinates defined by clouds of freely falling particles) and a preliminary statement of a singularity theorem, illustrated by a Bianchi I model. Lemaître argued that neither the anisotropy of space nor pressure can prevent the collapse to a singularity.

To avoid confusion with the other papers of historical importance by Lemaître, we will call the spherically symmetric ($\varepsilon=+1$) case of (2.12.1)–(2.12.4) the "Lemaître–Tolman" (L–T) model.

Lemaître derived eq. (2.12.3) and discussed its solutions in terms of the Weierstrass elliptic functions, presenting the results in a form suitable for numerical investigation. He contemplated the model in order to describe the development of condensations in a FLRW background. He noticed that different shells of matter may coalesce, which makes the solution "inadmissible", and observed that in such a situation the assumption $p=0$ becomes invalid. It is invalid, too, when the shells fall back onto the centre.

Lemaître's paper contains one consideration that would not be acceptable today. Let us note it as a historical curiosity. In a model with positive Λ, the "cosmological repulsion" caused by Λ is equal to the gravitational attraction towards $r=0$ at a certain distance r_e. At this distance, the comoving shell $r=r_e$ experiences no force acting on it. Lemaître assumed that r_e equals the size of a galaxy cluster, and from this assumption and observational data on the sizes and masses of galaxies and galaxy clusters he tried to estimate the age of the Universe and the duration of one cycle of a recollapsing model.

Synge (1934) presented a qualitative discussion and graphical representation of the solutions of (2.12.3) with $\varepsilon=+1$. The discussion is concentrated on the range of variability of $R(t,r)$.

Tolman (1934) derived eq. (2.12.3) with $\varepsilon=+1$ and discussed several properties of its solutions. He observed that the arbitrary functions in the model allow for choosing an arbitrary distribution of initial mass-density and initial velocity and that the model includes the Einstein static model as a subcase and is therefore suitable for describing deformations of the latter. The main result was the observation that the L–T model demonstrates the instability of the Einstein and FLRW models against the growth of inhomogeneities: ". . . at those values of r where the density in the distorted model is different from that in the Friedmann model, there is at least an initial tendency for the differences to be emphasized ... in cases where condensation is taking place ... the discrepancies will continue until we reach a singular state involving infinite density or reach a breakdown in the simplified equations." Note that this is a prediction of formation of condensations *and voids*.

Tolman also observed that the L–T model can describe several FLRW zones of different densities connected by inhomogeneous transition zones. See a few other quotations in Appendix B.

Bondi (1947) provided a physical interpretation of the coordinates of (2.12.1) in the case $\varepsilon=+1$, and presented a detailed derivation of (2.12.1)–(2.12.4) in the same case. He showed that the L–T model has the following properties:

1. $R(t,r)$ is the luminosity distance from the centre of symmetry to the observer at r, and at the same time it is the area-distance.
2. The active gravitational mass $M(r)$ (watch for notation – ours is in conflict with Bondi's) is not in general equal to the sum of rest-masses $M_0(r)$ comprising the dust cloud, but is corrected for the *total* energy (not only kinetic) of the shells so that $d/dr(M-M_0)(r)$ has the same sign as $E(r)$.
3. The arbitrary functions $E(r)$ and $M(r)$ are related to the initial conditions (distributions of energy- and mass-density, respectively) specified on any hypersurface $t=t_i(r)$.
4. The total energy of the shell $r=r_1$, $E(r_1)$ is related to the curvature of the space $t=$const in the neighbourhood of $r=r_1$. In fact, the relation is opposite to Bondi's conclusion: the sign of the curvature is the sign of $(-E)$. The curvature varies with r.
5. When ϵ is a function of time only, the L–T model reduces to the FLRW dust.
6. The redshift of light emitted at the centre and received at $r=r_1$ is a sum of contributions from the velocity of the source, the velocity of the receiver and the accumulated redshifts in the gravitational fields of the matter shells overtaken by the travelling ray. It does not depend directly on the cosmological constant.
7. Rapidly collapsing matter may force a light ray passing through it to move inward (this was in fact a prediction of black hole formation; see further comment in Section 3.4).
8. If $R_{,r}(r_1)=0$, then different shells of dust stick together at $r=r_1$. The model is then valid up to the smallest t_1 at which $R_{,r}(r_1,t_1)=0$ occurs, and for $t>t_1$ it is valid only in the region $r<r_1$.

9. Collapse back onto the centre leads to a singularity: either the dust particles fly through the centre and collide with the infalling particles, or they stick together at the centre to form a point mass.
10. If initially there is an empty region around the centre, then it will expand (see a further comment in Section 3.1).

Bondi thus initiated several lines of research, followed by other authors later.

Omer (1965) showed how (2.12.3) may be solved in terms of the Weierstrass elliptic functions and presented a graphical classification of solutions with respect to the values of $M(r)$ and $E(r)$.

Sakoto (1977) proved that in the L–T model with arbitrary Λ the "closed conformal infinitesimal transformation" (meaning a nontrivial conformal symmetry) exists for a nonzero matter-density only if the spacetime is homogeneous (and thus reduces to a FLRW model).

Ponce de Leon (1991b) derived the model (2.12.1)–(2.12.4) with $\varepsilon = +1$ as a starting point for a discussion of self-similar dust models.

Finally, Zecca (1991) derived (2.12.1)–(2.12.4) with $\varepsilon = +1$, and then discussed the solutions of (2.12.3) in terms of the Weierstrass elliptic functions, and the null geodesic equation for a collapsing model in the linear approximation; the small parameter was the distance (measured by R) from the final singularity $R = 0$. The discussion seems to indicate that a black hole forms before the collapse is completed. The L–T model with $\Lambda = 0$ was investigated in much detail in many other papers (see below), Zecca's paper does not, unfortunately, elucidate the influence of Λ on its properties.

The limit $\Lambda = 0$ of the L–T model has (so far) been rederived in at least 20 papers and books (see Box 2.4.2). It is the spherically symmetric limit of the Szekeres $\beta' \neq 0$ solutions. Some of the works listed in Box 2.4.2 indicate that their authors knew about Tolman's paper, but they either gave no reference to it or did not indicate that they had redone Tolman's work. They thus fulfilled the criteria adopted in this book for classifying a paper as a rediscovery.

The solutions of (2.12.3) with $\Lambda = 0$ are:

$$R(t,r) = -\frac{M}{2E}(1 - \cos\,\eta),$$

$$\eta - \sin\,\eta = \frac{(-2E)^{3/2}}{M}[t - t_0(r)] \quad \text{when } E(r) < 0,$$

(2.12.8)

$$R(t,r) = \{\tfrac{9}{2}M(r)[t - t_0(r)]^2\}^{1/3} \quad \text{when } E(r) = 0,$$

(2.12.9)

$$R(t,r) = \frac{M}{2E}(\cosh\,\eta - 1),$$

$$\sinh\,\eta - \eta = \frac{(2E)^{3/2}}{M}[t - t_0(r)] \quad \text{when } E(r) > 0,$$

(2.12.10)

where η is a parameter and $t_0(r)$ is an arbitrary function – the "bang time", which determines the local time of the initial singularity $t = t_0$; see Section 2.4.2. The

DATT I 1938
LANDAU - LIFSHITZ 1948
LANDAU - LIFSHITZ 1962
NOVIKOV 1963
BANERJEE 1967
PEEBLES 1967
BARNES - WHITROW 1970
CARR - HAWKING 1974
PAPAPETROU 1974
LIGHTMAN - PRESS - PRICE - TEUKOLSKY 1975
HENRIKSEN - DE ROBERTIS 1980
MAEDA - SASAKI - SATO 1983
SATO - MAEDA 1983
SATO 1984
KRISHNA RAO - ANNAPURNA 1986
LYNDEN-BELL 1987
MESZAROS - VANYSEK 1988
KRISHNA RAO 1990
LIU 1991
FENG - MO - RUFFINI 1991

Box 2.4.2. Rediscoveries of the L–T model with $\Lambda=0$.

solutions (2.12.8)–(2.12.10) have the same algebraic form as the FLRW solutions
with $p=0$, and the FLRW limit is given by (2.12.5)–(2.12.6). The subcase
M=const is a vacuum solution in each case; it is the Schwarzschild solution represented in the Lemaître coordinates, connected with a test cloud of dust. The sign
of E in this limit determines whether a given shell of the test cloud has its initial
energy smaller than, equal to or greater than the escape energy. The fact that the
FLRW and Schwarzschild solutions are both subcases of L–T was used in several
papers to consider patchworks composed of these solutions.

As can be seen from (2.12.1) and (2.12.4), if the equation $R_{,r}=0$ has solutions
$r=r_s$, then the locus of these solutions is in general singular: at $r=r_s$ the metric com-
ponent $g_{rr}(r_s)=0$, so $\det(g_{\alpha\beta}(r_s))=0$ and $\epsilon\xrightarrow[r\to r_s]{}\infty$. These are the shell-crossing singularities, at which different shells r=const collide and comoving coordinates become
inadmissible because they assign different values of r to the same points in spacetime. However, there will be no singularity at $R_{,r}=0$ if the solutions $r=r_s$ are time-
independent (i.e. comoving), while the functions $M_{,r}$ and $(1+2E)^{1/2}$ have at $r=r_s$
zeros of the same order as does $R_{,r}$. Then the points with $r=r_s$ are just the locus of
a local minimum or maximum of $R(t,r)$ at each constant t. Precisely this happens in
the closed FLRW model at the equator of each three-sphere t=const. In the coordinates of (1.3.1), the equator is at $r=(2/k)^{1/2}$; the coordinates of (1.3.8) and (1.3.9)

do not cover it. In a general L–T model, the set defined by $R_{,r}=0=M_{,r}=2E+1$ (if it exists) is the "neck" described by Barnes (1970, see Section 3.4) or the "bottleneck" of Suto, Sato and Sato (1984b, see Section 3.1); no light rays will pass through it when they reach it after the collapse has begun. The geometry of a L–T spacetime with such a neck was described in detail by Hellaby (1987), see Section 3.8.

On the basis of the solutions (2.12.8)–(2.12.10), a monumental body of literature was written in which the geometrical, physical and astrophysical implications of inhomogeneities in cosmological matter distribution were discussed. These papers, sorted by subject, will be described in Chapter 3. The following is a list of comments on the entries from Box 2.4.2.

Datt (1938) derived spherically symmetric dust solutions with $\Lambda=0$. Class I (this is our labelling, in Datt's paper this class is presented in section 5) is the L–T model (2.12.8)–(2.12.10). Class II (case e in section 6 of the paper) will be mentioned in our next section. Class III (Datt's section 7) is the spherically symmetric limit of the Szekeres $\beta'=0$ class and is described in our Section 2.6. Datt's section 6 contains more particular solutions: (a) is the Minkowski spacetime in spherical coordinates, (b) is the Minkowski spacetime in more elaborate coordinates, (c) is the flat FLRW model, (d) is a simple coordinate transform of (b).

The descriptions of the L–T model in the book by Landau and Lifshitz (1948 – exercise to section 96, 1962 – exercise to section 97, and all later editions) are just brief introductions. They are mentioned here because the book gives no references, and in the 1948 edition the solution is assigned to Datt. Later editions gradually extended the description.

Novikov (1963) derived (2.12.8) (i.e. only the case $E<0$) and specialized it so as to represent the closed FLRW model matched to the Schwarzschild solution. He concluded that an observer in the expanding FLRW region has a chance to send out a signal into the Schwarzschild region and receive an answer during the recollapse, provided the outer observer is not too far from the horizon. The observer in the Schwarzschild region has no chance of obtaining a response from the FLRW region to any of his signals.

Banerjee (1967) derived (2.12.8)–(2.12.10) and formulated the criterion for the occurrence of a shell-crossing singularity in the case $E=0$ (it is $M_{,r}/(2M)=t_{0,r}/(t-t_0)$). Next, Banerjee matched the L–T model to the Schwarzschild solution and derived the formula for the redshift of light emitted radially from the surface of the L–T cloud and received by an observer in the Schwarzschild region.

Peebles (1967) rederived (2.12.1)–(2.12.3) in the case $\Lambda=0$, $\varepsilon=1$ in a peculiar parametrization in which $2E=-r^2/\mathcal{R}^2(r)$ (this seems to force E to be negative) and presented the solution (2.12.8). However, the cases $E=0$ and $E>0$ can be recovered by merely renaming $r^2/\mathcal{R}^2=-2E$. Peebles' choice $2M(r)=Ar^3$, $A=$const, is just a specific choice of the coordinate r, permissible whenever $0\neq M_{,r}$ has a constant sign. Peebles used this solution to study the development of perturbations of the FLRW background introduced at an early time. He concluded that if an initially large perturbation were to decay so as to produce a nearly FLRW Universe now, then the initial expansion rate and density would have to be fine-tuned by $3R_{,t}^2/R^2 = \kappa\epsilon$. This makes smoothing out inhomogeneities by evolution improbable. The other conclusion is that small perturbations may develop into galaxies, but no numbers are given.

Barnes and Whitrow (1970) derived (2.12.3)–(2.12.4) for the case $\epsilon = +1$, $\Lambda = 0$ and observed that it is the same as the Newtonian equation of pressureless collapse.

Carr and Hawking (1974) derived the equations (2.12.3) and (2.12.8)–(2.12.10) as an intermediate stage in obtaining a self-similar dust solution. This paper will be mentioned once again in the following section.

In the book by Papapetrou (1974, pp. 99–103), eq. (2.12.3) is derived and interpreted and the solution (2.12.9) ($E=0$) is presented, but the physical discussion is partly incorrect. Papapetrou claims that the function $t_0(r)$ has no physical meaning (see Chapter 3 for counterarguments; see also Section 2.5, where its gradient is linked to the growing mode of perturbation of the FLRW background), and that $t = t_0(r)$ is the world-line of the centre of symmetry (this is the locus of the Big Bang, which may be timelike or spacelike or null; in the first case only may it be said to coincide with the centre of symmetry).

In the book by Lightman, Press, Price and Teukolsky (1975, pp. 98 and 460–461), the derivation of the $\Lambda = 0$ L–T model from equations corresponding to our Section 1.4 is one subject of an exercise, and the interpretation of the arbitrary functions is presented.

Henriksen and de Robertis (1980) derived the model and then used it for a study of structure formation in the Universe. This paper will be mentioned again in Section 3.1.

The papers by Maeda, Sasaki and Sato (1983), Sato and Maeda (1983) and Sato (1984) all contain derivations of (2.12.3) and (2.12.8)–(2.12.10). They are parts of an important series of papers that developed one of the most beautiful pieces of physics based on the L–T model: the description of formation of voids in the Universe; see Section 3.1.

Krishna Rao and Annapurna (1986) derived (2.12.3) from considerations similar to our Section 1.4 and presented the solution (2.12.9), then showed how it can be matched to the Schwarzschild solution.

Lynden-Bell (1987) derived (2.12.3) in passing, in an article devoted to accretion and collapse in Newtonian hydrodynamics.

Meszaros and Vanysek (1988) presented (2.12.3) and (2.12.8)–(2.12.10) as the model to replace the FLRW models for describing inhomogeneities in the Universe. The paper is notable for its astrophysical arguments against the cosmological principle (see a quotation in Appendix B).

The paper by Krishna Rao (1990) is a follow-up on Krishna Rao and Annapurna (1986) in which (2.12.8) and (2.12.10) are presented and discussed.

Liu (1991) derived (2.12.3) in order to compare this model with linearized perturbations of the FLRW background.

Feng, Mo and Ruffini (1991) derived (2.12.3) and calculated the influence of the inhomogeneity on the Hubble parameter and peculiar velocities of matter in the Universe. An elaborate set of numerical results is presented in the paper which emphasizes the distortions introduced into the interpretation of cosmological observations by the L–T inhomogeneity.

Stoeger, Ellis and Nel (1992) tested their observational cosmology program on the L–T model. The basis of the program is the introduction of coordinates connected with the observer's past null cone. The first problem then was to transform the L–T solution to such coordinates; this requires integration of the equations of null geodesics – an impossible task in this case. Therefore, the authors decided to rederive the model from the Einstein equations using the observational coordinates from the beginning. Stoeger *et al.*'s paper is thus not a rediscovery. The L–T metric in these coordinates is given (in the original notation) by:

$$ds^2 = A^2(\eta)(-dw^2 + 2dw\,dy) + C^2(w,y)(d\vartheta^2 + \sin^2\vartheta\,d\varphi^2), \qquad (2.12.11)$$

where $\eta \overset{\text{def}}{=} w - y$, $A(\eta)$ is an arbitrary function related to the observed redshift $z(\eta)$ by $A(\eta) = A_0/[1 + z(\eta)]$, $A_0 = $ const and C is given by:

$$C = A(\eta) \int W(y) dy, \qquad (2.12.12)$$

$W(y)$ being another arbitrary function; it is the analogue of the function $(1 + 2E)$ from the usual representation. The matter-density is:

$$\kappa\epsilon = \epsilon_0(y)/(AC^2), \qquad (2.12.13)$$

and $\epsilon_0(y)$ is an arbitrary function. The three arbitrary functions are connected by a rather complicated equation (see the Stoeger *et al.* paper, eq. (44b)) which is an image, in the observational coordinates, of the connection that would hold between ϵ_0, $E(r)$ and $M(r)$ in the comoving coordinates. Stoeger *et al.* explained how the functions are related to observable quantities, but the connection is not simple, and readers are advised to refer to the original paper. The function $z(y)|_{w=\text{const}}$ can in principle be read out from observations, but in practice the procedure would require cosmological parallax distances to be measured and this is unrealistic with present-day technology. For comparison, the FLRW metric in the observational coordinates is:

$$ds^2 = R^2(\eta)[dw^2 - 2dw\,dy - k^{-2}\sin^2(ky)(d\vartheta^2 + \sin^2\vartheta\,d\varphi^2)], \qquad (2.12.14)$$

where k is the curvature index and $R(\eta)$ is the scale factor.

Another paper by Lemaître (1933b) belongs in this section. In it, the author considered the Einstein equations for a spherically symmetric anisotropic fluid, that is, one that has different eigenvalues of the energy-momentum tensor in the radial direction and in the azimuthal directions. Adapted to our notation, the results are:

$$ds^2 = dt^2 - (R_{,r}^2/\cos^2 r)\,dr^2 - R^2(t,r)(d\vartheta^2 + \sin^2\vartheta\,d\varphi^2), \qquad (2.12.15)$$

where the $\cos^2 r$ arises by the choice of r, and:

$$R_{,t}^2 = -\sin^2 r + 2m/R + \tfrac{1}{3}\Lambda R^2, \tag{2.12.16}$$

$$\kappa \epsilon R^2 R_{,r} = m_{,r}, \tag{2.12.17}$$

$$\kappa p_\perp R^2 R_{,t} = -m_{,t}, \tag{2.12.18}$$

$$(p_{||} - p_\perp)R_{,r} = \tfrac{1}{2}Rp_{\perp,r}, \tag{2.12.19}$$

where $p_\perp$ is the radial pressure and $p_{||}$ is the azimuthal pressure. The set (2.12.16)–(2.12.19) is an extension of the approach presented in our Section 1.4 to anisotropic pressure (in the special case $\dot{u}^\alpha = 0$), and allows for a quick derivation of (2.12.1)–(2.12.4).

Siemieniec–Oziębło and Klimek (1978) were well on the way to deriving the subcase $p_{||} = p_\perp$ of the Lemaître (1933b) equations. They considered Einstein's equations in a spherically symmetric spacetime with a perfect fluid source and $p = p(t)$ in comoving coordinates. They would have rediscovered the $p \neq 0$ generalization of the L–T model, but in order to be able to present an explicit formula, they assumed that $p \ll \epsilon$ and derived an approximate solution.

2.13 Other subcases of the Szafron β′ ≠0 spacetime

Dodson (1972) considered the metric:

$$ds^2 = dt^2 - R^2(t)e^{-\lambda r}(dx^2 + dy^2 + dz^2). \tag{2.13.1}$$

This is a simple subcase of the Lemaître (1933b) class, and it becomes the general flat FLRW model when $\lambda = 0$. Dodson found and discussed in detail the solutions of the Einstein equations for (2.13.1), with the source being a fluid with anisotropic pressure, emphasizing the differences between (2.13.1) and a homogeneous model. Since this class of solutions is rather simple and no invariant definition of it was given in the paper, it is difficult to judge how generic the results are. Its inhomogeneity is generated by the anisotropy of pressure and vanishes together with the anisotropy.

The paper by Mitskievič and Senin (1981) was a preliminary step in obtaining the Senin (1982) solution (see Section 2.8). The authors obtained three solutions whose three-spaces $t = $ const had the local geometry of a torus. All three have FLRW limits and will be mentioned in this book, but only "Model 1" belongs to the Szafron family. Its metric is:

$$ds^2 = dt^2 - R^2(t)[dr^2 + (\sin r + A)^2(d\vartheta^2 + \sin^2 \vartheta \, d\varphi^2)], \tag{2.13.2}$$

where R obeys:

$$2RR_{,tt} + R_{,t}^2 + 1 = 0, \tag{2.13.3}$$

and A is an arbitrary constant. This solution is another simple subcase of the Lemaître (1933b) metric; compare eq. (2.12.15). The pressure anisotropy vanishes when $A=0$; the metric given by (2.13.2)–(2.13.3) then reduces to the closed FLRW dust. Note that the anisotropic fluids generating (2.13.1) and (2.13.2) have zero rotation, shear and acceleration, yet the metrics (2.13.1) and (2.13.2) are not FLRW because of anisotropy in pressure (see remarks in Section 1.3.2).

Bona, Stela and Palou (1987b) worked out the subcase of the Szafron $\beta' \neq 0$ subfamily in which the spaces $t=$const are flat (this is the counterpart of the solution from the same paper discussed in our Section 2.8). In this subfamily, it happens when $k(z)=0$, and the metric represented by (2.1.11)–(2.1.15) can then be parametrized so that in addition $h(z)=1$ and:

$$\Phi(t,z)=[C_1(z)+C_2(z)f(t)]^{2/3}f_{,t}^{-1/3} \qquad (2.13.4)$$

$$\kappa p(t)=\tfrac{2}{3}f_{,ttt}/f_{,t}-(f_{,tt}/f_{,t})^2. \qquad (2.13.5)$$

This solution reproduces the flat FLRW limit in its full generality when $C_1 C_{2,z}-C_{1,z}C_2=0$; the limit results in the form (1.3.16)–(1.3.18) with $k=0$ and $\varepsilon=f=W=1$.

Sussman (1992) investigated in detail the Bona–Stela–Palou (1987b) subcase, interpreted as a mixture of two fluids. The components of the mixture are: a homogeneous perfect fluid with $p=(\gamma-1)\epsilon$, whose dynamics is assumed to be the same as the FLRW dynamics with the same equation of state, and inhomogeneous dust. Since one of the equations is analogous to the Schrödinger equation, Sussman showed how the WKB technique can be applied to obtain approximate solutions with an arbitrary equation of state (it must then be assumed that $p(t)$ is slowly varying). Examples of exact solutions are presented and are in general specified in terms of hypergeometric functions. Sussman showed how the concept of a regular centre, defined for spherically symmetric spacetimes, can be generalized to the $\beta' \neq 0$ Szafron spacetimes: a regular centre exists if there exists a privileged observer for whom $\sigma_{\alpha\beta}(r_0)=0$ so that all his or her observations are isotropic. The K–S models and the $\beta'=0$ Szafron models generalizing them do not have a regular centre. Sussman also showed that a mixture of FLRW radiation with $k=0$ and inhomogeneous dust can be matched to a pure FLRW radiation with $k=0$, that is, the inhomogeneous dust density can vanish on a timelike hypersurface. A matching is possible for other FLRW backgrounds, but then the FLRW regions inside and outside have different dynamics. There exist analogues of the Big Bang and the shell-crossing singularities, and these are similar to the ordinary Szekeres singularities investigated by Goode and Wainwright (1982b). However, an additional singularity may exist in the mixture where the explosion of the homogeneous fluid component occurs. There, the density of the inhomogeneous dust diverges, too. In the Goode–Wainwright (1982b) classification it corresponds to a point singularity. Sussman discussed how different types of singularity may cause one of the densities to become negative during a period of evolution, and showed that parameters of the model can be chosen so that it evolves from an inhomogeneous Big Bang asymptotically towards FLRW, without shell-crossings. However, the homogeneous component has finite density and pressure at the Big Bang, which is difficult to understand. The work of Sussman is strictly analogous to the work of Lima and Tiomno (1988, see Section 2.8), where the $\beta'=0$ subfamily of the Szafron spacetimes was interpreted in the same way. However, details of the investigation differ in each paper.

Szafron (1977), in addition to having introduced the whole class, presented one particular example of a solution of the $\beta'\neq0$ subfamily. It is the exact counterpart of the Szafron–Wainwright (1977) solution from the $\beta'=0$ subfamily, and, just like the latter, was meant to be a test example for investigating the properties of a new kind of solution. The solution has the metric (2.1.1) with:

$$e^\beta = Q^{2/3}(t,z)e^\nu, \quad e^\alpha = e^{-\nu}(e^\beta)_{,z}, \tag{2.13.6}$$
$$Q(t,z)=C_1(z)t^{1-q}+C_2(z)t^q,$$

where e^ν is given by (2.1.13), q is an arbitrary constant and $C_1(z)$, $C_2(z)$ are arbitrary functions. Equation (2.1.15) still applies, but here $h=1$ and $k=0$. The pressure is:

$$\kappa p=\tfrac{4}{3}q(1-q)/t^2. \tag{2.13.7}$$

The (flat only) FLRW limit with a definite equation of state results when either $q=1/2$ or $C_1C_{2,z}-C_{1,z}C_2=0$; it is then of the same form as the FLRW limit of the Bona, Stela and Palou (1987b) solution. Szafron studied the asymptotic behaviour of the model as $t\to\infty$ and at the singularity $t=0$; the list of results is given in the paper. Szafron seems to imply that the solution (2.13.6)–(2.13.7) always has a symmetry group of dimension at least one, but this may be just an oversight in presentation.

Like the Szafron–Wainwright (1977) solution, the solution (2.13.6)–(2.13.7) is not the most general one with $p=p_0t^{-2}$. Both these solutions arise when $p_0<1/3$. When $p_0=1/3$, the solution is $Q=[C_1(z)+C_2(z)\ln t]t^{1/2}$ (this latter was mentioned by Bona, Stela and Palou 1987b); when $p_0>1/3$, the solution is $Q=[C_1(z)\cos(b\ln t)+C_2(z)\sin(b\ln t)]t^{1/2}$, where $b=\tfrac{1}{2}(3p_0-1)^{1/2}$. As mentioned in the comments on the Szafron–Wainwright solution (Section 2.8), this last case defines a recollapsing Universe with unequal cycles of expansion and recollapse.

The solution (2.13.6)–(2.13.7) is the subcase of the Bona, Stela and Palou (1987b) class defined by $f(t)=t^{2q-1}$.

Leibovitz (1971) derived a few spherically symmetric solutions, of which only one is within the scope of this book. The solution from his section 1 is static. The one from his section 2 has a source of unknown interpretation, and, if specialized to be a perfect fluid, it becomes just the flat FLRW dust. The two solutions from Leibovitz's section 4 have no FLRW limit; the matter-density becomes spatially homogeneous only when the source degenerates into the Λ-term. We shall consider here the solution from the author's section 3 which, transcribed to our usual notation, has the metric (2.12.1) with $\varepsilon=+1$, $f(\vartheta)=\sin\vartheta$ and:

$$R=[F(r)\tau(t)+G(r)]^{2/3}/\tau_{,t}^{1/3}, \tag{2.13.8}$$

where $F(r)$, $G(r)$ and $\tau(t)$ are arbitrary functions, and the pressure is:

$$\kappa p = -(\tau_{,tt}/\tau_{,t})^2 + \tfrac{2}{3}\tau_{,ttt}/\tau_{,t}. \qquad (2.13.9)$$

The function $\tau(t)$ is simply a way of parametrizing the pressure, convenient for integrating the Einstein equations. On closer inspection, (2.13.8)–(2.13.9) turns out to be the most general solution in the Szafron $\beta' \neq 0$ class with $k=0$ and spherical symmetry. Hence, the Leibovitz solution is the spherically symmetric limit of the Bona-Stela–Palou (1987b) solution (2.13.4)–(2.13.5). It is also the subcase with isotropic pressure and $E(r)=0$ of the Lemaître (1933b) class (the coordinates have to be transformed first so as to allow the limit $E=0$). It reproduces the flat FLRW model in its full generality when $F_{,r}G - FG_{,r} = 0$.

Bona and Stela (1987) considered a "Swiss cheese" model composed of the flat FLRW background into which spherical patches of an inhomogeneous medium were inserted and in which $p \neq 0$ everywhere. The solution they chose for the inhomogeneous patches is exactly the Leibovitz (1971) solution; their parametrization of $\tau(t)$ in (2.13.8)–(2.13.9) is different but equivalent.

Bona and Stela's paper contains some physical discussion. When $F(r)=$const, the inhomogeneous portion of the spacetime evolves asymptotically towards the Schwarzschild solution, while in the FLRW region the pressure tends to zero, that is, the whole configuration approaches the Einstein–Straus (1945 and 1946) "Swiss cheese". When $F(r)=r$ and $G(r)=\beta r\,[1-(r/r_0)^2]$, where β and r_0 are constants, the whole configuration tends asymptotically to a spatially homogeneous one, that is, the inhomogeneity in this case blends into the background.

In another paper, Bona, Stela and Palou (1987a) claim to have shown that (2.13.4)–(2.13.5) follows uniquely from the following assumptions: (1) the spacetime is spherically symmetric; (2) the source is a perfect fluid; (3) the hypersurfaces orthogonal to the fluid flow are flat. In fact, one more assumption is needed: geodesic motion of the fluid; otherwise some of the Stephani–Wolf (1985) solutions will be counterexamples (see Section 7.5).

Gautreau (1984) found the limit $\kappa p = -\Lambda$ of the Leibovitz (1971) solution. His paper is an important contribution to the problem of influence of cosmic expansion on planetary orbits, and will be discussed again in Section 3.3. Gautreau considered a condensation of fixed mass and comoving boundary embedded in an inhomogeneous spherically symmetric dust Universe. The properties of the condensation are not discussed; it is considered to be a model of a star in whose exterior gravitational field the planets move. The exterior gravitational field is assumed to be nonvacuum (part of a cosmological model). From the assumptions made about it one can see that it must be the L–T model or a subcase thereof. However, the solution is given in curvature coordinates and the transformation to comoving coordinates is nontrivial, therefore it will be presented here.

The metric (given at the end of Gautreau's section 4) is:

$$ds^2 = dt^2 - \psi^2 - R^2(d\vartheta^2 + \sin^2\vartheta\,d\varphi^2), \qquad (2.13.10)$$

where R is the radial coordinate, and ψ is the following differential form:

$$\psi = dR - \varepsilon[(2\mu + Z)/R + \tfrac{1}{3}\Lambda R^2]^{1/2}dt, \qquad (2.13.11)$$

where $\varepsilon = \pm 1$, Λ is the cosmological constant, μ is an arbitrary constant and $Z(t,R)$ is given by:

$$R^{1/2}\frac{\partial Z}{\partial t} + \varepsilon(Z + 2\mu + \tfrac{1}{3}\Lambda R^3)^{1/2}\frac{\partial Z}{\partial R} = 0. \qquad (2.13.12)$$

(Equation (2.13.12) is given in the paper only in the limit $\Lambda = 0$, but it is easy to reconstruct the above from the author's reasoning.) The function Z is related to the mass-density $\epsilon(t,R)$ by:

$$Z(t,R) = \kappa \int_0^R \epsilon r'^2 dr'. \qquad (2.13.13)$$

In order to transform (2.13.10) to comoving coordinates we observe that the form ψ, depending on two variables only, must have an integrating factor. Denoting this factor by $1/\chi$ we conclude that

$$\psi/\chi = dr \qquad (2.13.14)$$

is a perfect differential of a certain function $r(t,R)$. Now we choose (t,r) as the new coordinates. Then, from (2.13.12), it follows that in the new coordinates $\partial Z/\partial t = 0$, that is, $Z = Z(r)$, while (2.13.11) and (2.13.14) imply:

$$R_{,r} = \chi, \qquad (2.13.15)$$

$$R_{,t} = \varepsilon[(2\mu + Z)/R + \tfrac{1}{3}\Lambda R^2]^{1/2}, \qquad (2.13.16)$$

R now being a function of t and r. Equation (2.13.16) is now seen to be the subcase $E = 0$, $M = \mu + Z/2$ of (2.12.3), while (2.13.14) and (2.13.15) imply that, in the (t,r) coordinates, the metric component $g_{rr} = -R_{,r}^2$, as it should in a L–T model with $E = 0$. Equation (2.13.14) implies the integrability condition $(\chi^{-1})_{,t} = -\varepsilon\{[(2\mu + Z)/R + \tfrac{1}{3}\Lambda R^2]^{1/2}/\chi\}_{,R}$, and this is at the same time the integrability condition of (2.13.15)–(2.13.16).

The Gautreau solution is thus the limit $E = 0$ of the L–T model and at the same time the limit $\kappa p = -\Lambda$ of the Leibovitz (1971) spacetime.

Box 2.4.3 contains a substructure of interrelated solutions. Its largest element is the list of repeated discoveries of the same self-similar dust solution. A metric g is called self-similar (or homothetic) when there exists a vector field k such that:

$$\pounds_k g_{\alpha\beta} = 2g_{\alpha\beta} \qquad (2.13.17)$$

(the factor 2 on the right results from a conventional scaling). The generators of self-similarity form a Lie algebra, but it is easy to see that the difference of any pair of linearly independent generators k and l will generate an isometry, $\pounds_{k-l}g_{\alpha\beta} = 0$. Hence, the basis of any algebra of self-similar transformations can be chosen so that only one basis vector will generate (2.13.17) while all the others will be Killing vectors. The general theory of self-similar spacetimes was developed by Eardley (1974a); the

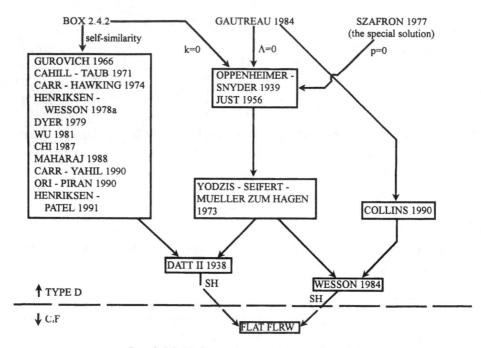

Box 2.4.3. Various subcases of the L–T model.

singularities in self-similar spacetimes were studied by Wu (1982). In particular, much attention was devoted to perfect fluid spacetimes in which the group of self-similarities is three-dimensional and acts simply transitively on hypersurfaces orthogonal to the fluid flow. Such groups can be classified in a way similar to the Bianchi classification (Eardley 1974a, Wu 1981), but the number of types is larger because the presence or absence of a generator of (2.13.17) provides an additional possibility in each type. In fact, the types VIII and IX do not exist with a generator of (2.13.17). Those so-called spatially self-similar spacetimes contain inhomogeneous solutions, but the inhomogeneity is of a very simple type. For work on a distinct but closely related class, readers interested in the subject may wish to consult the papers by Hewitt, Wainwright and Goode (1988) and Hewitt, Wainwright and Glaum (1991). In these, the authors qualitatively analysed the Einstein equations for perfect fluid metrics that have a two-dimensional symmetry group G_2, are homothetic with respect to the transformations generated by the generator collinear with the velocity field u^α, and whose sources obey the equation of state $p=(\gamma-1)\epsilon$. The Einstein equations in such a case reduce to an autonomous set of ordinary differential equations. In the first paper, Hewitt *et al.* presented one explicit solution, but it has no FLRW limit.

The authors who considered self-similar spacetimes often wanted to provide explicit examples of solutions, and in searching for them often turned to spherically symmetric metrics. Several of them considered spherically symmetric self-similar dust, thus ensuring that the example would belong to the L–T class. Self-similarity requires that $R(t,r)=rS(t/r)$ and $E(r)=$const in (2.12.1) so that $R_{,r}$ depends only on (t/r), $\Lambda=0$ and $M(r)/r=$const in (2.12.3). Since in a FLRW limit $2E=-kr^2$, it follows that the self-similar spherically symmetric dust solutions have $E=0$ in the FLRW limit and can reproduce only the flat FLRW dust.

The same solution, in various parametrizations, was obtained by the following authors:

Gurovich (1966) considered three solutions. The second of them is just a FLRW model, the third one does not allow the limit $\dot{u}^\alpha=0$. The first solution is the subcase of the L–T model specified above. Gurovich noted that this solution was not new.

The paper by Cahill and Taub (1971) is in large part devoted to the problem of the existence of spherically symmetric perfect fluid solutions obeying (2.13.17). As an example, the special L–T model is given.

Carr and Hawking (1974), in addition to deriving the case mentioned above, discussed qualitatively a self-similar spherically symmetric model obeying $\epsilon=3p$, without deriving an explicit solution.

Henriksen and Wesson (1978a) considered three spherically symmetric self-similar solutions. The first of them is static, the second one, with $\epsilon=p$, was discussed only qualitatively. The third one is the same as the others given above. In a few other papers, by Henriksen and Wesson (1978b), Wesson (1978b) and Wesson (1979), possible astrophysical implications of this solution were discussed. Wesson (1978a) did the same for the subcase $E=0$. The solution is also briefly discussed in the book by Wesson (1978c, his section 7.3); this last reference contains a rather large bibliography on cosmology, in which special attention is devoted to cosmological models in non-Einsteinian theories of gravitation. The paper by Henriksen (1989) should be mentioned in this connection. In it, the author argues that every L–T model with $\Lambda=0$ is self-similar in the following sense: one can always define a variable $\xi(t,r)$ such that (R/r) is a function of ξ only, but this property cannot in general be expressed in terms of an algebraic condition imposed on the Lie derivative of the metric, as in (2.13.17).

Dyer (1979) derived the solution and used it to investigate the properties of a hierarchical Universe.

Wu (1981) discussed kinematical and dynamical properties of self-similar cosmological models extensively, and devoted special attention to those that have $p=0$ and a G_3/S_2 symmetry group in addition. One of the solutions he obtained is the same as discussed here. An important implication of Wu's paper is that the plane- and hyperbolically symmetric self-similar dust solutions have no FLRW limit. The three solutions given for self-similar dust with a G_3/S_2 symmetry in the $\beta'=0$ subfamily are not within the scope of this book: the first and third of them are of Bianchi type I, the second is a vacuum solution.

Chi (1987) derived the subcase $E=0$, advertising it as a new solution, even though he cited Henriksen and Wesson (1978a) where the solution with arbitrary $E=$const was presented.

Maharaj (1988) supplied the explicit solutions for $0\neq E=$const and pointed out that they are all in the L–T class with $\Lambda=0$.

Carr and Yahil (1990) discussed spherically symmetric self-similar perfect fluid solutions in three ways: by qualitative analysis of the set of equations, by perturbations around the FLRW background and by numerical integration. Two examples of explicit solutions are given; one is static, the other is the usual one.

The contents of Carr and Yahil's paper were then repeated in Carr's lectures (Carr 1990); the other part of the lectures is concerned with the evaporation of black holes – a subject that is beyond the scope of this book.

Ori and Piran (1990) studied the collapse of a spherically symmetric self-similar perfect fluid obeying the equation of state $p = (\gamma - 1)\,\epsilon$. Most of their study is numerical, but in addition the authors showed by exact methods that a naked singularity necessarily appears in this situation. As an illustration, Ori and Piran derived the solution from Box 2.4.3 and identified it as a subcase of the L–T model.

Henriksen and Patel (1991) discussed in much detail the geometry of self-similar spherically symmetric dust spacetimes in various coordinate systems. They also discussed null geodesics in such spacetimes (not necessarily with a dust source) and self-similar collapse in FLRW and Kompaneets–Chernov (1964) spacetimes.

In this connection, the paper by Carter and Henriksen (1989) should be mentioned. The authors generalized the notion of self-similarity so that for a certain vector field k, $\pounds_k g_{\alpha\beta} = 2g_{\alpha\beta}$, $\pounds_k u^\alpha = au^\alpha$, where a is a constant. The traditional self-similarity results when $a = 1$. The L–T model with $E = 0$ provides an example of self-similarity in this generalized sense.

Another paper to be mentioned for completeness is that by Bogoyavlenskii (1977) in which spherically symmetric self-similar spacetimes were analysed qualitatively, without obtaining explicit solutions.

This brief overview obviously does not do justice to all the papers mentioned because each of them provided, sometimes extended, physical discussion. However, each leaves open the question of which properties are generic to inhomogeneous models and which are specific to self-similarity.

Box 2.4.3 contains a few other solutions. Oppenheimer and Snyder (1939) discussed the collapse to what was later called a black hole of a cloud of dust matched to the Schwarzschild solution. They ended up discussing the cloud of flat FLRW dust, but before specializing it in such a way they derived the subcase $\Lambda = 0 = E$ of the L–T model (their eqs. (13) and (19)–(22)). This can reproduce the flat FLRW dust, and is itself the subcase $E = 0$ of the solution represented in Box 2.4.2, the subcase $\Lambda = 0$ of the Gautreau (1984) solution, and at the same time the subcase $p = 0$ of the special Szafron (1977) solution. For further comments on the Oppenheimer–Snyder paper – see Section 3.4.

Just (1956) derived the same solution as Oppenheimer and Snyder. Just's paper was an introduction to a planned investigation of evolution of inhomogeneities superimposed on a FLRW background. Follow-up papers were indeed published (see Section 3.8). One statement from the paper was prophetic and can serve as a still valid memento:[17] "With the great simplicity of our method, it would not be

---

[17] Translated from the German by A.K.

surprising if it had been already given somewhere in the older literature; however the search for such an entry remained so far unsuccessful." If successful, the search would have identified the papers by Lemaître (1933a), Tolman (1934), Datt (1938), Oppenheimer and Snyder (1939), Bondi (1947) and Omer (1949).

Yodzis, Seifert and Müller zum Hagen (1973) showed by examples that naked singularities do occur in relativity. Their paper was one of the early counterexamples to the cosmic censorship hypothesis in its simplest form, and it will be discussed in Section 3.6. Their second example is not a solution of the Einstein equations, but a metric that generates an energy-momentum tensor of unknown interpretation. We shall be concerned here with the first example, which is the L–T model (2.12.1)–(2.12.3) with $\Lambda=0=E$ and $M/r=\mu=$const. Hence, it is the subcase $M/r=$const of the Oppenheimer–Snyder solution.

The special solution by Datt (1938), mentioned in Section 2.12 (case e from section 6 in the paper), denoted Datt II in Box 2.4.3, is the subcase $E=0$ of the self-similar solutions, and at the same time the subcase $t_0(r)=Cr$, $C=$const, of the Yodzis–Seifert–Müller zum Hagen (1973) solution.

Collins (1990) formulated a new criterion for uniform thermal histories that is equivalent to the old one. A perfect fluid spacetime with $p=p(\epsilon)$ and $\epsilon+p\neq0\neq\theta$ has flow-lines with uniform thermal histories if and only if there exist three linearly independent vector fields $\xi_{(a)}$, $a=1, 2, 3$, such that $\pounds_{\xi_{(a)}} u^\alpha=0=\pounds_{\xi_{(a)}} \epsilon$ for each vector field, for the velocity field u^α and the matter-density ϵ. Collins' paper is mostly devoted to identifying spacetimes with uniform thermal histories among those that are locally rotationally symmetric. Only two explicit solutions are given. The first one has no FLRW limit, although it is important for another reason: it is a family of solutions labelled by an arbitrary function of one variable, and its existence disproves the conjecture by Bonnor and Ellis (1986) that PUTH-obeying spacetimes are rare. The second explicit solution belongs to the L–T family with $E=0$ and is the subcase $2\mu+Z=De^{r/\xi}$ of the Gautreau (1984) solution, where D and ξ are constants.[18] Collins' coordinates are chosen so that $t_0(r)=r$.

Wesson (1984), motivated by the desire to find a simple inhomogeneous model, found a dust solution that is the limit $\Lambda=0$ of the Collins (1990) solution, and at the same time the subcase of the Yodzis–Seifert–Müller zum Hagen (1973) solution given by:

$$r_{\text{YSM}}=r_{\text{W}}^{3}, \quad R_0(r)=k_2^{3/2}r_{\text{W}}(\ln r_{\text{W}})^{2/3}, \quad t_{\text{YSM}}=-t_{\text{W}}-k_3/k_1,$$

where the subscript YSM refers to Yodzis–Seifert–Müller zum Hagen, and the subscript W refers to Wesson, $R_0(r)$ is the bang-time function in the YSM solution.

The Datt II (1938) solution and the Wesson (1984) solution both reproduce the flat FLRW dust in the spatially homogeneous limit.

---

[18] Collins considered $p=$const and Λ as separate constants. One can, however, assume $p=0$ in this case because it amounts to simply correcting the value of Λ. In this sense the Collins (1990) solution is a subcase of the Gautreau (1984) solution.

At this point, Box 2.4.3 is completely presented. There is one more special solution in Figure 2.4, the one by Møller (1975), who derived the special case $M/E=-r$ of the L–T solution with $\Lambda=0$ and discussed it as a model of stellar collapse. In fact, $M/E=-r$ is just a definite choice of the r-coordinate, but it excludes the case $E=0$, and so Møller's solution can reproduce only the open and the closed FLRW dust in the limit of spatial homogeneity.

2.14 Electromagnetic generalizations of the solutions of the $\beta'\neq0$ subfamily

In this collection of solutions, all cases follow from a single master equation found by Bronnikov and Pavlov (1979), who considered the Einstein–Maxwell equations for charged dust under the assumptions that the metric tensor and the electromagnetic tensor both have a G_3/S_2 group of isometries, and that the cosmological constant may be nonzero.

In order to reconcile the notations of the different papers discussed in this section, I have had to make changes in each one of them. The notation used here is loosely patterned after that of Vickers (1973), with modifications. The metric of a charged dust solution is:

$$\mathrm{d}s^2=e^{\gamma(t,r)}\mathrm{d}t^2-e^{\alpha(t,r)}\,\mathrm{d}r^2-R^2(t,r)[\mathrm{d}\vartheta^2+S_k{}^2(\vartheta)\mathrm{d}\varphi^2],\qquad(2.14.1)$$

where γ, α and R are to be found from the Einstein–Maxwell equations and $S_k(\vartheta)=\sin\vartheta$, ϑ or $\sinh\vartheta$ for spherical, plane and hyperbolic symmetry respectively. With these symmetries, and in these coordinates, the electromagnetic tensor can have at most two nonzero components, F_{tr} (the electric field) and $F_{\vartheta\varphi}$ (the magnetic field). Note that in the case of spherical symmetry, $F_{\vartheta\varphi}\neq0$ is due to a spherical distribution of magnetic monopoles – admittedly a nonrealistic source, but it is useful to know what the theory implies for it.

The Einstein–Maxwell equations reduce here to the following set:

$$e^{-\alpha/2}R_{,r}=\Gamma(r)-Q(r)Q_{,N}(r)/R,\qquad(2.14.2)$$

is the definition of e^α, where $\Gamma(r)$, $Q(r)$ and $N(r)$ are arbitrary functions and:

$$Q_{,N}=\frac{\mathrm{d}Q}{\mathrm{d}N}\equiv\frac{\mathrm{d}Q}{\mathrm{d}r}\bigg/\frac{\mathrm{d}N}{\mathrm{d}r},\qquad(2.14.3)$$

$$e^{-\alpha/2}\gamma_{,r}=2QQ_{,N}/R^2,\qquad(2.14.4)$$

$$e^{-\gamma}R_{,t}^2=2E(r)+2M(r)/R+Q^2(Q_{,N}^2-1)/R^2+\tfrac13\Lambda R^2,\qquad(2.14.5)$$

$$2E=\Gamma^2-\varepsilon,\qquad(2.14.6)$$

$\varepsilon=+1, 0, -1$ for spherical, plane and hyperbolic symmetry, respectively,

$$Q^2=Q_e{}^2+Q_m{}^2,\qquad(2.14.7)$$

where $[c^2 Q_e(r_0)/G^{1/2}]$ is the total electric charge (in electrostatic units) contained under the surface $r = r_0$, and $[c^2 Q_m(r_0)/G^{1/2}]$ is the total magnetic charge under the same surface. The electric and magnetic charge densities are, respectively:

$$4\pi G^{1/2}\rho_e/c^2 = Q_{e,r}e^{-\alpha/2}/R^2, \tag{2.14.8}$$

$$4\pi G^{1/2}\rho_m/c^2 = Q_{m,r}e^{-\alpha/2}/R^2, \tag{2.14.9}$$

while the energy-density of the dust is:

$$\kappa\epsilon = 2N_{,r}e^{-\alpha/2}/R^2. \tag{2.14.10}$$

The functions $\Gamma(r)$, $N(r)$, $M(r)$ and $Q(r)$ are connected by the equation:

$$\Gamma N_{,r} = (M + QQ_{,N}\Gamma)_{,r}. \tag{2.14.11}$$

Note that the ratios ρ_e/ρ_m, ρ_m/ϵ and ρ_e/ϵ are all independent of time. From the formulae given above one can see that the contribution of the magnetic charges to the energy-momentum tensor is indistinguishable from the contribution of the electric charges, and the assumption $Q_m = 0$ amounts to merely reparametrizing the solution of the Einstein equations. Such a reparametrization is equivalent to the duality rotation, and its admissibility results from the invariance of the electromagnetic energy-momentum tensor with respect to duality rotations. However, such a duality rotation does change the structure of the electromagnetic tensor which is:

$$F_{01} = Q_e e^{(\alpha+\gamma)/2}R^{-2},$$
$$F_{23} = Q_m S_k. \tag{2.14.12}$$

The electromagnetic field (2.14.12) defines the electric current by $F^{\alpha\beta}{}_{;\beta} = 4\pi j^\alpha$, where $j^\alpha = \rho_e u^\alpha$ and $u^\alpha = e^{-\gamma/2}\delta^\alpha{}_0$ is the velocity field of the dust and, in addition, it defines the magnetic current by $*F^{\alpha\beta}{}_{;\beta} = 4\pi j_m{}^\alpha$, where $j_m{}^\alpha = \rho_m u^\alpha$, and $*F_{\alpha\beta}$ is the dual electromagnetic tensor. Strictly speaking, then, (2.14.1)–(2.14.12) does not fulfil the Maxwell equations unless $\rho_m = 0$ (and consequently $Q_m = $const). Because of the electromagnetic field, the charged dust moves with the acceleration $\dot{u}^\mu = \frac{1}{2}e^{-\alpha}\gamma_{,r}\delta^\mu{}_1$.

Bronnikov and Pavlov preferred to use the coordinates (T, r) defined by:

$$e^{\gamma/2}dt = dT - W(T,r)dr. \tag{2.14.13}$$

The function $W(T,r)$ obeys $(We^{-\gamma/2})_{,T} = -(e^{-\gamma/2})_{,r}$, so that (2.14.13) is integrable. In such variables, the solution of (2.14.5) can be written as an integral representing $T(r,R)$. The paper contains a list of explicit solutions that can be obtained with various special subcases of the arbitrary functions of r, but the solutions are not discussed further; all have $\Lambda = 0$. On the basis of those solutions Bronnikov and Pavlov observed that in the case of spherical symmetry ($\varepsilon = +1$), all three types of model are admissible: the recollapsing one ("elliptic") with $E < 0$, the marginally bound one ("parabolic") with $E = 0$, and the ever-expanding ("hyperbolic") one. In the case of plane symmetry ($\varepsilon = 0$), only the two latter models are possible, and the parabolic one has $\gamma = 0$. Note that the plane symmetric parabolic model will not exist for neutral dust where $Q=0$ (eqs. (2.14.2)–(2.14.11)) were derived under the assumption $R_{,r} \neq 0$). In the case of hyperbolic symmetry ($\varepsilon = -1$), only the ever-expanding model exists.

The paper by Bronnikov (1983) extends some of the results of Bronnikov and Pavlov (1979). Bronnikov observed that shell-crossing singularities are not prevented by electric charge alone because there exist solutions with a charged dust source that develop shell-crossings (see comments on the Ori 1991 paper later in this section – shell-crossings are in fact inevitable when the charge-density is smaller than the mass-density, both expressed in the units in which $G=c=1$). Then, all cases that are free of singularities were enumerated (conditions for the absence of Big Bang were formulated in other papers, see below). Bronnikov showed how the solutions can be matched to the electrovacuum solutions that generalize that by Reissner and Nordström for Λ and for an arbitrary G_3/S_2 symmetry. Plane symmetric solutions were discussed in more detail; among other things, Bronnikov discussed their evolution and conditions for mirror-symmetry. Mirror-symmetric plane symmetric models are nonsingular only when $Q(0)=0$, $(M+QQ_{,N}\Gamma)|_{r=0}<0$ and $(\rho_e/\epsilon)^2 > G/c^4$.

Bronnikov showed that models with $Q=$const (i.e. dust moving in a source-free electric and/or magnetic field) necessarily develop shell-crossing singularities. This observation was later generalized by Ori (1991, see below). The specific model discussed at the end of Bronnikov's paper becomes vacuum in the limit $Q=0$.

In the limit of vanishing electromagnetic field, $Q=0$, the Bronnikov–Pavlov solutions go over into Ellis' (1967) case IIaii.

Shikin (1974) considered the Einstein–Maxwell equations for charged dust in free electromagnetic field. His result is the limit $\Lambda=0=\rho_m$, $Q_m=$const of (2.14.1)–(2.14.12), that is, it corresponds to the situation when the magnetic field is the exterior field produced by the distribution of magnetic monopoles of given symmetry. This was a new result when the paper was published. Shikin analysed the solutions of (2.14.2)–(2.14.7) qualitatively and found the range of values of R in different situations; in some cases R will never reach zero. The two explicit solutions presented have a singular limit $Q=0$, and so are not within the scope of this book. Finally, Shikin discussed how the G_3 symmetry group can be extended to larger groups. This part of the paper is a short exposition of his earlier results (Shikin 1972a). When $Q \neq 0$, the extension to a four-dimensional group produces a static solution, and no extension to a six-dimensional group is possible. In other words, this means that a FLRW geometry cannot accomodate electromagnetic field.

De and Ray (1983) found the plane symmetric limit ($\varepsilon=0$) with vanishing magnetic field ($Q_m=0$) of the Shikin (1974) spacetimes. By the time their paper was published, the result was not new, and the one explicit solution discussed in the paper becomes vacuum when the electromagnetic field is set to zero.

Chatterjee and Banerji (1980) found the limit $\rho_e=0$ ($Q_e=$const) of the result by De and Ray, that is, neutral plane symmetric dust moving in the exterior electric field produced by a charged plane. That result was not new, either. In fact, the authors considered the subcase $\rho_e=\rho_m=0=\Lambda=\varepsilon$, $Q_e=$const, $Q_m=$const of the

Bronnikov–Pavlov (1979) equation, that is, they allowed for a vacuum magnetic field produced by a magnetically charged plane, but, as explained above, our reinterpretation influences (2.14.12) only. In this limit, the acceleration is zero and the dust moves on geodesics, although the electromagnetic field contributes to (2.14.5) through the term $(-Q^2/R^2)$ and so influences the geometry of spacetime. The limit $Q=0$ of the Chatterjee–Banerji spacetime are the plane symmetric dust solutions of Box 2.4.1.

Chatterjee and Banerji also considered the corresponding spherically symmetric solutions, but ended up with an incorrect result: they did not notice that the Maxwell equations require Q to be constant when $\rho_e=0$.

Shikin (1972a, preliminary report is given in Shikin 1972b) considered the subcase $\rho_e=\rho_m=0=\Lambda$ (Q_e=const, Q_m=const) of the Bronnikov–Pavlov (1979) spacetimes, that is, neutral dust moving in the exterior electric and magnetic fields of the appropriate symmetry. From the point of view of eq. (2.14.5), however, Shikin's result can be more conveniently classified as the subcase $Q_m=0$, Q_e=const of the Shikin (1974) spacetimes, and we prefer such an interpretation. This class of spacetimes was new in 1972. In the further limit of plane symmetry, it reproduces the Chatterjee–Banerji (1980) spacetime. The paper is notable for its wealth of results and will be described here in more detail.

Shikin assumed that the orbits S_2 of the symmetry group G_3 may be either timelike or spacelike, and that the parallel magnetic and electric fields are orthogonal to S_2. When the S_2 are timelike, the metric is either static or has a static limit $Q=0$, that is, it has no FLRW limit. This case is elaborated in Shikin's paper, but it is not within the scope of this book.

In the cases of plane, hyperbolic or spherical symmetry, the algebra of the group G_3 is of Bianchi type VII_0, VIII or IX respectively. Each of these algebras can be extended to a four-dimensional algebra A_4 of a symmetry group either centrally (so that the fourth generator commutes with all of A_3) or noncentrally. For the central extension, the fourth generator may be timelike (this leads to a static spacetime) or null or spacelike. The paper elaborates only the last case. Then, for plane symmetry the central extension A_4 leads to a Bianchi I spacetime and allows for a further extension A_6 that produces the flat FLRW model. For hyperbolic symmetry, the central extension A_4 is a Bianchi type III model. For hyperbolic and spherical symmetries, the central extension A_4 cannot be extended to any A_6.

A noncentral extension A_4 of the plane symmetric case leads to a Bianchi V model. It can be further extended to an A_6 corresponding to the open FLRW model. For hyperbolic symmetry there are no noncentral extensions A_4, but there is an extension A_6 that leads to the open FLRW model. For spherical symmetry, there are noncentral extensions A_6 leading to all three FLRW models. This discussion of extensions is a description of the possible FLRW limits by a method dual to the one applied in this book.

Shikin noted that the case $R_{,r}=0$ requires a separate treatment, but concluded (incorrectly) that it leads to the K–S-like models with $\alpha=\alpha(t)$ in (2.1.1), thus overlooking the whole subfamily $\beta'=0$. Then he gave a complete list of explicit solutions of (2.14.5) with $\Lambda=0$ and $Q=$const; all the solutions can be expressed in terms of elementary functions in a parametric form. Using the explicit solutions Shikin observed that as long as $Q\neq0$ (and tacitly assuming $M>0$), the function $R(t,r)$ never assumes the value $R=0$ in any case, that is, the electric charge will prevent the Big Bang singularity. However, a shell-crossing singularity where $e^{\alpha}=0$ can occur, and it is in general nonsimultaneous. In the spherically symmetric case there exist solutions in which $R(t,r)$ oscillates forever between a finite minimum and a finite maximum (note the description of the Ori (1991) paper below – shell-crossings are in fact inevitable in this case and do not allow for a bounce, so the solutions for oscillating $R(t,r)$ are purely formal).

Note that to obtain the explicit solutions the property $Q=$const is essential. Then (2.14.4) implies $\gamma,_{r}=0$, so a transformation of t leads to $\gamma=0$. With $\gamma=0$, (2.14.5) decouples from the other equations and can be solved separately.

At about the same time Gorelik (1972) made an observation that explains the existence of models without a Big Bang on a simpler example. By investigating time-like geodesics in the Reissner–Nordström solution, Gorelik noted that the electric charge in the centre has a repulsive *gravitational* effect on neutral particles, that is, it creates effective antigravitation. Shikin (1972a) discovered the same effect in the cosmological context.

Vickers (1973) considered the spherically symmetric limit with zero magnetic field of the Bronnikov–Pavlov (1979) spacetimes, that is, charged spherically symmetric dust, with arbitrary Λ. This was a new result in 1973, and the presentation in Vickers' paper is most convenient for considering the L–T limit. Our presentation and notation are patterned after Vickers'. There are a few misprints in the field equations, but they do not influence the results. Vickers considered matching the charged dust sphere to the Reissner–Nordström spacetime, and this helps in understanding the physical meaning of the arbitrary functions. On the hypersurface $r=r_{b}$ where the solutions are matched, the following holds:

$$Q(r_{b})=e, \quad (M+QQ,_{N}\Gamma)(r_{b})=m, \tag{2.14.14}$$

where e and m are the Reissner–Nordström charge and mass parameters respectively. Equation (2.14.11) thus turns out to be the relationship between the active gravitational mass $m=M+QQ,_{N}\Gamma$ and the sum of the rest-masses N, with Γ measuring the gravitational mass defect for bound systems (when $\Gamma<1$) or gravitational energy excess (when $\Gamma>1$).

When $Q,_{r}\neq0$, the avoidance of the Big Bang singularity is not guaranteed. Vickers' discussion of the conditions for the avoidance of singularity is incomplete (he excludes the case $M<0$ for physical reasons) and does not distinguish between "no nonsingular solutions" and "no solutions at all". Therefore, the complete list of possibilities (for the case $\Lambda=0$) is given here.

(a) When $E<0$, singularity is avoided if and only if $M>0$ and $Q,_{N}^{2}<1$, but solutions will exist only if $M^{2}\geq2E(Q,_{N}^{2}-1)Q^{2}$, and they will be nonstatic only if the inequality is sharp (the last inequality is reversed in Vickers' paper). The sphere will then oscillate between a minimal and a maximal radius (we recall that this discussion ignores the possibility of shell-crossings, see below).

(b) When $E=0$, singularity is avoided if and only if $M>0$ and $Q_{,N}^2<1$. Collapse is then halted and reversed once and for all.

(c) When $E>0$, singularity is avoided if and only if either $\{M\geq0$ and $Q_{,N}^2<1\}$ or $\{M<0$, $Q_{,N}^2>1$ and $M^2>2E(Q_{,N}^2-1)Q^2\}$. Collapse is then halted and reversed as in case (b).

Note that with $M>0$ and $Q_{,N}=0$ the conditions are automatically fulfilled and Shikin's (1972a) conclusion follows; however, the condition $M^2\geq-2EQ^2$ in case (a) has to be assumed in addition (it is given in Shikin's paper).

From (2.14.3), (2.14.8) and (2.14.10) it follows that with $Q_m=0$, $Q_{,N}=Q_{e,r}/N_{,r}=c^2\rho_e/(G^{1/2}\epsilon)$. Note that with $M\geq0$, $Q_{,N}<1$ is a necessary and sufficient condition for nonsingular evolution, and this means that in order to prevent the Big Bang singularity, the electric charge density must be sufficiently *small* compared with the mass-density. A sufficiently large charge-density brings a positive contribution to the active mass that prevails over the electrostatic repulsion.

If the evolution of a nonsingular charged sphere is observed from a Reissner–Nordström exterior spacetime, then with $e^2\geq m^2$ the bounce can be fully visible. However, with $e^2<m^2$ the collapsing sphere will disappear under the outer horizon in the infinite future of the observer, and will re-emerge after the bounce into a different asymptotically flat region of the maximally extended Reissner–Nordström spacetime (see Graves and Brill 1960 for the extension of the Reissner–Nordström spacetime and Novikov 1966 and 1970 for the discussion of evolution of charged spheres). Vickers concluded by showing that there exists a nonsingular static solution for which $Q_{,N}^2=\Gamma^2=1$, $M=0$ and $e^2=m^2$ in the Reissner–Nordström exterior metric.[19]

Khlestkov (1975, see Box 2.4.4) basically just repeated the result of Vickers. He considered three cases when the Einstein–Maxwell equations can be integrated explicitly; these are: (1) $Q=$const – this is the spherically symmetric limit of the Shikin (1972a) solutions; (2) $N_{,r}=0$ – the limit $Q=0$ of this solution is a vacuum; (3) $Q^2/(M+QQ_{,N}\Gamma)=$const – this solution was new and nontrivial, but was not discussed further. Khlestkov observed that $Q=$const means that the electromagnetic field is generated by a single charge in the centre of symmetry. Finally, he derived

---

[19] The Vickers solution may have been obtained in 1968 by Bardeen. However, reports about that solution are available only in the form of a short abstract (Bardeen 1968) and a footnote (Ori 1991) suggesting that the result was never published in a more extended form. Attempts to reach the author for a comment were unsuccessful.

The problem of collapse and bounce of charged dust was also considered by Novikov (1966), who investigated the radial motion of a charged test-particle in the Reissner–Nordström spacetime and, on the basis of its equation of motion, similar to (2.14.5), verbally predicted that the motion of any spherical shell in a charged dust with $\Lambda=0$ would be described by the same equation, but with mass and charge depending on the radial coordinate. That prediction failed to account for the $e^{-\gamma}$ on the left-hand side of (2.14.5) and for the variability of $E(r)$.

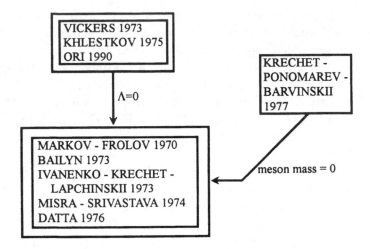

NOTE: THE SOLUTIONS OF VICKERS 1973 AND MARKOV + FROLOV 1970
WERE DISPLAYED IN FIGURE 2.4,
THE PURPOSE OF THIS BOX IS TO LIST ALL REDISCOVERIES OF THEM

Box 2.4.4. Generalizations of the L–T model with charged fluid source.

the equations of motion of charged test particles in this spacetime, but did not discuss the solutions.

Ori (1990) reobtained the Vickers class of spacetimes, but in coordinates in which all equations can be solved explicitly. The coordinates were R (the function from (2.14.1)–(2.14.11)), and the active gravitational mass m from eq. (2.14.14). In a slightly modified notation, the solution is:

$$ds^2 = \Delta F^2 dm^2 - 2\,(uF/u^R)dm\,dR + (u^R)^{-2}dR^2 - R^2(d\vartheta^2 + \sin^2\vartheta\,d\varphi^2), \quad (2.14.15)$$

where:

$$\Delta = 1 - 2m/R + Q^2(m)/R^2 - \tfrac{1}{3}\Lambda R^2, \quad (2.14.16)$$

$$u = \Gamma(m)(1 - QQ_{,m}/R), \quad (2.14.17)$$

$\Gamma(m)$ and $Q(m)$ being arbitrary functions, the same as in (2.14.1)–(2.14.11),

$$u^R = \pm(u^2 - \Delta)^{1/2} \quad (2.14.18)$$

is the single nonzero contravariant component of velocity, and:

$$F(m,R) = f_0(m) \pm p \int_{R_0}^{R} \mathscr{R}^{-3/2} r^3 dr \pm q \int_{R_0}^{R} \mathscr{R}^{-3/2} r^2 dr, \quad (2.14.19)$$

$f_0(m)$ being an arbitrary function, and:

$$p = \Gamma_{,m}, \quad q = \Gamma^{-1}(1 - \Gamma^2 Q_{,m}^2) - Q\frac{d}{dm}(\Gamma Q_{,m}),$$

$$\mathcal{R}=a+br+cr^2, \quad a=Q^2(\Gamma^2Q_{,m}^2-1), \tag{2.14.20}$$

$$b=2(m-\Gamma^2QQ_{,m}), \quad c=\Gamma^2-1.$$

The integrals in (2.14.19) can be calculated and the result represented as:

$$F(m,R)=f_0(m)\pm\tfrac{1}{2}c^{-2}(2qc-3pb)\ I$$

$$\pm c^{-2}\delta^{-1}\mathcal{R}^{-1/2}\left\{p[c\delta R^2+b(10ac-3b^2)R+a(8ac-b^2)]\right.$$

$$\left.-qc[(\delta-b^2)R-2ab]\right\}, \tag{2.14.21}$$

where $\delta=4ac-b^2$ and:

$$I=\int_{R_0}^{R}\mathcal{R}^{-1/2}dr, \tag{2.14.22}$$

the explicit result depending on the sign of c (see Ori's paper). Equation (2.14.21) applies when $\delta\neq0$; the result for $\delta=0$ is given in the paper as well. The only component of the electromagnetic field is:

$$F^{mR}=-F^{Rm}=Q/(FR^2), \tag{2.14.23}$$

and the matter-density is:

$$\kappa\epsilon=2(EFu^RR^2)^{-1}. \tag{2.14.24}$$

Ori emphasized the advantages of these coordinates: they allow us to write an explicit solution and are unique. This is a valuable result, but for most other purposes Ori's coordinates are not easy to use, in particular for the physical interpretation (almost impossible without reference to comoving coordinates) and for limiting transitions to other solutions (see discussion of this last point in his paper). The transformation to the comoving coordinates is effected by:

$$m=m(r), \quad R=R(t,r), \tag{2.14.25}$$

where $m(r)$ is an arbitrary function (in the approach used in this book, r is not uniquely determined), and $R(t,r)$ is defined by:

$$\frac{\partial R}{\partial r}=uu^RF\frac{\partial m}{\partial r} \tag{2.14.26}$$

(note that (2.14.26) is covariant with the transformations $r=r(r')$). Since $\partial m/\partial t=0$, it follows that $\partial r/\partial R=0$ and $u^r=0$ in the new coordinates. Equation (2.14.26) ensures that $g_{tr}=0$, $g_{tt}=e^\gamma=R_{,t}^2/[\Gamma^2-1+2(m-\Gamma QQ_{,m})/R+(\Gamma^2Q_{,m}^2-1)Q^2/R^2+\tfrac{1}{3}\Lambda R^2]$ and that g_{rr} is of the form given by (2.14.4). Note that in Ori's approach, both m and R are in general spacelike coordinates.

Ori then showed that for a collapsing and bouncing charged dust cloud, shell-crossings will inevitably occur before any shell reaches its minimal radius. He showed in addition how the solution can be transformed to the Schwarzschild coordinates and to the mass-proper time coordinates, and presented two special solutions: the marginally bound one with $\Gamma=1$, and the self-similar one with $Q/m, f_0$ and Γ all being constant. Ori was misled, by second-hand references, into believing that the L–T model was known before only in the case $\Lambda=0$.

Burlankov (1987) apparently intended to obtain the same result as Ori. His reasoning and the result are similar to Ori's. However, Burlankov did not provide any definition of the coordinates he used and did not explain his symbols in sufficient detail. Attempts to prove the equivalence of the results of the two papers were unsuccessful. Therefore, Burlankov's result cannot be accepted until details of his calculations are explained and verified.

The Vickers–Ori solution reproduces the L–T models in the limit $Q=0$, that is, zero charge.

In a later paper, the same author (Ori 1991) developed and presented in more detail the observation that shell-crossings are inevitable for charged dust. More exactly, the situation is as follows: if the charge-density is smaller (in units in which $c=G=1$) than the mass-density, so that bounce can occur, then the bounce will occur earlier for those shells that have collapsed to a smaller radius, and, as they are already re-expanding, they will collide with the shells that still keep collapsing. This means that the dust approximation becomes inapplicable in such a situation.[20] Ori studied the evolution of a charged dust sphere numerically, assuming that the shell-crossing singularity accumulates matter and charge (the shell "collision is completely inelastic"). The result is that a null singularity appears at the centre of the ball, proceeds towards the Reissner–Nordström singularity and closes off the entry into another asymptotically flat region. The singular shell eventually collects all the matter into it and crashes into the intersection of the null and R–N singularities. Hence, bounce may well turn out to be impossible. Presumably, the whole study was done for the case $\Lambda=0$, but the paper does not say this clearly.

A similar prediction (necessity of shell-crossings in a charged dust) was made by Raychaudhuri (1975), but in this paper it was studied in less detail.

The subcase $\Lambda=0$ of the Vickers (1973) solution has been obtained by a few authors (see Box 2.4.4), for the first time by Markov and Frolov (1970), and at that time it was a new solution. Markov and Frolov discussed matching the solution to the Reissner–Nordström spacetime and described the evolution of the resulting configuration as a pulsating sphere attached to the Einstein–Rosen bridge. Finally, they showed that the strong electric field in the throat will cause quantum production of particle-antiparticle pairs that will reduce the charge in the dust cloud asymptotically to the value $e_f \cong 137$, irrespective of the initial charge.

Bailyn (1973) rederived the Markov and Frolov solution in a slightly different parametrization, and found the conditions for the existence of oscillatory solutions equivalent to the condition (a) described above for the Vickers paper. Bailyn showed

---

[20] Solutions with charged fluid and nonzero pressure are known (see Sections 4.12 and 4.13) but they have zero shear and so do not generalize those considered here. Therefore they cannot be used to resolve the question of shell-crossings in the Markov–Frolov spacetimes.

that, under these conditions, $[Q^2-(2E+QQ,_N\Gamma)^2]$ must be a nondecreasing function of r in the interior of the charged sphere.

Ivanenko, Krechet and Lapchinskii (1973) rederived the Markov–Frolov result again, and, independently of Vickers and in fact with more clarity, showed that the Big Bang will be prevented if only $c^4\rho_e/(G\epsilon)<1$. They also related the arbitrary functions of the model to the standard initial conditions (initial densities of mass and charge, initial velocity), but the relation is rather complicated and is not discussed further.

Another rederivation was published by Misra and Srivastava (1974), and still one more by Datta (1976). Both papers considered matching the charged sphere to the Reissner–Nordström solution. Datta considered three special cases, but two of them have (in units with $c=G=1$) $\rho_e^2=\epsilon^2$, and the third one has $\rho_e^2>\epsilon^2$, so none admit a FLRW limit.

The Markov–Frolov class of solutions results from that of Shikin (1974) as the limit $Q_m=0$, $\varepsilon=+1$, that is, zero magnetic field and spherical symmetry.

De (1968) was well on the way to finding the Markov–Frolov class. He showed, by processing the Einstein–Maxwell equations, that nonstatic charged dust cannot be shearfree, and that static charged dust is unstable against perturbations of the balance between gravitational attraction and electrostatic repulsion. The Markov–Frolov solution demonstrates this instability explicitly. However, De's consideration was performed under the assumption that $\rho_e/\epsilon=$const. De's paper is remarkable for historical reasons – it was one of the earliest contributions to the subject.

A rather unusual generalization of the Markov–Frolov class was presented by Krechet, Ponomarev and Barvinskii (1977). They worked out the Einstein–Proca field equations, where the source was dust charged with meson charge, the carrier of the meson interaction being a massive vector field. Transcribed into our notation of (2.14.1)–(2.14.11), the result of the paper is that if the field carrier has mass μ (= const), then eq. (2.14.5) is modified to:

$$e^{-\gamma}R,_t^2=2E(r)+2M(r)/R+Q^2(Q,_N^2-1)/R^2-\tfrac{1}{3}(\mu^2/Q,_N^2)R^2, \qquad (2.14.27)$$

eq. (2.14.10) is modified to:

$$\kappa\epsilon=2N,_r e^{-\alpha/2}/R^2-2\mu^2/Q,_N^3, \qquad (2.14.28)$$

and eq. (2.14.8) is modified to:

$$4\pi G^{1/2}\rho_e/c^2=Q_{e,r}e^{-\alpha/2}/R^2-\mu^2\phi e^{-\gamma/2}, \qquad (2.14.29)$$

where ϕ is the electromagnetic scalar potential given by $\phi,_r=-Q_e e^{(\alpha+\gamma)/2}/R^2$. In comparing the KPB solution with our notation it should be observed that the authors used the opposite sign on the right-hand side of (2.14.2). This does not change $-e^\alpha=g_{11}$ in the metric.[21] The paper is partly a review of results known earlier; these are included in this book.

---

[21] One of their formulae is apparently in error, that is, their third equation under the label (8) which should read $\phi'=(e^{\nu/2})'/a$ instead of $\phi e^{-\nu/2}=1/a$ and only with such a correction can their equations reproduce (2.14.1)–(2.14.11) with $\Lambda=0=Q_m$ in the limit $\mu=0$. Their reference to Bronnikov and Pavlov (1976) is in error, too; they probably meant our reference to Pavlov and Bronnikov (1976).

Hamoui (1969) found the limit $\rho_e=0$ ($Q=$const) of the Markov–Frolov class, that is, spherically symmetric neutral dust with $\Lambda=0$ moving in its own gravitational field and in the vacuum electrostatic field of charges surrounding the centre of symmetry (this may in particular be a point charge at the centre). At that time it was a new solution. Hamoui wrote the field equations for a charged dust using the metric (2.14.1), but then considered only two subcases: $\gamma,_r=0$ and $\alpha,_t=0$. The second of these has no FLRW limit (unfortunately, the greater part of the paper is concerned with that second case). In the first case, it follows from the Einstein–Maxwell equations that either $F_{\mu\nu}=0$ or $\rho_e=0$ and $Q_e=$const. Note that $\gamma,_r=0$ means that $\dot{u}^\alpha=0$, and so the implication is that the dust can move on geodesics only when either there is no electromagnetic field or the dust is not charged.

Korkina and Chernyi (1976) found Hamoui's solution again and presented a physical discussion of it. They found that the charge at the centre will prevent collapse to a singularity and observed that this effect is not caused by electrostatic repulsion (the dust has no charge), hence the central charge must be exerting a repulsive gravitational action, just as was found by Gorelik (1972) in the Reissner–Nordström solution. Korkina and Chernyi showed how the solution can be matched to that of Reissner–Nordström and presented spacetime diagrams for both solutions. If the charge at the centre is small and $E<0$, so that the spacetime does not differ much from the closed FLRW spacetime, then the hypersurface of matching can be chosen so that it will be static in the Reissner–Nordström spacetime, but the dust inside it will be pulsating (collapsing and re-expanding forever).

The Hamoui solution is the spherically symmetric limit of the Shikin (1972a) class, and in the further limit $Q=0$ it reproduces the $\Lambda=0$ subcase of the L–T model from Box 2.4.2.

For completeness, the paper by Zecca (1993b) must be mentioned here. Zecca considered a spherically symmetric charged perfect fluid with the additional (tacitly assumed) condition $\dot{u}^\alpha=0$. His result, if correct, in the special case $p=0$ should coincide with the limit $\dot{u}^\alpha=0=\Lambda=Q_m$ of (2.14.1)–(2.14.12). In this case, $\gamma,_r=0$ and then $Q=$const, $\rho_e=0$ follow from (2.14.4), (2.14.7) and (2.14.8). However, Zecca's paper seems to imply that $Q,_r\neq0\neq\rho_e$ is possible when $p=0=\dot{u}^\alpha$, which is a contradiction with all other papers on this subject (whose calculations the author of the present book verified). Zecca probably did not consider some of the implications of the Einstein–Maxwell equations.

This completes Figure 2.4. Some more papers describing properties of the charged generalizations of the L–T model will be mentioned in Section 3.5.

2.15 Generalizations of the solutions of the $\beta'\neq0$ subfamily for viscosity

The relations between these solutions are shown in Figure 2.5.

Roy and Singh (1982) found six solutions with bulk and shear viscosity related to the Szafron subfamily with $\beta'\neq0$.

Roy and Singh's case 1 is a set of metrics having a G_3/S_2 symmetry, which are

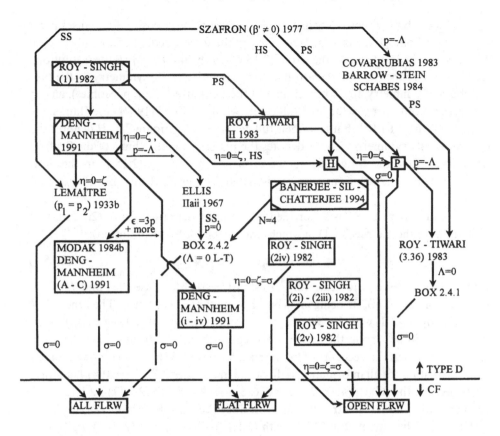

SYMBOLS USED:
η - shear viscosity,   ζ - bulk viscosity,   N - spacetime dimension,

H - hyperbolically ⎤ symmetric subcases of the Szafron (β' ≠ 0) subfamily,
P - plane   ⎦ so far not published separately

REFERENCES WITHOUT FRAMES ARE PERFECT FLUID SOLUTIONS FROM FIGURE 2.4,
(Roy - Tiwari (3.36) 1983 is not shown there because it is a minor point in a paper on a different
class of solutions)

Figure 2.5. Generalizations to Figure 2.4.

given by (2.12.1)–(2.12.2) with $E(r) = \frac{1}{2}Kr^2$, K=const. This special form of E simply
amounts to a definite choice of the coordinate r. Equation (2.12.3) is changed to:

$$2R_{,tt}/R + R_{,t}^2/R^2 - Kr^2/R^2 + 4\kappa\eta R_{,t}/R = 2l(t), \qquad (2.15.1)$$

where η is the shear viscosity coefficient and $l(t)$ is an arbitrary function; in the limit
of vanishing viscosity $2l(t)$ goes over into $-\kappa p(t)$. The formulae for matter-density
and pressure are:

$$\kappa\epsilon = (R_{,r}R_{,t}^2 + 2RR_{,t}R_{,tr} - 2KrR - Kr^2R_{,r})/(R^2R_{,r}), \qquad (2.15.2)$$

$$\kappa p = -2l(t) + \kappa(\zeta + 4\eta/3)(2R_{,t}/R + R_{,tr}/R_{,r}), \qquad (2.15.3)$$

where ζ is the bulk viscosity coefficient. The presence of viscosity thus causes pressure to be inhomogeneous. It does not change the form of eq. (2.15.2), but $R(t,r)$ is modified by the last term on the left of (2.15.1). In the limit $\zeta=\eta=0$ these solutions reproduce the subcases of the Szafron spacetimes with a G_3/S_2 symmetry (in the case of spherical symmetry, the isotropic limit of the Lemaître 1933b spacetimes), and in the further limit $\kappa p=-\Lambda$ they reproduce the whole of Ellis' (1967) case IIaii. They contain all three FLRW limits in their full generality. In Figure 2.5, H denotes the hyperbolically symmetric, and P denotes the plane symmetric limit of the Szafron spacetimes. Roy and Singh list all the explicit solutions of (2.15.1) in the limit $\zeta=\eta=p=0$, but the solutions are not discussed further (and were not new at that time).

Roy and Singh's case 2 is subdivided into five further cases. Cases (2i), (2ii) and (2iii) all have the metric (2.1.1) in which (transcribed into our notation):

$$e^\beta=\Phi(t,z)/H(x,y,z), \quad e^\alpha=\Phi\beta_{,z},$$
(2.15.4)

where $H=\sin(z+My)$ for case (2i), $H=\sinh(z+My)$ for case (2ii) and $H=\cosh(z+My)$ for case (2iii), M being an arbitrary constant in each case. The function Φ is defined by (2.15.1) with $K=M^2+1$, $r=z$. Comparison with (2.1.11)–(2.1.13) shows that M should be zero in the limit $\eta=0$ (because then $e^{-\nu}=H$ must have the property $H_{,yyy}=0$), but the necessity of taking such a value of M for $\eta=0$ is nowhere indicated. Possibly, not all information is disclosed in the paper. With $\eta=0=M$, all three cases become the subcases $A=B_1=B_2=0$, $C(z)=H$, $k=-z^2$ of (2.1.11)–(2.1.15); they then acquire plane symmetry and can reproduce the open FLRW model only.

Case (2iv) has the metric (2.1.1) with (2.1.11)–(2.1.13) and (2.1.15), $k=0$, $h=1$ and (2.1.14) modified to:

$$2\Phi_{,tt}/\Phi+\Phi_{,t}^2/\Phi^2+4\kappa\eta\Phi_{,t}/\Phi=2n/t^2+\tfrac{4}{3}(\kappa\eta)^2,$$
(2.15.5)

where n is an arbitrary constant. Because of $k=0$, it can reproduce only the flat FLRW model, and because of the specific time-dependence of the right-hand side of (2.15.5), the limit will have a specific equation of state. Because of $AC-B_1{}^2-B_2{}^2=1/4>0$, the solution can acquire spherical symmetry (when A, C, B_1, B_2 are all constant), but not hyperbolic or plane symmetry.

Case (2v) has the same metric as (2i)–(2iii), the same Φ as (2iv), and $H=ae^{My}+be^{-My}+c\sin(Mx+d)$, where M is an arbitrary constant and a, b, c, d are arbitrary functions of z. It reproduces the open FLRW model with a specific equation of state when $\eta=0=M=\sigma$.

Deng and Mannheim (1991) set out to find spherically symmetric solutions with zero acceleration and a viscous fluid source with nonzero shear. These assumptions guarantee that their class is the spherically symmetric limit of the Roy–Singh case 1. With $\eta=0$, their spacetime reproduces the isotropic limit of the Lemaître (1933b) class. The seven specific solutions discussed in Deng and Mannheim's paper arise by further *ad hoc* specializations of the functions. All of them have the equation of state $\epsilon=3p$. The first four have $k(z)=0$ in (2.1.14) and so can reproduce only the flat

FLRW limit with the radiative equation of state. All four models start from infinite density in a nonsimultaneous singularity and asymptotically approach the flat FLRW model both at infinite future and at spatial infinity. In all four cases, the singularity occurs first at spatial infinity and proceeds towards the centre. In case (ii) the singular set asymptotically approaches a world-tube of nonzero radius around the centre; in the other three cases it approaches the centre itself. In all four cases the centre is thus excluded from the spacetime. In cases (i), (iv) and in some instances of case (iii) the singularity is a Big Bang; in the other cases it is a shell-crossing singularity, and Deng and Mannheim seem to have overlooked this fact. All four cases obey the dominant energy condition.

The three remaining cases are referred to as A, B, C. Case A again has $k(z)=0$, and Deng and Mannheim study two possible choices of the bang-time function in it, systematically mistaking shell-crossing for a Big Bang. The first choice makes the *shell-crossing* simultaneous in all space. With the second choice, the singularity is completed in a finite time everywhere. The solution reproduces the flat FLRW model when $\sigma=0$.

Case B has $k(z)>0$ and reproduces the closed FLRW model when $\sigma=0$; case C has $k(z)<0$ and reproduces the open FLRW model when $\sigma=0$.

None of the seven cases has a nontrivial limit $\eta\rightarrow0$, so in each of them the resulting FLRW model formally has a nonzero shear viscosity coefficient. However, because shear itself vanishes in the limit, there is no contribution to energy-momentum from viscosity in that limit. The peculiar shapes of the initial singularity sets are connected with the specific choices of $t_0(r)$ and are independent of viscosity.

Modak (1984b) found a family of solutions that is within Deng and Mannheim's (1991) class, and is equivalent to the collection of cases A, B, and C of the latter. Modak's "simple special solution" is Deng–Mannheim's case A. Hence, the Modak solutions reproduce all three FLRW limits, but with a specific equation of state in each case.

Roy and Tiwari (1983), in a paper already mentioned in Section 2.10.3, considered plane symmetric solutions with a viscous fluid source. Their case II is just the plane symmetric limit of the Roy and Singh (1982) case 1, hence it is the plane symmetric counterpart of the whole Deng–Mannheim (1991) class. The subcase to which Roy and Tiwari give special consideration (their eq. (3.36)) is a plane symmetric perfect fluid solution with $\kappa p=-\Lambda$; it is therefore the plane symmetric limit of the case considered by Covarrubias (1983) and by Barrow and Stein–Schabes (1984). In the limit $\Lambda=0$ it reproduces the solutions from Box 2.4.1.

2.16 Other generalizations of the β′ ≠0 subfamily

Dandach and Mitskievic (1984) proposed an interesting way of generalizing the L–T model. They assumed that the metric in comoving coordinates has the form:

$$ds^2=dt^2-A^2(t,r,\vartheta)\,dr^2-B^2(t,r,\vartheta)\,d\vartheta^2-D^2(t,r,\vartheta)\,d\varphi^2, \qquad (2.16.1)$$

and supposed that it obeys the Einstein equations with a perfect fluid source. The equation $G_{01}=0$ then implies $(B_{,r}/A)_{,t}/B+(D_{,r}/A)_{,t}/D=0$, and a particularly simple solution of this is:

$$B_{,r}/A=f_1(r,\vartheta), \quad D_{,r}/A=f_2(r,\vartheta), \tag{2.16.2}$$

where f_1 and f_2 are arbitrary functions. For B and D, Dandach and Mitskievic then obtained equations that are identical in form to (2.1.14), except that in place of $k(z)$ there are arbitrary functions of r and ϑ, a different one in each equation. Unfortunately, the authors did not consider the following consequence of (2.16.2):

$$B_{,r}/f_1=D_{,r}/f_2 \quad (=A), \tag{2.16.3}$$

and the implications of (2.16.3) are quite strong (see below). Thus, the extended discussion in the paper must be considered unfinished.

Dandash (1985) investigated the same problem in the simpler case when A, B and D in (2.16.1) are independent of ϑ. However, he did not consider (2.16.3) either. This author (A.K.) has investigated the consequences of (2.16.3) for the Dandash metric in the simple case corresponding to $E=0$ in (2.12.1)–(2.12.3) (i.e. to $k=0$ in (2.1.14)). These are as follows: either the functions corresponding to M are constant and the Big Bang is simultaneous for both B and D, or the ratio of the two M-functions is constant and the two Big Bangs occur at the same moment. Dandash's paper is thus unfinished, too. Dandash studied a Bianchi I solution and a static solution separately, but they are not considered here.

Banerjee, Sil and Chatterjee (1994) found a generalization of the $\Lambda=0$ L–T model to N dimensions. They assumed that the metric has the $O(n)$ symmetry group (i.e. the group whose orbits are n-dimensional spheres, $N=n+2$) and that the Einstein tensor has the algebraic form of generalized dust, that is,

$$G_{\alpha\beta}=\kappa\epsilon u_\alpha u_\beta. \tag{2.16.4}$$

In notation adapted to ours, the metric of the solution is:

$$ds^2=dt^2-[1+2E(r)]^{-1}R_{,r}^2(t,r)dr^2-R^2(t,r)\,d\omega_n^2, \tag{2.16.5}$$

where $E(r)$ is an arbitrary function, $d\omega_n^2$ is the metric of an n-dimensional sphere of unit radius and $R(t,r)$ is defined by:

$$R_{,t}^2=2E(r)+2M(r)/R^{n-1}, \tag{2.16.6}$$

where $M(r)$ is another arbitrary function. The coordinates are comoving ($u_\alpha=\delta^0{}_\alpha$), and the matter-density is:

$$\kappa\epsilon=bM_{,r}/(R^nR_{,r}). \tag{2.16.7}$$

Banerjee *et al.* did not explain what b is. In the case $n=2$, their result reproduces the $\Lambda=0$ L–T model (see our eqs. (2.12.1)–(2.12.4)), then $b=1$. When $E=0$, eq. (2.16.6) can be solved explicitly with the result:

$$R = \left(\frac{n+1}{2}\right)^{2/(n+1)} \left\{2M \, [t-t_0(r)]^2\right\}^{1/(n+1)}, \tag{2.16.8}$$

where $t_0(r)$ is another arbitrary function. A simple explicit solution also exists in a five-dimensional spacetime (i.e. with $n=3$), with arbitrary E.

Banerjee *et al.* found several properties of their solution, most of which are natural consequences of the properties of the L–T model. In particular, they showed that the model can be matched to the corresponding N-dimensional generalization of the Schwarzschild solution (which follows from (2.16.5)–(2.16.6) when M=const, see Krasiński and Plebański 1980), that naked singularities can develop, both of the shell-crossing and of the central shell-focusing type, and that an open recollapsing model analogous to Bonnor's (1985b) exists (see Section 3.8).

This completes the overview of solutions of the Szekeres–Szafron family.

3

Physics and cosmology in an inhomogeneous Universe

In this chapter, those papers will be described which discuss physical and astrophysical implications of the various properties of inhomogeneous cosmological models. A great majority of the papers are based on the L–T model. Those considerations which are not based on any explicit solution are described in Section 3.9. Some more cosmological considerations, based on the McVittie (1933) solution, will be presented in Section 4.7. The papers are sorted by the subjects they discuss; each section is devoted to one subject; the sections are ordered in chronological order of the earliest contributions.

3.1 Formation of voids in the Universe

The first predictions that voids should form in an inhomogeneous Universe were formulated by Tolman (1934) and Sen (1934) on the basis of the L–T model. Tolman predicted it in just a casual remark (see quotation in Section 2.12), while Sen made a thorough study of stability of the static Einstein and the FLRW models with respect to the L–T perturbation, and concluded explicitly that "the models are unstable for initial rarefaction".

In a follow-up paper, Sen (1935) considered the influence of pressure on the stability. That investigation is based on the Einstein equations in a spherically symmetric perfect fluid spacetime, without invoking any explicit solution. Depending on the spatial distribution of pressure, stability may be restored or instability enhanced, but this observation is not developed further.

Bondi (1947) predicted the formation of voids in just one phrase: ". . . if originally there is a small empty region round O and if the matter nearest to O does not move inward at first . . . then it will never move inwards". (O is the centre of symmetry.)

These considerations were revived with more vigour in the late 1970s. A preliminary study was done by Occhionero, Vignato and Vittorio (1978). They used the L–T model with $\Lambda = 0$ to describe the formation of density maxima and minima in a distribution of matter that has initially spatially homogeneous density, but inhomogeneous energy distribution ($E(r)/r^2 \neq$ const). Their motivation was to describe the density variation in clusters of galaxies. They chose an initial energy distribution having one minimum and one maximum, and then showed that the density contrasts will increase with time while the maximum and the minimum will be comoving with matter as the matter expands. The results were produced numerically, but from the exact solution of the Einstein equations. Occhionero *et al.*'s paper is an illustration of the fact that the initial density distribution does not yet define a model; in addition, the complete initial data requires energy distribution.

Olson and Silk (1979) calculated the evolution of two L–T configurations numerically: in one, $E<0$ in the neighbourhood of the centre and E grows monotonically to a value $E_\infty>0$ at spatial infinity; in the other $E<0$ everywhere and $E\to0$ monotonically as $r\to\infty$. Olson and Silk found that dense cores develop, but holes around them do not form spontaneously: they appear at late times only if they were present in the initial data. The authors saw this as a theoretical difficulty: voids will not automatically form around galaxy clusters, but must already be contained in the Universe at the initial moment. This result was obtained under the assumption that $t_0=$const.

Henriksen and de Robertis (1980) studied basically the same problem as Olson and Silk (1979), but allowed for a rarefaction surrounding the condensations at the initial moment. They found that the region of lower density will propagate outwards and will be preceded by a density wave.[22]

This was probably the first appearance of this phenomenon in the literature. Details of the process are complicated and are usually studied numerically, but its mechanism can be explained quite simply. Let us take (2.12.9), that is, the L–T model with $\Lambda=0=E$, and consider the values $r=r_m$ at which the density given by (2.12.4) has its maxima or minima at a given instant t. These are among the solutions of the equation:

$$\frac{\partial}{\partial r}\epsilon(t,r)=0. \tag{3.1.1}$$

From (2.12.4) and (2.12.9) we find that (3.1.1) leads to:

$$(-2MM,_{rr}t_{0,r}+4M,_r^2t_{0,r}+2MM,_rt_{0,rr})(t-t_0)-2MM,_rt_{0,r}^2=0. \tag{3.1.2}$$

Note that as long as $M,_r\neq0\neq t_{0,r}$ (i.e. as long as the model is nonempty and the Big Bang is nonsimultaneous), every solution of (3.1.2) will depend on time, $r_m=r_m(t)$. This means that spatial maxima and minima of matter-density are in general not comoving with the dust, but propagate across the flow-lines as density waves. Details of their motion depend heavily on the shapes of $M(r)$ and $t_0(r)$ (and of $E(r)$ in the general case).[23]

Occhionero, Vecchia–Scavalli and Vittorio (1981a,b) showed that the L–T model can describe the formation of condensations with voids around them and halos around the voids. The implication of the first paper, where the cosmological

---

[22] The paper contains several references to a planned "Paper II", but that one was never published (private communication from R. N. Henriksen).

[23] Such density waves were predicted by Ellis, Hellaby and Matravers (1990) on the basis of linearized perturbations on a FLRW background. However, their existence was in fact demonstrated by examples in all the papers described further in this section.

constant was assumed zero, seems to be that the size of the voids resulting from the calculation is about 10 times smaller than the observed size. In the second paper, the cosmological constant was included and adjusted so that the cavities can become larger. They are then 1/2 the observed size.[24]

Occhionero, Santangelo and Vittorio (1982 and 1983), in two consecutive papers, arrived at a workable method of describing the evolution of holes in the Universe. In the first paper, they performed numerical experiments with different shapes of the function $E(r)$ in order to find an initial energy profile that would evolve into a suffi-ciently deep hole without producing too large a density contrast in the surrounding ridge. With their best choice (see papers), a matter distribution is obtained in which the density in the hole is 10% of the background density, and the maximal $\delta\epsilon/\epsilon$ in the ridge is 1.[25] In the second paper, the authors showed that a lower-density region in a L–T Universe will have the tendency to expand faster than the Hubble flow of the background, pushing a wave of high matter-density in front of its boundary. Occhionero *et al.* suggested that the observed large-scale matter distribution in the Universe resulted from such a process. The new feature with respect to older works by Occhionero *et al.* was that here the authors did not assume a concentration of matter inside the low-density region, and this seems to have been the key to the success of their new description.

Working apparently independently of Occhionero and coworkers, Sato and coworkers arrived at the same idea and picked up where the former had left off. The consecutive papers develop the idea in more and more detail.

Sato (1982) considered linearized perturbations of a flat FLRW background and suggested that the "cellular structure" observed in matter distribution in the Universe might have resulted from several expanding voids. The shells surrounding different voids would collide after some time and would flatten against each other, forming "pancakes". The model used (separate perturbations in different places of a FLRW background) becomes inapplicable as soon as the first shells collide; the rest of the story is speculation, but it points towards a direction for further research.

Maeda, Sasaki and Sato (1983) studied (within the L–T framework) the evolu-tion of a configuration where a low-density FLRW region (i.e. a void) surrounds the centre of symmetry, is surrounded by a L–T transition region, and that one is in turn surrounded by another FLRW region with higher density; the outer FLRW region has positive curvature and recollapses. Maeda *et al.* showed that the rim of the void moves on a timelike hypersurface, and so can be intersected by light rays at all times. If the void occupies less than half the volume of space, then maximal density in the transition region is higher than the density of the unperturbed FLRW background. If it occupies more than half the volume of space, then minimal density in the transi-tion region is lower than in the void. The void has a tendency to expand forever, but

[24]   The reference in this paper, to "Occhionero, Vecchia–Scavalli, Vittorio (1981b), in preparation (Paper II)", could not be located.
[25]   One reference invoked in the paper, "Hoffman, Salpeter, Wasserman, in IAU Symposium 104", does not exist.

is eventually swallowed up in the final singularity of the surrounding positive-curvature FLRW region.

Sato and Maeda (1983) pointed out that Newtonian studies imply spherical symmetry to be a stable property in the expansion of voids, that is, initially nonspherical voids become more spherical during expansion. This is in contrast to nonspherical collapse, where spherical symmetry is an unstable property. Sato and Maeda also showed that the rim of an expanding void should, in a finite time, become a shell-crossing singularity. The equation of motion of the shell, in the Newtonian limit, is the same as the equation of a spherical blast wave created by an explosion, derived by Sedov (1959).

Maeda and Sato (1983a) investigated the expansion of a shell of zero thickness and finite surface density of matter in a dust medium having different (and spatially homogeneous) densities on the two sides of the shell. This was a way around considering the shell-crossing singularity as an actual entity. Maeda and Sato derived the equations of motion of the shell and the equation for mass-accumulation in the shell, and then solved the equations in the Newtonian approximation for those cases when the outside Universe has negative, zero or positive curvature. They concluded that the Newtonian approximation is unsuitable for asymptotic considerations.

Again Maeda and Sato (1983b) integrated numerically the equations derived in the preceding paper, in their full relativistic form, for all three types of FLRW background. If the background is flat ($k=0$), then shells situated inside the particle horizon accelerate their expansion towards a constant velocity, which is independent of their initial size. Shells larger than the particle horizon expand slowly until the horizon passes them, and then accelerate to the same limiting velocity. If the background is open, then the expansion of the shell is initially accelerated, but it subsequently slows down and freezes into the motion of the background. (See the remarks on the paper by Tomimura 1978 in Section 2.12; Tomimura discovered the same effect in the plane symmetric counterpart of the L–T model.) If the background is closed, the motion of the shell is always accelerated and tends to the velocity of light at the recollapse time. The dependence of the enlargement of the void on the time of its formation is represented on graphs. In general, the earlier the formation time, the larger the enlargement, but enlargement is increased in models with higher background density. In the closed background, the void may encompass the whole space before the final singularity is reached. This will happen with appropriately chosen initial conditions, for example, the initial size of the void.

The papers by Sato *et al.* described above are extensively and very clearly summarized by Sato (1984) himself. This article is recommended as further reading to those readers who are interested in the subject.[26]

Note that Sato *et al.*'s results imply that the configuration considered by Einstein and Straus (1945 and 1946) is unstable, see Section 3.3.

---

[26] The paper by Qadir and Wheeler, to which reference is made in this article and in some other articles, exists only in preprint form (private communication from A. Qadir).

The investigation was later continued in two papers by Suto, Sato and Sato (1984a,b) that are in fact beyond the scope of this book, because they are based on numerical integration of the Einstein equations for a spherically symmetric metric with nonzero pressure in the source. In the first paper, the source is a perfect fluid, in the second one, the source is a mixture of ideal gas and radiation. It is found that small-size voids will be smoothed out by dissipation, but sufficiently large voids will survive. The pressure gradients prevent, as was expected before, the formation of a shell-crossing singularity. Suto *et al.* found numerically that a peculiar bottleneck forms at those values of r where $E(r) = -\frac{1}{2}$. This effect was first identified and described by Barnes (1970) for the L–T model. A light ray arriving at the neck after collapse has set in will not be able to pass through it. Suto, Sato and Sato discovered that the effect persists with nonzero pressure gradients.

Hoffman, Salpeter and Wasserman (1983) calculated numerically the possible evolutions of voids in backgrounds with $\Omega \leq 1$ (Ω is the density parameter of the FLRW cosmology). Specifically, they considered observational consequences of various initial conditions. They concluded that deep holes (i.e. with large contrast) can form only when Ω is close to 1, and are always surrounded by dense thin ridges. The holes and ridges can form honeycomblike structures even in the absence of dissipation. The velocities of the ridge and of matter in the hole can differ from the Hubble flow by 25–30% which leads to errors of the same magnitude in estimating the size of the holes. Voids cannot in general form around overdensity cores. The time evolution of matter-density in the holes is nearly the same when the hole is alone in the Universe and when there is a lattice of holes. Hoffman *et al.* considered the influence of these structures on the temperature distribution of the microwave background radiation, and arrived at the conclusion that with $\Omega \ll 1$ the temperature variations would be "considerably larger" than the observational limits; with $\Omega = 1$ they are "only slightly larger". This question will be considered in our Section 3.7.

Zhu (1983) investigated the problem whether a void expands faster or slower than the surrounding Universe. However, the investigation involves a plain error. The author stated that with $M = M(r)$ the void has a constant mass, while for $M = M(t,r)$ it has varying mass. In fact, the second case does not obey Einstein's equations, and so does not provide any information. With $M = M(r)$, Zhu proves the rather trivial fact that a parabolic void in a parabolic background will have a smaller Hubble parameter than the background has.

Lake and Pim (1985) studied the evolution of voids by modelling them with a Schwarzschild vacuole matched to a surrounding FLRW region through a surface matter distribution. The most important results were as follows:

1. With the initial epoch, the Schwarzschild mass and the initial size of the void all given, there is a minimum initial growth rate below which the void collapses.
2. With the initial epoch and the growth rate given, there is a minimal initial size of the void below which the void will collapse.

3. Increasing the Schwarzschild mass slows down the expansion of voids, and above a certain maximum the void will collapse.

Note how these results complement those of Sato *et al.* in implying the instability of the Einstein–Straus configuration. Lake and Pim stress the contrast between the conclusions given above and some of those reached by Maeda and Sato (1983). However, there is no contradiction between them: Maeda and Sato assumed that the void contains a low-density but nonvacuum FLRW medium, while Lake and Pim took the void to be perfectly empty with a point-mass at the centre. It is not strange that such different configurations behave differently.

Pim and Lake (1986) continued the study of Lake and Pim (1985). They considered the void to be filled with a FLRW radiation, and the background to be the flat FLRW spacetime. The conclusions given above are then modified as follows:

1. The photon density in the void now plays the same role as the Schwarzschild mass did before. The minimal initial growth rate is now lower than for a vacuum void, and voids grow more readily.
2. Increasing the photon density in the void helps the void to expand. It is even possible, with appropriate initial data, that the radiation pressure will halt the collapse of the void and change it into expansion. With a vacuum void, collapse was irreversible once it had begun.

At late times voids grow as fast as the particle horizon. Hence, the observed voids must be much younger than the Universe.

In another paper (Pim and Lake 1988), the authors extended the study to the case of open and closed backgrounds. The radiation pressure in these cases also turns out to help the voids to expand. If the Universe is closed, then the blackbody radiation within voids would have to have a significantly higher temperature than in the background. Small voids grow faster, but they do not grow at all when they are too small.

The problem of formation of voids was also studied by Bonnor and Chamorro (the two papers contain an overview of observations of structure in the Universe). In the first paper (Bonnor and Chamorro 1990) the authors considered the following configuration: Minkowski spacetime around the centre matched to the L–T model with $\Lambda=0$ across a sphere, the L–T model asymptotically going over to a FLRW model far from the void. The conclusions of the paper are: (1) no such void can exist in a Universe with positive spatial curvature; (2) a nonexpanding void can exist only in a zero-curvature background; (3) an expanding void may exist only in an exterior region of negative spatial curvature. It should be observed that Bonnor and Chamorro take the time-variability of $R(t,r_0)$ (where r_0 is the comoving outer radius of the void) as the indicator of expansion. This is the proper geometric radius of the void. In fact, what is cosmologically relevant is the expansion or not of the void with respect to the background matter, that is, whether the rim of the void is comoving or not. Also, the assumption that the interior of the void is not only of low density, but is simply empty and flat, is a very strong constraint on the model.

Bonnor and Chamorro find that the zero-curvature exterior of a void that is non-expanding (in their sense) has its Big Bang in the infinite past on the boundary of the void, and at progressively later times at greater distances. If $M \geq 0$ and $M_{,r} \geq 0$, then there will necessarily be a region of shell-crossing and negative matter-densities around the boundary of the void, but "these unphysical features can be confined to early times". Next, Bonnor and Chamorro considered an expanding void, but under the assumption of a simultaneous Big Bang in the background. Here, the arbitrary functions can be chosen so that there will be no shell-crossings at any time anywhere, and the authors gave an explicit example of such a model. The authors concluded that, at face value, their results imply that the existence of voids excludes the closed FLRW model. However, conclusions may be different if the void is not assumed completely empty (and, in fact, they were shown to be different by Occhionero *et al.* and Sato *et al.*).

In the next paper (Bonnor and Chamorro 1991), the authors extended their investigation. The void was assumed to be nonempty, but the boundary of the void and the boundary of the high-density envelope of the void were forced (by assumptions about initial data) to be comoving with the dust. (This assumption kills several of the interesting effects described by Occhionero *et al.* and Sato *et al.*, and has in fact been shown in the latter papers to be an unstable configuration.) Outside the high-density envelope the spacetime is assumed to be FLRW. Under these assumptions, an upper bound on the density in the void follows if the background is hyperbolic. In an elliptic background, the average density in the void with a simultaneous Big Bang must obey $\bar{\epsilon} > 8 \cdot 10^{-30} \mathrm{g/cm^3}$. If our Universe is recollapsing, then such a density is not appreciably below the average cosmic density, $10^{-29} \mathrm{g/cm^3}$, and so the void would "hardly be noticeable". This problem can be solved satisfactorily if it is assumed that matter in the voids is very much older than outside, that is, that the Big Bang in the void occurred earlier (the age of the Universe should be 10^{10} years outside and 10^{11} years in the void). Another solution would be to make the void parabolic and the background elliptic, but then shell-crossings would appear, which the authors want to exclude.

Chamorro (1991) developed this approach further. He considered a central FLRW region of low density, surrounded by a L–T transition region, which is in turn surrounded by an exterior FLRW region of higher density. Both boundaries between the regions were assumed comoving, all three regions were elliptic (recollapsing), and the initial data was chosen so that no shell-crossings ever appear. The main point of the paper was a demonstration on an explicit example that initial data exist that do not lead to shell-crossings and surface layers of matter. With such data, if the Big Bang is simultaneous across all three regions, then $\epsilon_v/\epsilon_b \geq 0.68$, where ϵ_v and ϵ_b are matter-densities in the void and in the background respectively. This ratio is too high compared with observational implications. When the condition of simultaneity is relaxed, the density ratio can be made as small as necessary, but then matter in the void must be older than that outside, and the L–T transition region must be stratified: close to the void, the matter must be older than in the void, and close to the outside FLRW region the matter must be younger than in the background.

As can be seen, while Sato *et al.* decided to approximate the shell-crossing singularity by a nonsingular layer or to prevent it by including pressure gradients numerically, Bonnor and Chamorro chose to exclude the singularity by assumptions about initial conditions. This results in difficulties with adjusting the model to observational data.

Meszaros (1991) considered the $\Lambda=0$ L–T model with shell-crossings as a model for structure formation. His main results were as follows:

1. A sufficient condition for shell-crossing in the L–T model was derived, stronger than that of Hellaby and Lake (1985a,c).
2. On a specific example of the L–T model it was shown that it is possible to choose initial conditions so that the model starts as a homogeneous one and after the time equal to the age of the Universe develops a void of size 10–100 Mpc surrounded by a shell-crossing singularity.
3. If the inhomogeneity of the L–T model is small enough, then the perturbative theory applies to it.

The second point implies that it is possible to have a L–T model which develops into a configuration similar to the present Universe but whose inhomogeneity leaves no imprint on the background radiation. The first point seems to be just a misunderstanding – Meszaros apparently overlooked the progress made by Hellaby and Lake in understanding the difference between a shell-crossing singularity and a local extremum of the function $R(t,r)$ which is a "neck" in the space – see comments on the papers by Hellaby and Lake (1985a,c) and by Hellaby (1987) in our Section 3.8.

The same author (Meszaros 1993) studied L–T voids with all possible combinations of the following properties:

1. Within the void, there is either the Minkowski vacuum, or a L–T medium with $E<0$, $E=0$ or $E>0$.
2. In the envelope of the void, the L–T model has $E<0$, $E=0$ or $E>0$.
3. The FLRW background (outside the envelope) has $k<0$, $k=0$ or $k>0$.
4. Shell crossings in the L–T region are present or absent.

There are 72 such combinations altogether. Some of them can be rejected as self-contradictory (e.g. shell-crossings necessarily occur), some others can be rejected on observational grounds (e.g. if matter in the void has $E>0$, while it is known that its density is less than 10% of the background density, then the background density comes out definitely too high). Only 14 combinations survive the scrutiny (see paper). The main result of the paper can be stated as follows: "the discovery of voids is an observational support that either the Universe is open or there is a shell-crossing" (Meszaros 1993). The author then showed that a perfectly empty void with static edges is unstable against the introduction of mass at its centre.

3.2 Formation of other structures in the Universe

Under this heading we will consider papers aimed at describing galaxies, clusters of galaxies and other phenomena which can be treated by similar mathematical methods.

Formation of galaxies (then called "nebulae") was considered as early as 1933 by Lemaître (1933c). Using (2.12.15)–(2.12.19) with $m_{,t}=0$ and $\Lambda \neq 0$ he showed that the initial mass distribution may be set up so that a region of finite comoving size r_0 around the centre will recollapse while the region $r > r_0$ will keep expanding forever. In this particular situation, the curvature of the space is positive everywhere and the expansion of the outer region is caused by the cosmological repulsion. Lemaître stated that this local instability may provide the mechanism for the formation of the "nebulae".

The next paper by Lemaître (1934) is descriptive rather than mathematical. It elaborates the prediction that an inhomogeneous model of the Universe should contain collapsing regions spread over an expanding background. Assuming that the expanding background is FLRW, Lemaître presented an astrophysical study relating the parameters of the observed "nebulae" to the hypothetical collapsing regions in the FLRW background.

Tolman (1934) and Sen (1934) both concluded that the L–T model implies the instability of the static Einstein and the FLRW models against the formation of condensations. This was a good starting point for considering galaxy formation within the exact theory. However, several years later, galaxy formation became the subject of perturbative calculations and developed into an almost separate science.

Bonnor (1956) was the first to make a quantitative prediction concerning formation of galaxies based on the L–T model. He considered the following configuration: a FLRW dust tube around the centre of symmetry, surrounded by a L–T transition zone, and that one surrounded by another FLRW dust region so that at any $t = t_0 =$ const, the density in the outer FLRW region is different than that in the inner FLRW region. The boundaries of the FLRW regions were assumed comoving.[27] If both FLRW regions have positive spatial curvature and the density in the inner FLRW region is higher than that in the outer region, then the inner region will start to recollapse earlier than the background and will form a condensation. Bonnor assumed that the condensation has the mass of a typical galaxy, that is, it contains $N \cong 3 \cdot 10^{67}$ nucleons, and that it formed at $t_i \cong 1000$ years after the Big Bang. Then the following problem arises: if such a condensation formed as a statistical fluctuation in a homogeneous background, then the initial density contrast is $\delta\epsilon/\epsilon = |\epsilon_c - \epsilon_b|/\epsilon_b \cong N^{-1/2} \cong 10^{-34}$, where ϵ_c is the density in the condensation and ϵ_b is the background density. However, in order to develop into a galaxy of typical density, the initial perturbation at t_i would have to be of the order $\delta\epsilon/\epsilon \cong 10^{-5}$. On

---

[27] The whole configuration may be described as a single L–T model with appropriately chosen shapes of $E(r)$ and $M(r)$.

the other hand, if a perturbation of the order of 10^{-5} is to arise as a statistical fluctuation, then it can involve only 10^{10} particles.

If the outer FLRW region has negative curvature while the inner FLRW region has positive curvature, then the initial perturbation has to be about 10 times larger than in the preceding case. If both FLRW regions have negative curvature, a galaxy cannot form at all by gravitational condensation. The introduction of the cosmological constant does not help in a model which begins from a Big Bang. Hence, two possibilities are left: either $\Lambda \neq 0$ and the Universe begins as an instability in the Einstein static Universe in the asymptotic past (then there is an arbitrary amount of time available for the statistical fluctuations to grow) or there exists a mechanism for producing large perturbations.

This much is contained in Bonnor's paper. It seems that one more possibility should be considered: the initial fluctuation may have a smaller mass than the galaxy into which it will develop, and may accrete matter after it forms. The boundary of the condensation would then have to be assumed noncomoving, and its evolution should be followed by the methods developed for investigating the voids (see Section 3.1). No such consideration has so far been published.[28]

The calculations based on perturbative schemes narrowed the gap between observations and theory, but the qualitative problem still remains unsolved: either the initial perturbations were larger than statistical fluctuations, or else there was not enough time for them to grow into galaxies.

Novikov (1964a) used a similar configuration to that of Bonnor to describe another process that is in principle observable. The arbitrary functions in the model can be chosen in such a way that the resulting spacetime consists of a flat FLRW background with several empty holes in it, each described by a Schwarzschild solution. Around the centre of each hole there is a flat FLRW region that has its local Big Bang later than the background. Direct emission of particles from such "cores of delayed expansion", or emission of light in consequence of collisions between matter emerging from the core and matter falling into it, might be observable. Novikov suggested this as a mechanism for the emission of energy from quasars. Such an explanation is no longer considered valid, but the theoretical possibility remains that regions of delayed Big Bang might be observable (see Eardley, Liang and Sachs 1972, and a few papers discussed below). The idea of "lagging cores" of Big Bang seems to go back to Ambartsumyan (1964), but he did not back it up with any calculations.

Neeman and Tauber (1967) discussed Novikov's (1964a) idea in more detail, using Bonnor's (1956) configuration. The results were as follows:

1. If the delayed core is described by a FLRW model with $p = 0 = \Lambda$, then neither $k = +1$ nor $k = -1$ can be fitted to yield the right order of magnitude of the radius of the core so that it can describe a quasar.

---

[28] This suggestion was made by S. Bażański, private communication.

2. For a FLRW model with $\epsilon = p$ and $\Lambda = 0$, the size comes out right, but the core is expanding too fast and so cannot be "lagging".
3. Introduction of the cosmological constant makes it possible to fit theoretical results to observations in both cases 1 and 2, but then the cosmological constant in the core would have to be different than that in the background Universe.

Neeman and Tauber then discussed a pulsating core, but that discussion is not based on any exact solution.

Kantowski (1969) argued that the L–T model might be appropriate for describing the density variations in large clusters of galaxies. The author showed that the spatial distribution of density at the present time can be fitted to observations, but did not attempt to calculate the time evolution. Then Kantowski hypothesized that the Coma cluster might be a portion of the L–T model that is already recollapsing, and that other clusters might exist where the recollapse has already advanced to the stage of galaxy collisions.

Tomita (1969a) calculated numerically the evolution of condensations in a flat FLRW background under the assumption that the boundary of the condensation is comoving. The calculation did involve some approximations resulting from the assumption that the initial density contrast is small, but it is otherwise based on the L–T model (the small density contrast was, however, larger than statistical fluctuation). Tomita found that the density contrast of a condensation grows, the central region becomes gravitationally bound at the time τ given by $H_0\tau = 0.05$ (H_0 is the present value of the Hubble parameter, at present $\tau \cong \frac{2}{3}$) and grows in size, while a trough surrounding the condensation develops. If a small-scale condensation forms in the middle of a large-scale one, then the growth of the density contrast in the smaller condensation is faster than it would be in an isolated small condensation. The trough around the small condensation becomes a local minimum of density, but the density in it grows with time.

In the next paper (Tomita 1969b), the author considered the possible influence of the cosmological constant on the process. He assumed the background to be of the Lemaître type (that is, with a long stagnation epoch) and found that in such a model the rate of growth of condensations during the stagnation period is very small, so that the efficiency of the condensation process is not increased much in this case.

Trevese and Vignato (1977) studied the development of clusters of galaxies in more detail. They took the arbitrary function $E(r)$ to be:

$$E(r) = -\gamma r^2 \exp[-(\beta r)^{\delta(r)}], \tag{3.2.1}$$

where $\delta(r) = \delta_0 + \delta_1 \exp[-(\beta r)^{-\lambda}]$, β, γ, δ_0, δ_1 and λ being arbitrary constants. This has the properties that (I) it approximates the recollapsing FLRW model for small r, (II) it approximates the flat model when $r \to \infty$, (III) no shell-crossings occur. With such initial data, Trevese and Vignato calculated the matter distribution within a cluster of galaxies and its time evolution. They concluded that the resulting density distribution and velocity dispersion are in good agreement with the observational data.

Moffat and Tatarski (1992) applied the L–T model with $E=0$ to the description of structure formation in the Universe.[29] They calculated the formula for redshift of light emitted by a distant source and observed at the centre and showed how the redshift can be decomposed into a part due to expansion and a part due to the difference in the gravitational potential. By adjusting the bang-time function one can accommodate all objects with the "age problem" into the model, at the price of assuming that matter away from the centre is older than that at the centre. Moffat and Tatarski stressed that in order to determine the evolution of the model it is more appropriate to specify the initial density distribution on the past light cone of the observer, instead of doing it on a $t=$const hypersurface. They showed numerically that if the density distribution on the $t=$now hypersurface has the form of a Gaussian curve, then the redshift has a maximum of about 0.4, and the density distribution along the light cone has a minimum followed by a maximum. However, if the same initial density distribution is assumed on the past light cone of the observer, then redshift behaves similarly as in FLRW, and the distribution of ρ on $t=$now has a minimum. In the second case, sources with a large redshift would be emitting their light much earlier than in a FLRW model. Finally, for several density distributions along the light cone, Moffat and Tatarski calculated the bang-time functions numerically. These are instructive, but may be specific to the assumed case $E=0$. The redshift vs. time graphs are very similar to the FLRW ones. The authors apparently failed to notice the difference between the Big Bang at $R=0$ and the shell-crossing singularity at $R_{,r}=0$, but otherwise their paper is a very important contribution, and very relevant for future observational cosmology. It also lists several references on observations of large-scale structure.

Kurki–Suonio and Liang (1992) used the L–T model to demonstrate the ambiguity in reading out the spatial distribution of matter from redshift measurements. The observations provide the density of matter in the redshift space, $\hat{\epsilon}(z)$, and this is related to the functions of the L–T model by:

$$\hat{\epsilon}(z)=\frac{2H_0{}^3M_{,r}}{\kappa(1+z)z^2R_{,tr}},$$ (3.2.2)

where H_0 is the local present value of the Hubble "constant" at the location of the observer (the centre of symmetry) and z is the observed redshift; the values of t and r in (3.2.2) refer to the time and position of emission. Equation (3.2.2) connects one observed quantity, $\hat{\epsilon}(z)(1+z)z^2$, to two unknown functions, $M(r)$ and $R(t,r)$, and obviously cannot determine both. Kurki–Suonio and Liang investigated numerically the dependence of these functions on $\hat{\epsilon}(z)$ when various shapes of $t_0(r)$, $E(r)$ and initial density distribution $\epsilon(t,r)$ are assumed. They also assumed no shell-crossings

---

[29] The authors quoted the COBE measurement as a confirmation of isotropy of the microwave background radiation. This may be one of the historically last such quotations; COBE detected anisotropies in the background radiation soon afterwards (Mather *et al.* 1993).

$(R_{,r}>0)$ and redshifts monotonically growing with distance $(R_{,tr}>0)$. The results, greatly abbreviated, are as follows:

1. If $t_0(r)$=const, then the amplitudes of overdensity in the redshift space are larger than the amplitudes of proper density by 40% or more. The overdensity regions in proper density have larger sizes than their images in the redshift space.
2. When $M(r)/r^3$=const and $E(r)/r^2$=const, that is, when the bang-time function is the only factor generating inhomogeneity, overdensities in the redshift space translate into *underdensities* in proper density, of larger size and much larger amplitude.
3. Since the two effects influence observations in opposite directions, they can momentarily cancel each other, producing a homogeneous density in the redshift space while the Universe is highly inhomogeneous. Conversely, the Universe may be momentarily very smooth in density, but appear highly inhomogeneous in the redshift space because of the effects of inhomogeneous velocities. To illustrate this last possibility, Kurki–Suonio and Liang fitted the velocity function in the L–T model so that the proper density is constant on the past light cone, while the density in redshift space is the same as in one of the deep-redshift observational surveys. The proper density at the present time then has smaller amplitudes in the peaks, and less regular distribution of the peaks. The authors concluded that using the FLRW relation between redshift and comoving distance to describe inhomogeneities is "fundamentally self-inconsistent". This may be the most important point of this whole book. They, further, said: "What we see on the past light cone are only momentary 'images'".

Kurki–Suonio and Liang's paper contains several graphs and spacetime diagrams that illustrate the conclusions more clearly. In particular, the spacetime diagrams show very pictorially how the bang-time nonsimultaneity generates decaying inhomogeneities while variations in initial energy generate growing inhomogeneities, even though these processes are not discussed in the paper.

3.3 Influence of cosmic expansion on planetary orbits

Early studies of this problem, based on McVittie's solution, were performed by McVittie (1933) and Järnefelt (1940a and 1942). They will be described in Sections 4.6 and 4.7. From the physical point of view they were not quite correct because McVittie did not define the radius of the orbit in any invariant or measureable way, while Järnefelt allowed the small parameter in the perturbative calculation to depend on time.

The first formally correct study of the problem of expansion of orbits was carried out by Einstein and Straus (1945, with an erratum 1946). They showed that the Schwarzschild solution can be matched to any FLRW model. This implies that the planetary orbits are (in this configuration) not influenced by the expansion of the Universe. The Schwarzschild mass m is related to the FLRW mass integral M_0 by:

$$m = M_0 r_0^3/(1+kr_0^2/4)^3 \overset{\text{def}}{=} \mu(r_0),$$ (3.3.1)

where r_0 is the radius of the Schwarzschild vacuole in the FLRW coordinates of (1.3.1). If the FLRW model is flat ($k=0$), then (3.3.1) simply means that the FLRW mass removed from the vacuole equals the Schwarzschild mass put in. The geodesic radius of the vacuole, $R(t) \int_0^{r_0} r \, (1+\frac{1}{4}kr^2)^{-1} dr$, expands together with the Universe.

Einstein and Straus' result was for many years taken as the general implication of Einstein's relativity theory. Note, however, that the matching condition (3.3.1) need not be fulfilled if the Einstein–Straus configuration is taken only at a single moment $t=t_0$ as an initial condition for a L–T model. Then, the results of Occhionero *et al.* and of Sato *et al.*, described in Section 3.1, imply that if $m<\mu(r_0)$, then the boundary of the vacuole will expand faster than the FLRW background. If $m>\mu(r_0)$, then initial conditions may be set up so that the vacuole will start to collapse (see the paper by Lake and Pim, 1985, our Section 3.1). This indicates that the Einstein–Straus configuration is unstable against perturbations of the condition (3.3.1), that is, it is an exceptional situation.

Only two other studies of this problem were published. Noerdlinger and Petrosian (1971) investigated the motion of galaxies in galaxy clusters using McVittie's (1933) solution. Their paper will be described in Section 4.7. The other study was by Gautreau (1984). He based his study on the subcase $E=0$ of the L–T model that he derived in the curvature coordinates (see Section 2.13). In these coordinates, quantities depending only on R are defined in terms of the curvature radius of the orbits of the symmetry group; these do not participate in the cosmic expansion and therefore R of any single orbit can be used as a standard of length.

In the curvature coordinates, the solution (2.13.10)–(2.13.13) takes the form:

$$ds^2 = A\tau_{,T}^2 dT^2 - A^{-1} dR^2 - R^2(d\vartheta^2 + \sin^2\vartheta \, d\varphi^2),$$ (3.3.2)

where:

$$A = 1 - (2\mu + Z)/R + \tfrac{1}{3}\Lambda R^2,$$ (3.3.3)

and $\tau(T,R)$ is defined by:

$$\tau_{,R} = -\epsilon\sqrt{1 - A}/A,$$ (3.3.4)

$\varepsilon = \pm 1$. The function Z is given by (2.13.12), and $\mu=$const is the mass of the star. The metric (3.3.2) applies only outside the star, that is, for $R \geq R_b=$const>0, the metric for the inside of the star is not considered. By investigating the equations of timelike geodesics in such a metric, Gautreau showed that circular orbits do not exist. This is in fact a purely Newtonian phenomenon: in the model (3.3.2)–(3.3.4), the smoothed-out cosmic matter-density extends throughout the planetary system, and, as a result of cosmic expansion, matter streams out of every sphere $R=$const. Hence, each planet moves under the influence of a gravitational force that is decreasing with time, and so the orbit must spiral out. Gautreau derived the Newtonian formula for the rate of change of orbital radius:

$$\frac{dR}{dt}=8\pi R^4 H\bar{\rho}/(2\mu),\tag{3.3.5}$$

where R is the orbital radius, H is the Hubble parameter, $\bar{\rho}$ is the mean cosmic density of matter and μ is the mass of the star. The effect is thus greater for orbits of greater radius; for Saturn it is:

$$\left(\frac{dR}{dt}\right)_S=6\cdot10^{-18}\text{m/year}.\tag{3.3.6}$$

This is obviously unmeasureable (one proton diameter per 1000 years). For a star at the edge of the Andromeda galaxy the effect would be:

$$\left(\frac{dR}{dt}\right)_{gal}=1100\text{ km/year}.\tag{3.3.7}$$

From the "practical" point of view Gautreau's result thus implies that planetary orbits do not react to the expansion of the Universe. However, it is important to know that in principle the effect is nonzero. In the Einstein–Straus approach, it was exactly zero. As explained above, the model of Einstein and Straus is unstable against the perturbations of (3.3.1), and so is less realistic than the model of Gautreau.

3.4 Formation of black holes in the evolving Universe

The precursor of all the papers discussed here was the paper by Oppenheimer and Snyder (1939). These authors discussed the collapse of a FLRW dust cloud matched to the Schwarzschild solution. They found that the collapse to the Schwarzschild horizon $r=2m$ takes a finite proper time for each infalling particle, but an infinite amount of time for a static distant observer in the Schwarzschild region who would see the star gradually redden. For a comoving observer on the surface of an object of the mass of the Sun, but rarefied so that the initial density equals that of water, the time needed to reach the horizon would be of the order of one day. As the horizon is approached, light can escape outwards within a cone around the radial direction that becomes progressively narrower and closes completely at the horizon.

The novel idea in the Oppenheimer–Snyder paper was to study the black hole in the process of its formation. The elaborate theory of black holes developed from the 1960s onwards is based on stationary vacuum solutions, that is, it describes black holes that have always existed and are observed from afar. Oppenheimer and Snyder were the first to take a look at the collapsing matter from inside. The L–T model allows for an even more sophisticated approach (see below).

The evolution of the Oppenheimer–Snyder configuration was described with more clarity by Szekeres (1972).

Bondi (1947) observed that if matter in the L–T model is collapsing with a great velocity so that $[1+2E(r)]^{1/2}+\partial R/\partial t<0$, then along a light ray emitted away from the centre, with the tangent vector k^μ, $k^\mu R_{,\mu} < 0$, that is, the ray is forced to move inward. A necessary condition for this is $R < 2M$, that is, a sufficiently high matter-density over a sufficiently large region. With hindsight, this was a prediction that black holes would form under certain conditions.

Raychaudhuri (1966) considered collapse in a L–T model. The model was arranged so that $E(r) < 0$ for $r < a$, $E(r) = 0$ for $r = a$ and $E(r) > 0$ for $r > a$. Then each shell with $r \leq a$ begins to recollapse at a certain time. Raychaudhuri claimed that light rays emitted radially from the recollapsing region after the moment of maximum expansion will never reach any observer comoving with $r > a$ (this moment of maximum expansion is different on every sphere). However, this result is incorrect, see below.

Barnes (1970) developed the approach of Raychaudhuri (1966) further. Barnes formulated the conditions for the absence of shell-crossings ($M_{,r} \geq 0$ and $M_{,r}/M - 3E_{,r}/(2E) \geq 0$ for $E > 0$, $M_{,r} \geq 0$ and $E_{,r} \leq 0$ for $E \leq 0$) and corrected Raychaudhuri's condition for the formation of a black hole. The condition $R_{,t}(t_i, r_i) < 0$, that is, collapsing motion of the shell $r = r_i$, is a necessary, but not sufficient, condition for the ray emitted radially from $r = r_i$ at $t = t_i$ to become trapped. The sufficient condition is:

$$R(t_i, r_i) \leq 2M(r_i); \tag{3.4.1}$$

this means that the emitting shell must have fallen within its Schwarzschild radius. For collapsing matter $R_{,t} \leq 0$, and so $(2M/R)_{,t} \geq 0$, that is, once (3.4.1) holds for $t = t_i$, it will continue to hold for all $t > t_i$. Hence, a radial ray that meets a shell obeying (3.4.1) anywhere along its path will become trapped and will never escape beyond $r = a$. A sufficient condition for a radial ray to escape beyond $r = a$ is $2M(a)/R(t_i, r_i) < 1$. The trapping or escape of rays emitted at different moments is illustrated by numerical examples in the paper. For each direction, there is one and only one ray that reaches $r = a$ asymptotically in the future.

Demiański and Lasota (1973) reproduced Barnes' (1970) results by a slightly different method.

Polnarev (1977) repeated Barnes' (1970) and Demiański–Lasota's (1973) considerations on a simpler example: the Big Bang was assumed simultaneous and the region outside the matter that eventually falls into the black hole was assumed parabolic ($E(r) = 0$). The results are in agreement with the two preceding papers; the process of collapse is discussed in more detail in this paper.

Papapetrou (1978) considered the same configuration as did Polnarev.

Liang (1974) presented a more elaborate discussion and spacetime diagrams for the process of black hole formation, but with fewer details of the calculation. In the same paper, Liang considered plane symmetric dust collapse and showed that if the dust density is positive everywhere, then such a solution cannot be matched to any static plane symmetric vacuum solution – the exterior spacetime is then the Kasner solution, and the collapse terminates in a spacelike singularity. If the dust density is negative, then collapse is stopped and reversed, the exterior spacetime is then the static Taub solution (the plane symmetric counterpart of the Schwarzschild solution, compare the papers by Horsky, Lorenc and Novotny (1977) and Novotny and Horsky (1979) described in our Section 2.12). The cylindrically symmetric dust collapse was studied perturbatively, and that consideration is outside the scope of this book.

3.5 Collapse in electromagnetic field

Before exact solutions for this problem were considered (see Section 2.14), Novikov (1966 and 1970) discussed this process qualitatively, by investigating observations of a collapsing charged sphere from the Reissner–Nordström spacetime. The 1966 paper was mentioned in Section 2.14, in the 1970 paper Novikov showed that a static portion of charged matter located between the Cauchy horizon and the singularity in the Reissner–Nordström spacetime can prevent the emergence of the collapsing cloud into another asymptotically flat region if the charges of the cloud and of the additional portion are properly balanced. What Novikov did not say, though, is that the additional matter would have to be created out of the singularity at a definite moment in time, and then disappear into the singularity after a finite time.

Raychaudhuri and De (1970) investigated the properties of charged dust without solving the Einstein–Maxwell equations and without assuming any symmetry. The results obtained were as follows:

The charge density ρ, the electric field E^α, the magnetic field H^α, the vorticity vector $w^\alpha = (-g)^{-1/2} \epsilon^{\alpha\beta\gamma\delta} u_\beta u_{\gamma,\delta}$ (where ϵ is the Levi-Civita symbol), the acceleration vector $\dot{u}^\alpha$, the shear $\sigma^{\alpha\beta}$ and the expansion θ are related by:

$$E^\alpha{}_{;\alpha} + E^\alpha \dot{u}_\alpha - 2H^\alpha w_\alpha = 4\pi\rho, \tag{3.5.1}$$

$$H^\alpha{}_{;\alpha} + H^\alpha \dot{u}_\alpha + 2E^\alpha w_\alpha = 0, \tag{3.5.2}$$

$$2H^\alpha \sigma_{\alpha\beta} - \tfrac{1}{3}\theta H_\beta - u^\alpha(H_{\beta,\alpha} - H_{\alpha,\beta}) + (-g)^{-1/2} g_{\beta\mu} \epsilon^{\alpha\mu\lambda\sigma}(u_\alpha \dot{u}_\lambda E_\sigma + u_\lambda E_{\sigma,\alpha}) = 0. \tag{3.5.3}$$

If the dust has zero electric conductivity, then in addition:

$$\epsilon \dot{u}^\alpha = \rho E^\alpha, \tag{3.5.4}$$

$$2E^\alpha \sigma_{\alpha\beta} - \tfrac{1}{3}\theta E_\beta - u^\alpha(E_{\beta,\alpha} - E_{\alpha,\beta}) - (-g)^{-1/2} g_{\beta\mu} \epsilon^{\alpha\mu\lambda\sigma}(u_\alpha \dot{u}_\lambda H_\sigma + u_\lambda H_{\sigma,\alpha}) = 0. \tag{3.5.5}$$

If the magnetic field is zero, then the electric flux through any surface element bounded by dust particles is a constant of the motion. For $w_\alpha = H_\alpha = 0$, E_α is collinear with $(\rho/\epsilon)_{,\alpha}$ and if $p = 0$ in addition, then the motion cannot be shearfree. The paper presented more formulae, among them the generalization of the Rainich conditions to a nonvacuum case.

Bekenstein (1971) investigated spherically symmetric charged perfect fluids, again without solving the Einstein–Maxwell equations. The results of the paper include: a generalization of the Lemaître–Podurets–Misner–Sharp approach (see our Section 1.4) to the case of charged matter, the equation of hydroelectrostatic equilibrium, and another derivation of the Christodoulou formula $M=m_0+e^2/(4m_0)$, expressing the active gravitational mass M of a black hole in terms of its irreducible mass m_0 and charge e. Bekenstein showed that with nonzero pressure, charge will not necessarily prevent the Big Bang singularity. This is so because the positions of the horizons are not static when $p\neq0$, and the inner horizon itself is collapsing. The reversal of the matter collapse can occur only inside the inner horizon; whether shells of collapsing matter manage to overtake the collapsing horizon depends on the details of the equation of state.

Gertsenshtein and Stanyukovich (1974) criticized Novikov's (1966) approach. They argued that the re-emergence of the charged ball of matter into another asymptotically flat sheet of the Reissner–Nordström spacetime should "be preserved also when the charge tends to zero", while the L–T solution does not have this property. The whole argument is a misunderstanding, and the paper is mentioned here only for completeness, because it discusses the $\Lambda=0$ L–T model.

Pavlov (1976) wrote the set of first integrals found by Vickers (1973) and then presented the Reissner–Nordström solution in the same coordinates. He expressed the constants in the R–N solution through the boundary values of the functions in the interior solution. The main achievement of the paper is supposed to be a new coordinate system in which the Vickers equations can be expressed through "quadratures". This progress is of dubious value since the solution is still defined by differential equations that cannot be integrated in any simple way (though each is solved by calculating a single integral). Moreover, the physical interpretation of the new coordinate T is nowhere explained. The author's motivation was apparently the same as that which led Ori (1990) to his result, see Section 2.14.

Pavlov and Bronnikov (1976) investigated the Vickers models as reformulated by Pavlov (1976) for the occurrence or absence of singularities. Some of the solutions are singular, others are not. However, the meaning of the singularities is unclear because the interpretation of the coordinates is not specified. This consideration was refined later (Bronnikov and Pavlov 1979, Bronnikov 1983, see Section 2.14).

Lake and Nelson (1980) proved, without solving Einstein's equations, that a charged fluid cannot collapse to zero volume if the following conditions hold: (1) the weak energy condition; (2) the absence of shell-crossings; (3) the absence of heat flux; (4) a finite volume of the fluid is matched to the Reissner–Nordström metric. It is not assumed that the fluid is perfect, and the R–N exterior has to obey $2mr_\Sigma\geq e^2$ where r_Σ is the curvature radius of the collapsing body's surface. This result generalizes the one established for spherical charged dust using the almost-explicit Vickers (1973) solution. Note that the collapse is halted only if the charge is smaller than a certain upper limit, just as for dust.

Singh and Abdussattar (1983) showed that the trajectories of charged dust particles in their own gravitational field are geodesics in another Riemannian space.

The connection coefficients $\Gamma^\alpha{}_{\beta\gamma}$ in the new space are expressed in terms of the Christoffel symbols $\left\{ \begin{matrix} \alpha \\ \beta \ \gamma \end{matrix} \right\}$ and the metric $g_{\beta\gamma}$ of the original space by:

$$\Gamma^\alpha{}_{\beta\gamma} = \left\{ \begin{matrix} \alpha \\ \beta \ \gamma \end{matrix} \right\} + \mathscr{E}^\alpha g_{\beta\gamma}, \qquad (3.5.6)$$

where:

$$\mathscr{E}^\alpha \stackrel{\text{def}}{=} (\rho/\epsilon) F^\alpha{}_\beta u^\beta. \qquad (3.5.7)$$

Singh and Abdussattar showed that the new connection is metrical, but no explicit formula for the corresponding metric could be given.

Stein-Schabes (1985) studied the gravitational collapse of dust in the presence of electromagnetic field, using a subcase of Hamoui's (1969) solution (defined by $E(r)=0$ and $M(r)$ being chosen so as to assure that $\epsilon(r=0)$ and $R_{,t}(r=0)$ are bounded, and $\epsilon(r=0)>0$). It is spherically symmetric dust moving in the electrostatic field of a single charge at the centre. The author showed that no Big Bang singularity will occur, while shell-crossings are possible (these results were not new by then, see Section 2.14). The shell-crossings are locally naked.

Solutions with a shearfree charged perfect fluid source were studied at length by Sussman (1987, 1988a,b), see Sections 4.4, 4.6, 4.7 and 4.12 of this book.

3.6 Singularities and cosmic censorship

The investigation of singularities in cosmological exact solutions of the Einstein equations was initiated by Eardley, Liang and Sachs (1972). They proposed a definition of geometry of the singular hypersurface in the case when the singularity is "velocity dominated", that is, the curvature terms in the Einstein equations in the vicinity of the singularity become negligible compared with the terms involving the expansion rate. In this case, one can factor out the time dependence of the asymptotic metric, and the time-independent three-metric g_{ab} that remains, no longer singular, is defined to be the metric of the singular hypersurface. The extrinsic curvature of this hypersurface is then uniquely defined, too. This definition is illustrated on the Ellis (1967) plane symmetric solutions IIaii, which Eardley *et al.* rederived, and on the L–T solutions (2.12.8)–(2.12.10).

Later, the L–T models became one of several test-laboratories for the cosmic censorship hypothesis. We stress that this section is not meant to be an overview of problems of cosmic censorship, but only an overview of applications of the L–T model as evidence or counterexample to various formulations of the hypothesis. Another frequently used test-laboratory is the Vaidya (1943, 1951 and 1953) solution, but papers discussing the Vaidya solution are not included in this book.

The first to use a subcase of the L–T model for discussing cosmic censorship were Yodzis, Seifert and Müller zum Hagen (1973). As can be seen from Section 3.4, if a black hole is formed, then its apparent horizon (the envelope of the region of trapped surfaces) first appears after some time along each world-line, and then it grows to encompass more and more world-lines. A shell-crossing singularity, if it is present at all, usually exists for a part of the lifetime of the Universe only: it can either keep emitting matter for some time and then disappear, or come into existence at a certain moment and then keep swallowing up matter. For the naked singularities, the second situation is relevant. Yodzis, Seifert and Müller zum Hagen showed on a definite example that initial conditions can be chosen in a L–T model so that the shell-crossing singularity shows up earlier than the apparent horizon along some world-lines. In this case, if the collapsing ball of dust is matched to the Schwarzschild spacetime outside, then the singularity is visible from spatial infinity for a finite period of its existence. From the Schwarzschild spacetime, however, the singularity will be seen for an infinite period of the local observer's proper time (see the Penrose diagrams in the paper).

In the second part of the paper, Yodzis *et al.* showed that pressure will not always prevent the singularity from being visible. That part of the paper is based on a metric whose energy-momentum tensor has no clear interpretation, and so it is not included in this book. Yodzis *et al.* concluded that naked singularities are not unconditionally ruled out in the Einstein theory, but they can be ruled out under some additional assumptions. For example, forming naked singularities might be an unstable property that will disappear when the initial data or the equations of state are modified.

Banerjee (1975) showed that the prediction of a naked shell-crossing singularity in the L–T model can be deduced from his earlier paper (Banerjee 1967).

Ellis, Maartens and Nel (1978) showed that many qualitative features of the Universe may be accounted for by a static model with an ever-existing active singularity at the antipode to the observer. The features include the redshift-distance relation and the microwave background radiation. In their final appendix Ellis *et al.* show that the L–T model can be arranged so that it has a permanent singularity. The article is most valuable for its demonstration of how the same observational data can be interpreted on the basis of two very different models.

Later research on cosmic censorship concentrated on the search for conditions that would exclude naked singularities.

Eardley and Smarr (1979) considered various slicings of spacetimes that are usable in numerical relativity. They took the collapsing L–T model with $\Lambda=0=E$ as a test example, and observed that it can contain a new kind of singularity which they called shell-focusing. If $L=\lim_{r\to 0+}[t_0(r)/M(r)]=0$, then the final singularity is spacelike in the neighbourhood of $r=0$ and is hidden within the event horizon. If $L=\infty$, then part of the singularity is null and visible from afar. Depending on the detailed shapes of the functions $t_0(r)$ and $M(r)$, it may be visible from a finite distance only or from the null infinity. If $0<L<\infty$, then each of these cases may occur. Thus the shell-focusing singularity is another counterexample to the simplest formulation of the cosmic censorship hypothesis.

A brief and simple presentation of the idea described above was given by Eardley (1979).

The paper by Seifert (1979) is a brief presentation of the (then) current status of research on cosmic censorship.

Szekeres (1980) described, at an elementary level, the singularities that may occur in various solutions (those of Schwarzschild, Weyl, Szekeres 1975a, L–T). His section on the Szekeres solutions is just an abbreviation of his 1975b paper (see our Section 2.4.2). For the L–T model, Szekeres discussed the subcases when $E=0$, $M=M_0 r^3$, M_0=const, and the bang-time function has two shapes: $t_0(r)=1/(1+r^2)$ (a "hump singularity") and $t_0(r)=-1/(1+r^2)$ (a "trough singularity"). In the case of the "hump singularity" the null geodesics leaving it are initially vertical ($dt/dr\rightarrow\infty$ as $t\rightarrow t_0$), and so the light from such places should be infinitely blueshifted, unlike in FLRW, where the null geodesics are initially horizontal and the light is infinitely red-shifted. The blueshift would have a strong influence on the background radiation, and would exist even far from the top of the "hump". The presence of this effect indicates an instability of the infinite initial redshift of the FLRW models. In the case of the "trough singularity", a shell-crossing singularity appears after the Big Bang. The null geodesics leaving it are horizontal, but the singularity itself is not simultaneous, and so photons leave it with a finite redshift. Since portions of the singularity are timelike, some photons will never leave it, but will hit its other part. Compare these conclusions with those of Hellaby and Lake (1984; see below).

Hellaby and Lake (1984, with an erratum 1985b) investigated the influence of the initial singularity on redshifts/blueshifts of light rays from the Big Bang. Infinite blueshifts will be absent only if the Big Bang is simultaneous. The stronger require-ment of no finite redshifts from the Big Bang is then satisfied automatically, if only comoving emitters are considered. For noncomoving emitters, the redshift is finite along a single direction for each emitter (and the same is true in the FLRW models), so this does not lead to infinite energies reaching any observer. Simultaneous Big Bang implies no decaying mode of inhomogeneity (see Section 2.5). The only models that have simultaneous Big Bang in the past and are either asymptotically homogeneous in the future or have a simultaneous Big Crunch in the future, are the FLRW models. Hellaby and Lake derived their results by developing the equations of null geodesics and the components of the null geodesic vector k^α into power series in R in the vicinity of the Big Bang.

Christodoulou (1984a, an abbreviated version in 1984b) showed that within the family of the L–T models with $E<0$, $M=-rE$ and $\Lambda=0$ there exists an open set of initial data for which both the weak and the strong cosmic censorship postulate are false. The weak form says that regular initial data cannot develop into singularities that are visible from infinity. The strong form says that such data cannot develop into singularities that are even locally visible. A singularity appears after a finite time at the centre of the L–T cloud considered by Christodoulou and is visible from future null infinity. The light rays from the singularity are infinitely redshifted. Christodoulou's result is complementary to that of Eardley and Smarr (1979): Christodoulou considered initial data which exclude $E=0$; Eardley and Smarr assumed $E=0$; both results are qualitatively similar.

Newman (1986a) discussed the problem of strength of the naked singularities in the L–T model. The discussion is rather technical, so its results are described here in simplified language in the hope that expert readers will pick up the message, while nonexpert readers will look up the original paper. Imagine a nonspacelike geodesic and a set of spacelike vectors orthogonal to it and parallelly propagated along it. If the geodesic is timelike, the set will be three-dimensional; if it is null, the set will be two-dimensional. The limiting focusing condition (LFC) means, roughly, that the volume (for timelike geodesics) or area (for null geodesics) of the parallelepiped spanned on the set of vectors is decreasing as the singularity is approached. The strong LFC means that the volume/area tends to zero at the singularity. The conclusions of the paper are as follows: (1) Radial null geodesics hitting or leaving a shell-crossing singularity in a L–T model do not obey LFC. (2) Radial null geodesics hitting or leaving a shell-focusing singularity in a L–T model obey LFC, but not strong LFC. The first result implies that a shell-crossing singularity is weak and thus is not a conclusive counterexample to cosmic censorship. Newman remarked that Christodoulou (1984a,b) in fact considered a time-symmetric subcase of a L–T model, not a general one.

Another paper by the same author (Newman 1986b) is a review of the history and development of the concept of cosmic censorship and its counterexamples. It may be considered as an earlier, but more detailed and expert, version of this section of our book.

The paper by Eardley (1987) is another review, but much more concise. Eardley showed how the existence of a shell-focusing singularity implies the breakdown of cosmic censorship. Then he proposed that the naked singularity is due to hydrodynamic description of dust: the dust shells are not allowed to go through the centre or to rebound, but are forced to stick together forever after the collision. Eardley hypothesized that if these limitations were relaxed, then no naked singularities would exist. Part of his paper is devoted to the existence of naked singularities in the Vaidya solution, but that part is outside the scope of this book.[30] Eardley pointed out that the singularity results when the collapse towards the centre is sufficiently *slow*, otherwise a horizon would form.

Newman and Joshi (1988) showed that the shell-crossing and shell-focusing naked singularities in the L–T model found in the preceding papers are special cases of a theorem that states, roughly, that "... the evolution of non-singular initial data on a suitable partial Cauchy surface cannot give rise to globally naked singularities which are both gravitationally strong, in the sense of refocusing of Jacobi fields, and severely causally disruptive in the sense of obscuring a region of spacetime from observers at $\mathscr{I}^{+}$." (This is a quotation from the paper.)

Waugh and Lake (1988) found that the study by Newman (1986a) excludes self-similar configurations (through his assumptions about energy-density), and discussed

---

[30] The reference "T. Maithreyan, D. Eardley, ITP preprint (1986)" mentioned in the paper could not be located.

the self-similar L–T model with $E=0=\Lambda$ from the point of view of cosmic censorship. They found that a shell-focusing singularity may be formed in it, which will be globally naked and strong. It is strong in the sense that $\lim_{\lambda \to 0}(\lambda^2 R_{\alpha\beta}k^\alpha k^\beta) \neq 0$, where k^α is the tangent vector to a null geodesic hitting the singularity, λ is the affine parameter on the geodesic with $\lambda=0$ at the singularity and $R_{\alpha\beta}$ is the Ricci tensor of the L–T spacetime.

Gorini, Grillo and Pelizza (1989) identified another subcase of the L–T models that was not captured by Newman (1986a) and showed that it develops a globally-naked shell-focusing singularity that obeys the strong limiting focusing condition, so it is to be taken as a serious counterexample to (one particular formulation of) the cosmic censorship postulate.

A wider class of L–T models with such a singularity was found by Grillo (1991), who showed by an example that a $\Lambda=0$ L–T model with $E<0$ (recollapsing) and with a certain definite choice of $E(r)$, $M(r)$ and $t_0(r)$ will form a locally naked shell-focusing singularity which is strong, that is, which satisfies the strong limiting focusing condition. A valuable contribution of Grillo's paper is the explicit calculation of the position of the apparent horizon (Barnes 1970, Demiański–Lasota 1973 and Eardley–Smarr 1979 only showed that it will exist).

Lemos (1991a) studied the self-similar L–T model with $\Lambda=0$ for the occurrence of a strong globally naked singularity. He showed that the singularity will appear whenever a certain constant parameter defining the timetable of the initial singularity is large enough. The other paper from the same year (Lemos 1991b) is very much a copy of the 1991a paper.

In a subsequent paper (Lemos 1992a), the author proved that the Vaidya solution can be understood as the limit $E(r) \to \infty$ of the L–T model with $\Lambda=0$. This was shown by performing an approximate coordinate transformation (up to $1/E$) on the L–T solution. The main part of the paper was devoted to demonstrating that the Vaidya and L–T solutions with $\Lambda=0$ have several identical properties from the point of view of cosmic censorship, for example, stability of the solutions and the existence of naked singularities and their strength. Compare this to the similar result by Gleiser (1984) concerning the quasi-spherical Szekeres models (third paragraph above our eq. (2.4.16)). The other paper (Lemos 1992b) is a short draft of the same study.

Hellaby (1994) studied the limiting transition from the L–T model to the Vaidya solution in more detail. The main result of the study was that in order to allow the Vaidya limit, the source in the L–T model must be a portion of dust with a vacuum sphere around the centre.

Clarke and O'Donnell (1992) verified whether the shell-crossing singularity in the L–T model can be removed by assuming that on the other side of the dust caustic there are three superimposed flows of dust (produced by the dust existing before flying through the singular set). The result was that a self-consistent extension of the spacetime through the caustic exists and that there exist coordinates in which the Einstein equations are satisfied in the distributional sense.

Dwivedi and Joshi (1992) identified a class of the functions $M(r)$ which generate a strong naked singularity at the centre. The initial density distribution in this class has the property that its first and second derivatives by r are zero at $r=0$. Most of the earlier papers considered special cases of this class.

In another paper, the authors (Joshi and Dwivedi 1993) showed that among the L–T variety of models one can find a subset in which there will be an at least locally naked strong curvature singularity at the centre. The class includes the configurations of Eardley and Smarr (1979) and of Christodoulou (1984). A continuous family of nonspacelike curves can be emitted from the singularity. Compare the paper by Królak *et al.* (1994, see our Section 2.4.2) in which it was shown that the Big Bang singularity in the $k(z)=0$ subcase of the $\beta' \neq 0$ Szekeres (1975a) solutions is a naked strong curvature singularity as well.

Szekeres and Iyer (1993) discussed the occurrence of locally naked singularities in more general spherically symmetric spacetimes. In passing, they showed again, in a rather simple way, that the shell-crossing singularity in the L–T model with $\Lambda=E=0$ is locally naked, and that in double null coordinates the metric components behave like C^1-functions across it. The spacetimes considered in the paper are assumed to have "power-law" singularities, occurring for metrics of the form:

$$ds^2 = dt^2 - [t_0(r)-t]^{2a}f^{\,2}(t,r)dr^2 - [t_0(r)-t]^{2b}g^2(t,r)(d\vartheta^2 + \sin^2\vartheta\, d\varphi^2), \quad (3.6.1)$$

where $t_0(r)$ is an arbitrary function, a and b are arbitrary constants and $f(t,r)$ and $g(t,r)$ are functions that are regular and nonvanishing at $t=t_0(r)$. Szekeres and Iyer did not assume any specific source in the Einstein equations, and so (3.6.1) is not in fact a solution. The L–T model with $\Lambda=E=0$ results when $-a=b/2=1/3$, $g(t,r)=r$ and $f(t,r)=(t_0+2rt_{0,r}/3-t)$; the flat FLRW dust results when, in addition, $t_0=$const. The conclusions of the paper are as follows: (1) in such spacetimes, the dominant energy condition can be satisfied in a neighbourhood of a locally naked singularity (this contradicts the strong cosmic censorship hypothesis), but (2) if the singularity is not a shell-crossing, then the dominant energy condition will only be satisfied if either one of the eigenvalues of pressure, p_r or $p_\perp$, becomes negative when approaching the singularity, or else the source is asymptotically an extreme fluid ($|p_r| \to \epsilon$ or $|p_\perp| \to \epsilon$ at the singularity).

Dwivedi and Joshi (1994) showed that for every configuration C containing a strong naked singularity there exists, within the collection of spherically symmetric spacetimes obeying the weak energy condition, a continuous family of equations of state containing the configuration C. Hence, the existence of the singularity is a consequence of the initial conditions rather than of any specific equation of state.

Unnikrishnan (1994) commented on the earlier papers in which the L–T model was used as a counterexample to the cosmic censorship hypothesis. In those papers, Unnikrishnan said, it was shown that a strong naked singularity may simply exist in the model, but it was not verified from what kind of initial data such a configuration might possibly have evolved. Unnikrishnan claims that the initial data leading to a strong naked singularity are, in one way or another, already singular: the dust density at the centre, $\epsilon(t,r=0)$, is either nondifferentiable or infinite. When the initial density is finite and differentiable at the centre, it can only evolve into one of the following configurations:

1. a weak shell-focusing singularity;
2. a strong shell-focusing singularity hidden within an event horizon.

Neither of these would be a genuine counterexample to cosmic censorship. However, P. S. Joshi, T. P. Singh and I. H. Dwivedi (private communication from PSJ) do not agree with Unnikrishnan's critique and plan to publish a rebuttal.

At the moment of closing this book, this was where the matter stood. The current status of the cosmic censorship hypothesis was reviewed by Clarke (1993), and, more extensively, by P. S. Joshi (1993). The latter reference contains a detailed and expert discussion of the implications of the singularities in the L–T model for cosmic censorship, and an overview similar to this section.

3.7 Influence of inhomogeneities in matter distribution on the cosmic microwave background radiation

The first to consider the L–T model from this point of view were Raine and Thomas (1981). They considered a large-scale but small-amplitude condensation in the path of light rays of the microwave background reaching an observer. The rays were assumed to be emitted in a FLRW region and received in another FLRW region. The equations of null geodesics were integrated numerically in order to calculate the temperature at the reception point and compare it to the temperature at endpoints of rays propagating all the way through a FLRW medium.The temperature variation was calculated as a function of the direction of observation. The condensation was assumed to have the diameter of 1000 Mpc, amplitude 1.5% and to be centred at the redshift of 3. The plot of ΔT vs. direction was drawn for different values of parameters in the background FLRW region to detect the influence on these curves of (1) the deceleration parameter; (2) the distance between the observer and the centre of the lump; (3) the density contrast in the lump. Raine and Thomas concluded that with $q_0=0.35$ the anisotropy observed in the background radiation can all be accounted for by assuming it to be a dipole, while with $q_0=0.25$ the anisotropy can be reconstructed by mixing the dipole and the quadrupole components in the ratio 85%:15%. If $q_0=0.02$, as favoured by measurements of deuterium abundance, then the velocity of the Local Group would imply that the density contrast is $\approx 15\%$ – "unacceptably large" as Raine and Thomas say. The paper opened up a new line of research which led to very illuminating results later.

The paper by Saez and Arnau (1990) was a preliminary report on a project aimed at a more thorough investigation along the lines of Raine and Thomas (1981). It presents numerically calculated initial profiles of velocity and the corresponding graphs of $\Delta T/T$ vs. the direction of observation for three different initial density profiles.

Paczyński and Piran (1990) used the L–T model to calculate the dipole and quadrupole moments of anisotropy in the cosmic microwave radiation. The L–T model they used had a preassumed connection between the arbitrary functions, $M_{,r}(r)=3r^2[1+2E(r)]^{1/2}$, and a simultaneous Big Bang. For the calculations, the definite form $2E(r)=-r^2(1-r^2)/(1+r^2)$ was used. In addition, the model was assumed to be filled with a test radiation field with the following ratio of photon density to baryon density: $S(r)=n_\gamma/\epsilon=1+a_S r^2/(1+r^2)$, where a_S is taken either to be zero or $a_S=-0.5$. By essentially the same method as used by Raine and Thomas (1981, see above) and Panek (1992, see below), Paczyński and Piran then calculated the dependence of the temperature of the received radiation on direction, and calculated the dipole and quadrupole components of the inhomogeneity. The dipole component for $a_S=-0.5$ is much larger than that for $a_S=0$ (a_S influences the temperature through $n_\gamma \propto T^4$, $\epsilon \propto T^3$). However, stricly speaking, a_S is not a parameter of the L–T model, but an arbitrary element in the calculation. Paczyński and Piran were aware of the *ad hoc* nature of their choice of parameters, and the paper was meant to point out a new possibility, not to estimate any actual parameter of the Universe.

Panek (1992) made the first really comprehensive study of the problem, with the results being clearly related to observations. He stressed the need for going to the nonlinear regime and using more realistic models of structure than Swiss cheese or perfectly empty voids. He used the Raine and Thomas (1981) method and considered three kinds of inhomogeneity in the path of the light rays:

1. Voids, with four different initial amplitudes and similar initial density profiles. Their evolution as found by Panek is very similar to that found by Sato *et al.* (see our Section 3.1).
2. Great attractors, with two different initial profiles of density.
3. Clusters of galaxies. The initial density profiles (two) in this case differ from those in point 2 only by the size and amplitude of the inhomogeneity.

The anisotropies in the background radiation caused by these structures are found by integrating the null geodesics equations backwards in time from the present time to the assumed emission epoch (the same for all rays). The observed temperature is $T=T_E/|k_t(t_E)|$, where T_E is the emission temperature and k_t is the time component of the vector tangent to the ray (k_t (today) is assumed to be -1). Since the reception times for all rays are the same, and the emission times for all rays are the same, the temperature changes arise because the rays penetrating the void spend more time reaching the void's boundary than they would if there were no void. Accordingly, they are more redshifted when they reach the observer and their temperature is lower. If the Universe has the density parameter $\Omega=1$ and the void is placed at $100\,h^{-1}$ Mpc from the observer, with the diameter $60\,h^{-1}$ Mpc, then it will cause a decrease in the temperature of the radiation of about $3\cdot10^{-7}$ ($=\delta T/T$) at an angular scale of about $15°$. Smaller voids produce smaller fluctuations; the same holds for more distant voids. In the calculation, the dipole component and its associated quadrupole component of the temperature fluctuation were subtracted (they are caused by the perturbation of the Hubble flow generated by the inhomogeneity).

The perturbation of the "great attractor" type causes a temperature decrease, too (the photon escaping from the condensation has to climb out of a potential well deeper than the one it fell into – due to the evolution of the condensation in time). For a great attractor placed at $42\,h^{-1}$ Mpc from the observer and of size $30\,h^{-1}$ Mpc, the relative amplitude of the temperature variation is about $2 \bullet 10^{-6}$ over the angular scale of $10°$.

The galaxy cluster perturbation is less massive, but much denser at the centre than the great attractor. For a cluster of size $\simeq 42\,h^{-1}$ Mpc at $100\,h^{-1}$ Mpc from the observer, the temperature fluctuation is between 1 and $6 \bullet 10^{-6}$ (depending on the density profile) on the angular scale of $10°$ again.

Detection of such temperature anisotropies is not possible at present, and may never be possible. Fluctuations of the order of 10^{-5} are expected to be produced by inhomogeneities existing at the moment of last scattering, and those of the order of 10^{-4} are observed, due to the Sunyaev–Zeldovich effect (scattering of background photons in hot centres of galaxies). The former may completely obliterate the effects of inhomogeneities encountered by the light rays on their paths later.

Arnau, Fullana, Monreal and Saez (1993) performed another study by the same method. They computed numerically the dependence of the temperature contrast on the direction of observation for the microwave background radiation for a model consisting of a FLRW background, with an arbitrary density parameter, and with a localized L–T perturbation superimposed on it. The calculation was performed for seven cases of overdensity perturbations and ten cases of underdensity perturbations, with different background density parameters and different density contrasts. The paper had the character of a preliminary study, and in large part was devoted to the description of the numerical code and tests performed on it.

More sophisticated tests of the code were performed by Saez, Arnau and Fullana (1993), who showed that the code can be used to fit a model with a few free parameters (describing the density profile of the condensation, the background density parameter and the Hubble parameter) to the observed characteristics of the Great Attractor or of the Virgo cluster, and then to calculate the anisotropy of temperature of the background radiation caused by these condensations.

Arnau, Fullana and Saez (1994), using the code described in the preceding papers, calculated the temperature anisotropies in the background radiation produced by models of the Great Attractor with different density and velocity profiles, and allowing for different density parameters of the FLRW background. It turned out that the effect of $\Omega < 1$ is rather strong. In the paper, Arnau *et al.* probed the effects of:

1. Various velocity profiles in the attractor model with a fixed background density parameter and fixed distance from the observer to the condensation.

2. Various background density parameters with a fixed velocity profile and the distance adjusted so as to produce the largest effect.
3. Various distances to the condensation with a fixed velocity profile and a fixed background density parameter.

The result was that the maximal anisotropy to be expected is up to $3 \cdot 10^{-5}$ (when $\Omega = 0.15$), at the angular scale of $10°$. Note that this is larger by an order of magnitude than Panek's result (1992, see above) obtained for $\Omega = 1$, but still barely within the error bar of the most recent measurements (Mather *et al.* 1993). Moreover, such anisotropies are comparable to those expected from inhomogeneities at last scattering, and so have a chance to be detected.

The papers by Panek and by Arnau *et al.* have one more important implication that these authors did not formulate. The high degree of isotropy of the cosmic microwave background radiation has been frequently used as an argument in favour of the FLRW models, and the "reasoning" was that inhomogeneities in matter distribution would leave an imprint on the radiation.[31] The results of the said papers show that no trace of such an imprint could be expected until the precision of measurements of temperature anisotropies reached the level of 10^{-6}, and this has happened only very recently (Mather *et al.* 1993).

3.8 Other papers discussing the L–T model

In this section, those papers will be reviewed whose contributions do not fit into any of the large groups mentioned in Sections 3.1–3.7. The aim of this section is to list all, even trivial, contributions so that readers can use the following as a bibliographical reference for the L–T model.

Omer (1949) derived perturbative formulae for several observable quantities in the L–T model that apply in a small neighbourhood of the observer located at the centre of symmetry. They represent: (1) mass within the comoving radius r; (2) number of galaxies per unit sky area brighter than a given magnitude, as a function of redshift and comoving distance; (3) redshift as a function of comoving radius; (4) equations of null geodesics. The only qualitative observation in Omer's paper is that if the Big Bang begins earlier at locations close to the centre and proceeds outwards, then the more distant regions are younger and denser.

Just (1960) found that the L–T solution is labelled by two arbitrary functions of r and showed how it can be specialized to a FLRW model. Then he showed that the solution can be used to describe an arbitrary number of spherical condensations with comoving boundaries superposed on a FLRW background. Just apparently suggested that the condensations necessarily have comoving boundaries, but this fact results from the assumption that the background permanently remains FLRW.

---

[31] References are deliberately omitted to avoid personalized conflict.

Just and Kraus (1962) related the functions $E(r)$ and $M(r)$ to the initial conditions, thus repeating one of Bondi's (1947) results. Then they showed that in the configuration considered by Just (1960, see above) the total mass of each condensation must equal the mass that a homogeneous FLRW medium filling the same volume would have. This implies that each condensation must be surrounded by a depression in which the density is smaller than in the background (Just and Kraus cite observations that have revealed the presence of such depressions). They seem to imply that this is a general property of the L–T model, but in fact it is only a consequence of the assumption specified in the preceding paragraph. The papers by Henriksen and de Robertis (1980), Occhionero *et al.* and Sato *et al.* described in Section 3.1 make it evident that such a configuration is unstable. Just and Kraus justify the requirement of comoving boundaries of condensations by the observations that without it "we cannot prove . . . that the cluster models remain non-overlapping and at rest with respect to the expanding universe if they fulfill these conditions initially". In this way, Just and Kraus missed the results of Section 3.1.

Novikov (1962a) showed that, with some choices of the arbitrary functions in the L–T model, adding some amount of rest-mass to a source may decrease the gravitationally active mass. The L–T model used is specified numerically, and the effect is only demonstrated on a definite amount of mass added. In fact, the model is a FLRW region going over into the Schwarzschild spacetime through an interpolating L–T region. The interpretation proposed is this:[32] "The (negative) potential energy of gravitational interaction between the medium added and the one present previously is greater than the energy corresponding to the mass of the medium added. This leads to a decrease of the total gravitational mass." The effect is an example of what happens beyond the "neck" described by Barnes (1970, see Section 2.12, remarks following eq. (2.12.10)). By another numerically specified example Novikov subsequently showed that an infinite amount of added rest-mass may leave the gravitational mass still finite.

In another paper, the same author (Novikov 1962b) introduced the definitions of the R- and T-regions in spherically symmetric spacetimes (these are the regions in which the gradient of the coefficient of the metric $(d\vartheta^2 + \sin^2\vartheta \, d\varphi^2)$ is spacelike and timelike, respectively), and presented the following two theorems for the $\Lambda = 0$ L–T model: I. If $\lim_{r \to r_0} R(t_0, r) = \infty$ for a certain t_0 (the r_0 may be infinite), and for $r > r_1$ the matter-density $\epsilon \geq A/8\pi R^2$ where $A = \text{const} > 1$, then there exists a r_2 such that the points with $r > r_2$ and $t = t_0$ are in the T-region (this means that with sufficiently high density, the spheres $t = \text{const}$, $r = \text{const}$ will not be static to any observer – comment by A.K.). II. For every $r = r_2$ there exist such t_1 and t_2 (dependent on r_2) that the points with (t, r_2) where $t_1 < t < t_2$ are in the T-region (this means that each matter particle remains in a T-region for a period of time – in the vicinity of the Big Bang singularity – comment by A.K. again). Novikov's paper contains, in addition,

---

[32] Translated by A. K.

several theorems concerning general spherically symmetric spacetimes which will be presented in Section 3.9.

Ginzburg and Ozernoi (1964) applied the $E=0=\Lambda$ subcase of the L–T model matched to the Schwarzschild solution to a study of the role of magnetic field in gravitational collapse. The magnetic field was assumed to be a test field in the L–T/Schwarzschild spacetime. However, after presenting the model very briefly, Ginzburg and Ozernoi transformed it to curvature coordinates by approximate calculations and performed the whole work on the approximate result. Thus the paper is cited for completeness only.

Kolesnikov and Stanyukovich (1965), while considering the Einstein equations for spherically symmetric spacetimes with a perfect fluid source in curvature coordinates, presented a prescription for obtaining the L–T model with $\Lambda=0$ in these coordinates. The prescription consists of solving a linear partial differential equation and calculating an integral, and is similar to the one used later by Gautreau (1984, see Section 2.13). The same consideration was presented by the same authors in another paper (Kolesnikov and Stanyukovich 1966), and the transformation between comoving coordinates and curvature coordinates in the L–T model was again considered by Stanyukovich (1969).

Gertsenshtein (1966) argued that recollapse in the L–T model with $E<0=\Lambda$ should not be treated as a final singularity terminating the existence of the model, but that the dust should fly through the centre and keep oscillating. However, Gertsenshtein did not propose any solution to the mathematical difficulties of continuing a solution through a singularity.

Similar considerations were presented by Pachner (1966a,b). The two papers listed as 1966a presented the explicit solutions of (2.12.3) with $\Lambda=0$ in detail.

The other paper by Pachner (1967b) is a summary of the earlier papers by the same author. It contains one false statement. Pachner argues that the volume of a recollapsing L–T cloud of dust never goes to zero unless the model degenerates into FLRW. This is so only if one accepts the notion that dust particles fly through the centre and continue to move. Otherwise, the singularity accumulating matter *has* zero volume and swallows the whole cloud in a finite time.

Another paper by the same author (Pachner 1967a) is a short review of properties of spherically symmetric perfect fluid spacetimes in which the L–T model with $\Lambda=0$ is derived and briefly described.

Papapetrou and Hamoui (1967) studied the subcase of the $\Lambda=0$ L–T model in which the dust forms a layer between a Minkowski vacuole and the Schwarzschild space, and in a certain moment $t=T$ is compressed into a shell of zero thickness. Papapetrou and Hamoui show that the evolution of such a configuration for $t>T$ is determined by initial data at $t<T$, and on this basis they argue that a shell-crossing is not really a singularity.

Nariai and Tomita (1971) mentioned the L–T model in an appendix to a paper discussing the formation of proto-galaxies by perturbative calculations. Properties of the L–T model were very briefly compared and contrasted with the properties of a solution found by Nariai (1967a,b, see our Section 4.6).

Bonnor (1972) used the L–T solution with $\Lambda=0=E$ and a suitably chosen $M(r)$ to model the hierarchical density distribution $\epsilon\cong(\text{distance})^{-1.7}$, reportedly observed by de Vaucouleurs (1970). Bonnor showed that this is inconsistent with the observations of redshift vs. luminosity distance.

Markov and Frolov (1972) applied the Schwarzschild and Reissner–Nordström solutions to considerations about the minimal possible size of particles. The $\Lambda=0$ L–T solution is invoked only in order to introduce the Lemaître coordinates in the Schwarzschild solution, and this paper is mentioned only for historical completeness.

Stanyukovich and Sharshekeev (1973) used the L–T model for exactly the same purpose as did Markov and Frolov (1972). The main subject of the paper is a study of the Schwarzschild spacetime.

Misner, Thorne and Wheeler (1973, p. 859) presented the L–T model briefly, and concentrated on playing down the importance of the shell-crossing singularity: it produces negligible tidal forces on neighbouring dust particles, and "it is an idealization that gets smeared down to finite density by finite pressure". The second point still awaits a demonstration on any exact solution.

Lund (1973a) considered canonical quantization of the motion of spherically symmetric dust (as described by the L–T model) and showed that in the quantum theory the relative probability of the collapse ending in a singularity is zero. More exactly, the set of initial data leading to a singularity is of measure zero in the set of all possible initial data.

The same author (Lund 1973b) showed that, from the point of view of hamiltonian quantization, the G_3/S_2 subcases of the $\beta'=0$ Szekeres solutions have just two degrees of freedom. This paper thus in fact belongs to our Section 2.6, but its message would be lost there more readily than here.

Bonnor (1974) showed that the set of the L–T models evolving asymptotically towards a FLRW spacetime is labelled by one arbitrary function of the radial variable. This is meant to be a counterexample to the statement by Collins and Hawking (1973), based on Bianchi models, that only a subset of measure zero of all models can evolve towards FLRW. Note how Bonnor's result follows from the approach of Goode and Wainwright (1982b, see Section 2.5): a L–T model can evolve towards FLRW only if the growing mode of perturbation is absent, that is, when $\mathcal{M}=\text{const}$ (or $k=0$, which leads to $\mathcal{M}=\text{const}$ anyway; the $\mathcal{M}$ here is the one of G–W). This leaves the L–T model with only one arbitrary function of r, and the remaining perturbation of the FLRW background is decaying.

Eardley (1974b) used the special L–T model with $E=0$, $M(r)=r$ to illustrate the hypothetical process of accretion of matter onto a white hole. The accreted matter would be stopped by that ejected from the white hole at a "blue sheet"; the matter accumulated there would convert the whole setup into something observationally indistinguishable from a black hole. Detailed calculations were not presented, but were promised to be presented in a future paper. The paper did not invoke any real astrophysical situation.

Korkina and Chernyi (1975) discussed the special relativistic and Newtonian limits of the L–T model with $\Lambda=0$. The case $E<0$ has no special relativistic limit; in the case $E=0$ the limit is the Minkowski spacetime in the ordinary spherical coordinates; for $E>0$ the limit is the Minkowski spacetime in the Robertson coordinates. The Newtonian limit always exists, and it shows that E is the analogue of the energy of the r-shell of matter while M is the analogue of the mass within the shell (Korkina and Chernyi were not the first to observe this last point).

Mavrides (1976a,b) used the subcase $\Lambda=0=E$ of the L–T model to construct the following configuration: a flat FLRW dust with greater density and later-occurring singularity matched to the Schwarzschild spacetime, and that one matched to a flat FLRW dust with smaller density and earlier Big Bang. This was used to model de Vaucouleurs' observation (also called the "Rubin–Ford effect") that H (the Hubble constant) in the local supercluster of galaxies is 100 km/(s•Mpc) while outside $H=50$ km/(s•Mpc). The second paper is a more detailed elaboration of the first one; the same problem was then discussed in yet another paper (Mavrides and Tarantola 1977).

Papapetrou (1976a,b) used the L–T model with $\Lambda=0=E$ to construct a perturbation of the flat FLRW model consisting of the interior Schwarzschild solution surrounded by the vacuum Schwarzschild solution, surrounded in turn by the flat FLRW region. In the first paper the configuration was constructed; in the second one it was investigated for the change in the redshift distribution that it would cause. No qualitative conclusions were drawn.

Miller (1976) described at length how a Universe with the L–T geometry and with $E=0=\Lambda$, $t_{0,r}<0$ comes into being by gradually emerging from the Big Bang singularity. In such a configuration (the Big Bang first occurring far from the centre and proceeding inwards), one can assign mass to the singularity and interpret it as the mass to be emitted from it. The mass of the singularity is simply the function $M(r)$ from (2.12.3) calculated at $r=r_0(t)$, where $r_0(t)$ is the inverse function to the bang time $t_0(r)$; it is well behaved on the singularity. When the Big Bang reaches $r=0$, the singularity may either disappear or may continue to exist with a negative mass and still emitting matter. In the second case it becomes a permanent timelike singularity. In order to illustrate this point, Miller discussed the Schwarzschild solution with negative mass, the Vaidya solution and the self-similar spacetimes of Cahill and Taub (1971). These three cases are not within the scope of this book, the last one because it is defined by a set of equations and not by an explicit solution.

Fennelly (1977) showed that in the L–T model with $\Lambda=0=E$, the functions and the observer's position can be adjusted so that the observed anisotropy of the Hubble constant can be reproduced.

Silk (1977) studied the L–T model with $\Lambda=0$ from the point of view of perturbations of the FLRW models. The main conclusions were as follows: a gradient of the bang-time function $t_0(r)$ generates decaying modes in density contrast and curvature contrast; a gradient of the initial energy distribution generates the corresponding growing modes. The growing modes have zero density contrast and a finite curvature contrast at the initial singularity, respectively, while the decaying modes both become infinite at the singularity. In an ever-expanding model, energy gradients produce frozen-in finite contrasts in both quantities at the asymptotic future. Some of these observations were known earlier; the connection between $t_0(r)$, $M(r)$ and the growing and decaying modes was later found by Goode and Wainwright (1982b) to hold for the Szekeres models; see Section 2.5.

Gertsenshtein (1977) used the $\Lambda=0$ subcase of the L–T model (represented in the proper time-curvature coordinates) to discuss the closed FLRW model matched to the Schwarzschild spacetime. The exact purpose and results of the paper are rather obscure. Gertsenshtein seems to want to imply that collapse does not lead to a singularity but is followed by re-expansion. The re-expansion is supposed to be observable from outside. From the paper it is not clear that the interior solution is FLRW, the author calls it "self-similar".

Lake and Roeder (1978) derived the full L–T model only in order to specialize it to the FLRW model matched to the Schwarzschild spacetime. This served as a background for a study of the distinction between exterior and interior vacuum solutions. In passing, it was shown that "T-solutions" exist which arise as the degenerate case $R_{,r}=0=1+2E$ of the L–T model; see our Section 2.6.

Olson (1980) calculated the analogue of the Newtonian matrix of tidal forces in a L–T spacetime and observed that it has the same algebraic form if Newtonian time, radius and mass are identified with L–T's comoving time, curvature radius and "effective gravitating mass" (Bondi 1947) respectively.[33]

Peebles (1980, section 87) presented a short derivation of the L–T model in order to use it as a basis for considering linearized perturbations of a FLRW background.

Samoilov (1981) showed that in the L–T model with $\Lambda=0=E$ matched to the Schwarzschild solution, coordinates can be chosen so that the R-region is not covered by a single coordinate chart, apparently contrary to the statement by Novikov (1964b, see Section 3.9). The consequences of this result are not explained.

Zeldovich and Grishchuk (1984) provided a formal proof of the apparently obvious fact that if a L–T model contains a region with $E>0$ surrounded by a space with $E<0$, then shell-crossings (caustics) must appear. However, see the remarks on the paper by Hellaby and Lake (1985a) given below – they provided counter-examples. The paper by Zeldovich and Grishchuk inspired a few other authors to investigate in more detail the geometry of such L–T models in which $R(t,r)$ has local extrema with respect to r; see in particular the paper by Hellaby (1987, discussed below).

Lake (1984) showed that a shell-crossing singularity in a L–T model can be assigned an intrinsic metric (induced by (2.12.1) through $R_{,r}=0$) and a surface

---

[33] Further use of the result was predicted for a paper "in preparation" by Olson and Wasserman, but that paper was never published (information from D. Olson).

energy tensor that obeys a three-dimensional analogue of the strong energy condition. Thus, a shell-crossing should not be considered as another variety of Big Bang for which this method would not work.

Bonnor (1985a) showed that a "physically acceptable" closed L–T model must always recollapse. It is closed when its spatial volume is finite, and it is physically acceptable when the density is everywhere positive, has no shell-crossing singularities and no comoving surface layer singularities.

In another paper, the author (Bonnor 1985b) demonstrated that the L–T model with a simultaneous Big Bang, $M = M_0 r^3$ and $E(r) = -E_0 r^2/(1+r^2)$, where M_0 and E_0 are constants, has infinite spatial volume even though $E < 0$ and it recollapses. Far from the centre, the space $t =$ const in this model is approximately conical and does not close up.

Laserra (1985) presented a simple description of the L–T model with $\Lambda = 0 = E$ and of some of its subcases. The paper does not seem to contain any new information and is registered here for completeness only.

Hellaby and Lake (1985a, with an erratum 1985c) considered the conditions for no shell-crossings. The reason for the investigation was that the usually assumed conditions $R_{,r} > 0$ everywhere, and $M_{,r} > 0$, are too restrictive: with appropriately chosen $M(r)$ and $E(r)$, the locus of $R_{,r} = 0$ may be simply a local extremum of the function $R(t,r)\big|_{t=\text{const}}$ (see the remarks in the second paragraph after our eq. (2.12.10)). Hellaby and Lake gave a detailed list of conditions for no shell-crossings, with the possibility of extrema of $R(t,r)$ fully taken into account (see paper). Then they presented counterintuitive examples:

1. A model with $E > 0$ (negative curvature) everywhere that has a globally closed space.
2. A model with $E < 0$ (positive curvature) everywhere that has an open space.
3. A model that consists of a region of negative curvature placed between two regions of positive curvature, and yet it does not develop shell-crossings because the positive-curvature regions do not recollapse. This contradicts a statement by Zeldovich and Grishchuk (1984, see above). Hellaby and Lake showed that at an extremum of $R(t,r)$ where $R_{,r}(t,r_e) = 0$, the equation $1 + 2E(r_e) = 0$ need not be fulfilled, but only $E_{,r}(r_e) = 0$, and this resolves the apparent contradiction. However, with $1 + 2E(r_e) \neq 0$ a surface layer of matter is present at $r = r_e$.

Meszaros (1986) considered the implications of the assumption that the observed part of the actual Universe is described by a finite-volume portion of the L–T model with a simultaneous Big Bang. There are two chances of confirming this assumption observationally: (1) by observing the outer surface of the L–T sphere; (2) by detecting the "cosine anisotropy" in the matter-density and in the Hubble parameter. There is no observational evidence for point (1). As for point (2) Meszaros notes the following: I. Observations of matter-density are inconclusive. II. The Hubble parameter and the temperature of the microwave background radiation do reveal cosine anisotropies. They are usually explained away as resulting from the motion of the solar system around the galaxy. However, the directions and magnitudes of the velocities calculated from the two effects are inconsistent. This inconsistency is a problem in a FLRW model, but it can be accounted for on the basis of a L–T model. Unfortunately, the precision and interpretation of the observations are questionable. The quadrupole anisotropy of the background radiation is much simpler to explain in a L–T model than in a FLRW model. Meszaros concludes that there are more arguments for the L–T models than against them, but present observations are inconclusive, and so "the usual Friedmann model need not necessarily be rejected yet".

Bonnor and Ellis (1986) discussed the postulate of uniform thermal histories and those L–T models that obey it but are spatially inhomogeneous. The conclusions of the paper in fact follow from the later study by Bonnor and Pugh (1987) of the Szekeres models from the same point of view, see our Section 2.5, the paragraphs following eq. (2.5.39).

Sannan (1986) applied Hawking's definition of mass to a L–T model, with parameters chosen so as to ensure a rapid falloff of the inhomogeneity with distance (the definition makes sense only in this case). Sannan found the mass to be positive and constant in time, as expected. The calculation was performed by resorting to perturbations. The paper is rather technical; those readers who are interested in further details are advised to look it up.

Hellaby (1987) presented an illuminating discussion of the L–T models by comparing them with the Schwarzschild spacetime represented in the Lemaître coordinates (Hellaby calls them "Novikov coordinates"). The latter results directly from the former in the limit of vanishing matter-density. The comparison reveals the meaning of the "neck" at $R,_r = 0 = M,_r = 1 + 2E$ in the L–T model: it is analogous to the "throat" in the Kruskal coordinates. However, in the L–T case the parts of the spacetime on opposite sides of the neck do not have to be isometric to each other. In particular, on one side there can can be a Big Bang in part of the volume and a stream of matter from past infinity elsewhere, all ending up in a Big Crunch; while on the other side there can be a Big Bang everywhere followed by collapse to a Big Crunch in part of the volume, the other part expanding forever. A "string of beads" of finite volume, connected by necks to its neighbours, is also possible. This model starts out as a collection of disjoint bubbles that coalesce into a single space, then pinch off during recollapse. Each bubble has two white holes at antipodes just after its birth, and two black holes just prior to recollapse. Geodesics may pass through the necks, but the flow-lines of the dust source cannot; the locus of $R,_r = 0$ must be comoving. Hellaby proved that the apparent horizon coincides with the event horizon only in vacuum (when $M,_r = 0$, that is, in the Schwarzschild limit). A detailed discussion of horizons shows that the communication between the two sides of the neck in the presence of matter is no better than in the vacuum (Kruskal) case: no light ray emitted at $\mathscr{J}^-$ in one sheet can hit $\mathscr{J}^+$ in the other sheet. The considerations are illustrated with several spacetime- and Penrose diagrams. The paper is readable at many levels of expertise and is very strongly recommended to the readers of this book.

Goicoechea and Martin–Mirones (1987) derived the analogue of Mattig's magnitude-redshift formula for the L–T model, and emphasized the difficulties of decomposing the observed deceleration parameter into a large-scale cosmological contribution and a local component due to an inhomogeneity placed in the direction of the Hydra-Centaurus complex at $z=0.1$. The observed deceleration parameter contains multipole components that can in principle be deciphered for small values of shear, but not for large values.

Gribkov and Soloviev (1987) discussed the L–T model in curvature coordinates, with the additional requirement that the L–T region is matched to the Schwarzschild solution. They mainly discussed the dependence of the time of collapse to a singularity on the initial density distribution. No qualitative physical conclusions were presented.

Bona (1988) discussed conformal symmetries of spacetimes having a G_3/S_2 symmetry. For this book, the following result (theorem) is meaningful: a dust spacetime admitting a spacelike G_3/S_2 symmetry group will admit a conformal vector invariant under the G_3 if and only if it is either (I) FLRW or (II) self-similar of the Cahill–Taub (1971) class or (III) of the Kantowski–Sachs (1966) class.

Hellaby (1988) showed that if a portion of the L–T model is replaced by a subset of a FLRW model with the same spatial volume evolution, then the density of matter in such a model may differ by up to 30% from the average density in the same L–T region. This result highlights the lack of an adequate theory of averaging in cosmology, see Chapter 8 of this book. It is in fact a problem for the future because the accuracy of cosmological observations is still too small to detect such discrepancies.

Bartnik (1988) used the subcase $E=0=\Lambda$ of the L–T model to construct an example of a spacetime that does not admit a constant mean curvature hypersurface. The construction is Schwarzschild matched to FLRW.

Lynden–Bell and Lemos (1988) derived the L–T models with $\Lambda=0$ just in passing, as an appendix to a paper whose main subject was a Newtonian self-similar collapse. They presented a concise discussion of the formation of naked singularities or apparent and event horizons. The solution is also displayed in curvature coordinates (the same as in Gautreau 1984, where the case $E=0\neq\Lambda$ was considered).

Lemos and Lynden–Bell (1989) presented the L–T equation in the case $\Lambda=0$, and then integrated it under the tacitly inserted assumptions $E/M=$const and $t_0=$const (simultaneous Big Bang). Then, for this special case, they derived the condition for the formation of the apparent horizon. A very short review of papers discussing self-similar models is given. This discussion constitutes the last section of a paper devoted to self-similar solutions in Newtonian hydrodynamics.

Rindler and Suson (1989) considered determining the functions in the L–T model from the results of astronomical observations. The discussion concerned determination "in principle" and seems to be a simplified version of the discussion by Omer (1949).

The paper by Gorini, Grillo and Pelizza (1990) is a short description of the L–T models in very general terms, and does not seem to contain new results.

Liu (1990) interpreted the functions $E(r)$ and $M(r)$ in terms of the standard hydrodynamical initial conditions. This work was in fact done by Bondi (1947) for the first time, and was repeated on a few other occasions afterwards.

Korkina (1991) showed that the de Sitter solution can be matched to the L–T solution across a $t=$const hypersurface. On the hypersurface of matching, the matter-density is constant, that is, the L–T model that is born out of the de Sitter model starts its evolution with a spatially homogeneous density. The pressure suffers a jump which Korkina interprets as a phase transition.[34] Definite examples of such matching were discussed and it was shown that, in particular, the de Sitter solution may change over into a FLRW model at some time.

Tomita (1992) compared the L–T model and its plane symmetric counterpart (Ellis IIaii 1967) to linearized perturbations of the FLRW background. He found that irregularities in spatial curvature drive density perturbations whose growth factor is the square of the growth factor of linearized adiabatic peturbations.

Ribeiro (1992a) introduced the expressions for various astrophysically relevant quantities in the L–T model, such as observer area distance, luminosity distance, mass-density along the light cone, number density of galaxies along the light cone (these should be compared with those derived by Omer, 1949). Then Ribeiro introduced a fractal distribution of matter by assuming that the number of objects N_c and luminosity distance d_l to the surface of the sphere containing the N_c objects are connected by $N_c = \sigma d_l^D$, where σ is a constant and D is the fractal dimension of the distribution. Then, a detailed discussion of the equations of radial null geodesics follows. Ribeiro discussed at length the matching between the L–T and FLRW spacetimes, although this is in fact trivial if one remembers that FLRW is a subcase of L–T. One important (even if not quite new) conclusion emerged from the discussion: the matching is possible only if the L–T mass contained inside the boundary sphere is the same as the FLRW mass removed from there. Thus, any under- or overdensity must be compensated for by over- and underdensity, respectively.[35] Finally, Ribeiro discussed how the assumed distribution can be numerically checked for agreement with observations. The actual carrying out of the numerical procedure was postponed to a follow-up paper (Ribeiro 1993a, see below).

In a sequel paper (Ribeiro 1992b) the author calculated the density vs. distance function along the past light cone in the $k=0$ FLRW (i.e. Einstein–de Sitter) model. This function is not constant, of course, but even so it does not seem to agree with the observations. Ribeiro stressed that cosmological models should be tested in precisely this way. The Einsten–de Sitter model tested in this way does not possess any features of a fractal distribution of mass.

In the third paper (Ribeiro 1993a) the author showed that the density vs. distance relation measured along the past light cone in the open and the closed FLRW models is not homogeneous either, nor is it fractal. However, the arbitrary functions

---

[34] The jump could presumably be avoided with a general, that is, $\Lambda \neq 0$, L–T model.
[35] At the same time, this is the necessary condition for comoving boundaries of voids.

$E(r)$, $M(r)$ and $t_0(r)$ in the L–T model *can* be chosen so that $N_c = \sigma d_l^D$. Ribeiro found several examples of such functions by numerical experiments. In most of the examples, either the Hubble law is contradicted (the distance to the light source fails to be proportional to its redshift) or the value of the Hubble constant is lower than astronomers currently believe it should be. The only example that yields acceptable parameters is $2E(r) = \sinh^2 r$, $2M(r) = \alpha r^p$, $t_0(r) = \ln(e^{\beta_0} + \eta_1 r)$. If $\alpha = 10^{-4}$, $p = 1.4$, $\beta_0 = 3.6$ and $\eta_1 = 1000$, then the implied fractal dimension is $D = 1.3$, the Hubble constant is $H = 61$ km/(s•Mpc), and $\sigma = 5.4 \cdot 10^5$. Even better agreement with observations follows when $\eta_1 = 0$; then $D = 1.4$ and $H = 80$ km/(s•Mpc). Ribeiro emphasized at length that no positive observational evidence of spatial homogeneity of the Universe is available. Should it be available, it would actually rule out the FLRW models because observations made in them along the past light cone reveal inhomogeneity (the cone intersects hypersurfaces of different densities).

The subject of the three papers by Ribeiro is discussed at more length in a review paper by the same author (Ribeiro 1993b).

Zecca (1993a) rederived the L–T model with $\Lambda = 0$ in the Newman–Penrose variables, then derived the Dirac equation on the background of the model, and solved it approximately assuming that $R_{,t} \cong 0$. This assumption is fulfilled in the vicinity of the maximal value of $R(t,r)$ in the recollapsing model. Under this assumption, the Dirac equation may be solved by the method of separation of variables.

Burnett (1993) proved the following theorem: "There exists an upper bound to the lengths of timelike curves in any Tolman spacetime that possesses S^3 Cauchy surfaces and whose energy density is positive." This is meant to be the ultimate form of the statement that a closed L–T model must recollapse. As Burnett observed, earlier papers on the subject did not prove the existence of the said upper bound. An explicit formula for the bound in terms of the initial data on S^3 is given in the paper.

3.9 General theorems and considerations

Finally, let us quote some general results, proved without recourse to explicit solutions, that place the L–T models in a broader context. This section only presents a selection of such results and does not pretend to be complete or even representative.

Novikov (1962b) proved the following theorems for a general spherically symmetric spacetime: (1) There exists a solution for which the volume of a finite mass goes through a nonsingular minimum and the minimum is simultaneous for the whole space. (2) The same is possible for an infinite mass. (3) A solution exists in which a volume minimum occurs at small distances from the centre, and, at the same moment, a volume maximum occurs at large distances from the centre. (4) If $M(r) = \int \epsilon \, dV$ grows with r faster than the affine radius $l = \int_0^r (-g_{11})^{1/2} dr'$, and does so for a sufficiently large interval of l, then the configuration cannot be static. (5) If $M(r)$ grows faster than $l(r)$ on a sufficiently large interval of monotonic growth of R, then the volumes of different shells cannot go through an extremum simultaneously (R is the coefficient of $(d\vartheta^2 + \sin^2\vartheta \, d\varphi^2)$ in the metric). (6) If the surface of a spheroidal ball is in a T-region and is shrinking, then, in a finite proper time, it will shrink to a point; if the surface is in a T-region and is expanding, then it started its expansion from a point. Finally, Novikov calculated the position of the boundary between the R- and T-regions in a FLRW model and concluded that the radio-sources observed must lie in the T-region. Consequently, they seem to have emerged from a point. This seems to be an early version of a singularity theorem, though formulated rather shyly and immediately followed by warnings that the real Universe was more anisotropic in the past than it is now, and so the FLRW models did not necessarily apply to it earlier. However, the data (at the time) allowed one to conclude that the Universe has been expanding for the last $5 \cdot 10^9$ years. The properties listed above reveal what should be expected from exact solutions generalizing the L–T model for pressure with nonzero spatial gradient.

The 1964b paper is a continuation of the earlier one (Novikov 1962b), and partly repeats the older results. The new results are as follows. (1) In the Schwarzschild solution the boundary between the R- and T-regions is the $r = 2m$ event horizon, and such boundaries also exist for the Kottler (i.e. Schwarzschild with $\Lambda \neq 0$) and Reissner–Nordström solutions. (2) The Schwarzschild solution is a subcase of the L–T solution given by $M =$ const, in this case the comoving coordinates of the L–T model go over into coordinates connected to congruences of freely falling particles in the Schwarzschild spacetime. Such coordinates cover the T- and R-regions simultaneously. Novikov discussed the solutions of (2.12.3) in this case, interpreted them and gave references to earlier papers where this system was used (here he quoted Lemaître 1933a for the system with zero velocity at infinity). This part of the paper is a very elegant and historically important contribution to the theory of the Schwarzschild solution.

The other paper (Novikov 1964c) mostly repeats the results of the papers already mentioned in this book.

Ryan (1972) studied the Einstein equations for the following metric numerically (in the original notation except for the signature):

$$ds^2 = e^{\alpha(t,\vartheta)}dt^2 - e^{-2\Omega(t)}e^{2\beta(t,\vartheta)}[(1+\beta_\vartheta{}^2)d\vartheta^2 + \sin^2\vartheta(d\varphi^2 + \sin^2\varphi\, d\psi^2)], \quad (3.9.1)$$

where $\Omega(t)$, $\alpha(t,\vartheta)$ and $\beta(t,\vartheta)$ are functions to be determined from the field equations. The discussion was meant to be a survey of possibilities offered by inhomogeneous models (studied on this particular example), and so was very much in the spirit of this book, although no explicit solutions were invoked. The main results of the paper are that such models can evolve from an inhomogeneous initial state towards homogeneity, and they make it possible to study the change of topology of spatial slices.

Gold (1973) described verbally several processes that might be going on in a Universe with density inhomogeneities of large spatial size. The description is based on a half-intuitive, qualitative consideration of the possible effects caused by such inhomogeneities. Recalling the papers by Barnes (1970) and by Demiański and

Lasota (1973), one can even recognize a description of the formation of a black hole in the expanding Universe ("There would be a system of 'nesting Universes' in which each is perhaps only slightly smaller than the one in which it nests"). Neither the L–T model nor any other explicit solution was used in this discussion.

In a series of papers, Dautcourt (1980, 1983a,b and 1985) developed a new approach to integrating the Einstein equations which is better suited to incorporating observational results into cosmological models. In this approach, the initial data are given on the past light cone of a single event in the spacetime, and then the Einstein equations are integrated off that cone to the future. The first paper (Dautcourt 1980) is a short exposition of the ideas of the other papers. In the second paper (Dautcourt 1983b) the author gave a prescription for integrating the Einstein equations in vacuum and for dust with the initial data given on the past light cone of an observer. The Einstein Universe and the FLRW models were used as the only illustration. In the third paper (Dautcourt 1983a) the author described the ideal observations needed to determine the initial data. In practice, several of those observations are beyond the limits of current technology. In the last paper (Dautcourt 1985), the author noted that the observable quantities introduced in the preceding papers depend on the four-velocity of the observer, and proposed a preferred set of coordinates with a simple relation to the observable quantities and a simple transformation law under a Lorentz rotation of the observer. Dautcourt's approach was later developed further by Ellis and coworkers; the paper by Stoeger, Ellis and Nel (1992) mentioned in Section 2.12 (the paragraph around eq. (2.12.11)) is one entry in that series.

In another series of papers, Adams *et al.* (1982, 1985 and 1987) introduced a formalism for describing the propagation of gravitational waves on spatially homogeneous backgrounds of the Bianchi types I–VII. The first step in the formalism is to transform the Bianchi metric to the form:

$$ds^2 = e^{2a}(dt^2 - dz^2) - e^{2b}h_{ij}\sigma^i \otimes \sigma^j, \qquad (3.9.2)$$

where $i, j = 1, 2$ and σ^i are differential forms in x^1 and x^2, and then to replace all the functions of time (a, b and h_{ij}) by unknown functions of t and z. In the first paper, several properties of such spacetimes were discussed on the general level (e.g. in which types the polarization, "$+$" or "$\times$", may be preserved in time). The only explicit solution discussed was a vacuum one. In the second paper, the consideration was worked out in more detail for the Bianchi types I, III, V and VI, but the explicit solutions discussed were still only the vacuum ones. In the last paper, Adams *et al.* considered the propagation of gravitational waves with the "$+$" polarization on a Bianchi I dust background, but only the background solution was presented exactly; the full solution was considered perturbatively.

Partovi and Mashhoon (1984) calculated various cosmologically relevant parameters for two types of a spherically symmetric inhomogeneous spacetime: one with a perfect fluid source and a barotropic equation of state, the other with a dust source.

The quantities are calculated in terms of successive powers of the redshift around the values in a FLRW background. It was found that the leading terms in z are identical to the FLRW terms, while the next following terms are formally modified, but in such a way that the inhomogeneous corrections cannot be properly read out from the formulae, and may lead to assigning incorrect values to the standard cosmological parameters. The conclusion is that up to those following terms these inhomogeneous models are observationally indistinguishable from the FLRW models.

Burnett (1991) proved that in a spherically symmetric spacetime with a Cauchy hypersurface of topology $S^1 \times S^2$ whose source satisfies the non-negative pressures and dominant energy conditions, there exists an upper bound to the lengths of timelike curves. Some L–T models are illustrations of this theorem. Part of the paper is devoted to the existence of such an upper bound in the K–S spacetimes, but those results are not directly connected with this book.

Brauer (1991) considered the "Existence of finitely perturbed Friedmann models via the Cauchy problem" (the paper's title). The results are in short:

1. If the matter variables and metric variables are in a suitable function space, then "neighbouring" solutions exist for all three FLRW models.
2. In general, there exist no one-parameter families of solutions of the evolution equations with a C^ω dependence on the parameter and a C^k dependence on the coordinates that converge to a FLRW spacetime for a certain $t \rightarrow t_0$.
3. For $k=0$ and $k=-1$, the problem is linearization stable (i.e. every solution of the linearized problem is tangent to a curve of exact solutions in the space of solutions); for $k=+1$ linearization stability requires additional restrictions.

The suitable function space for point 1 is H^k – a space of tensor fields whose norms are square integrable, and whose derivatives of order $\leq k$ have norms that are square integrable on every compact subset of the manifold of the solution. In point 2 the one-parameter families can only be C^k-functions of the parameter. The additional restrictions for $k=+1$ in point 3 require a lengthier elaboration and will not be quoted here. How these results are illustrated by the exact solutions of this book will be left as an exercise for the reader.

Brauer and Malec (1992b) considered spherically symmetric (finite!) perturbations superimposed on a finite-size region of the $k=0$ FLRW Universe. The perturbation was assumed to have, on the initial hypersurface, the special form g_{ij} (perturbed)$=$ $\phi^4(r)g_{ij}$ (FLRW), where g_{ij} is the spatial three-metric. Both the L–T and the Stephani–Barnes models are in this set. Brauer and Malec did not consider any explicit solutions. The theorems are sufficient conditions and necessary conditions for the existence of trapped surfaces. The assumptions are chosen from the following collection:

A1. The density perturbation is non-negative.
A2. The trace of the second fundamental form of the initial data surface remains unperturbed ($\delta K^a_{\ a} = 0$).
A3. There is no perturbation of the matter current.

The theorems proved are the following:

(I)   (A1, A2, A3)$\Rightarrow$(the condition $\delta M > L + S(\hat{\rho}/6\pi)^{1/2}$ ensures that the sphere of area S and radius L is trapped; δM is the mass of the perturbation and $\hat{\rho}$ is the density of the FLRW region on the initial hypersurface) (this is a sufficient condition).

(II)   (A2, A3)$\Rightarrow$(if $\delta M < L/2 + S (\hat{\rho}/6\pi)^{1/2}$, then the sphere is not trapped) (this is a necessary condition).

(III) (A1, A2, A3)$\Rightarrow$($\delta M = \int_V \rho_V dV \leq L + R_0$, where δM is the mass of the inhomogeneity inside the sphere, L is the proper radius of the sphere and R_0 is its areal radius).

(IV)   (A1, A2, A3)$\Rightarrow$(if $(\hat{\rho}/6\pi)^{1/2} \geq 1/4\pi R_0 + L/8\pi R_0^2$, where all the quantities refer to the same sphere S, then S cannot be trapped) (this is the condition of absence of large trapped surfaces).

(V)   (A1, A2)$\Rightarrow$(if at a sphere S:

$$\delta M - \int_V \delta J_b n^b dV \geq \tfrac{7}{6}L + L^2(8\pi\hat{\rho}/3)^{1/2}, \tag{3.9.3}$$

then there exists a trapped surface inside S) (this is a sufficient condition). Here δJ_b is the perturbation of the matter current, and n_a is the outward unit normal vector to the initial data hypersurface.

Another paper by the same authors (Brauer and Malec 1992a) is a slightly extended version of the 1992b paper, with exactly the same results.

The conditions for forming trapped surfaces in a spherically symmetric distortion (of the same kind) in the closed FLRW background were elaborated by Malec and O'Murchadha (1993). The necessary conditions in the case of the open FLRW background were formulated by Brauer, Malec and O'Murchadha (1994).

This completes the part of the book that is concerned with the L–T models and related solutions. The solutions considered in later parts of this book will mostly not be related to the preceding ones in any way other than having the FLRW models as a common subset.

4

The Stephani–Barnes family of solutions

4.1 Definition and general properties

This family has a simple invariant definition: it is a collection of all perfect fluid solutions for which $\sigma = \omega = 0 \neq \theta$. Already at this point, it can be seen that the common subset of these solutions and those of the Szekeres–Szafron family will consist of the FLRW models only: the perfect fluid solutions that would belong to both families at the same time must have $\sigma = \omega = 0 = \dot{u}^\alpha$, and so, by the criterion of Section 1.3.2, must be FLRW.

In comoving coordinates, the metric of this family is:

$$\mathrm{d}s^2 = D^2 \mathrm{d}t^2 - V^{-2}(t,x,y,z)(\mathrm{d}x^2 + \mathrm{d}y^2 + \mathrm{d}z^2), \tag{4.1.1}$$

where:

$$D = F(t)\, V_{,t}/V, \tag{4.1.2}$$

$F(t)$ is an arbitrary function, and V is determined by the equation:

$$w_{,uu}/w^2 = f(u), \tag{4.1.3}$$

where $f(u)$ is an arbitrary function. The variable u is related in simple ways to the coordinates x, y, z, and the function $w(t,x,y,z)$ is related in simple ways to the function V, but several cases have to be considered separately, and in each case the relation is different. Therefore we will present the details later on. With (4.1.2) and (4.1.3), all the nontrivial Einstein equations are fulfilled; the remaining ones just define matter-density and pressure (the formulae will follow later). The Weyl tensor is proportional to $f(u)$; with $f(u) = 0$ the solutions become conformally flat, but not necessarily FLRW ($\dot{u}^\alpha$ may be nonzero in the limit). The function $F(t)$ is related to the scalar of expansion $\theta = u^\alpha{}_{;\alpha}$ by:

$$\theta = 3/F(t), \tag{4.1.4}$$

hence, in this family, $\theta = \theta(t)$ in comoving coordinates.

The solutions split into two subfamilies; one is of Petrov type D, the other is conformally flat. The type D solutions necessarily have a G_3/S_2 symmetry; the conformally flat solutions in general have no symmetry at all and so are not contained in the former as a limit. As cosmological models, these solutions have an undesirable property: with each of the two equations of state that are cosmologists' favourites, dust ($p = 0$) and a barotropic perfect fluid ($p = p(\epsilon)$), they become trivial. With $p = 0$, they necessarily become FLRW; with $p = p(\epsilon)$ they become either FLRW or the

Wyman (1946) solution or the Collins–Wainwright (1983) solution; neither of the last two solutions is a reasonable cosmological or stellar model (they will be described in later sections of this book). Other than $p=0$ or $p=p(\epsilon)$, no-one so far has had any idea of a cosmologically relevant equation of state. To allow nontrivial solutions of this family, the equation of state would have to involve a third thermo-dynamic function (e.g. temperature or entropy), or would have to explicitly involve spatial coordinates as parameters. Without any equation of state being imposed, the metrics contain arbitrary functions of time (arising as integration "constants" in the solutions of (4.1.3)), and do not define any time-evolution of initial data. Therefore, most of the solutions of this family have been discussed from a general geometric or analytic point of view, and very little is known about their application to cosmology apart from the fact that they are exact perturbations of the FLRW models. Many of them have been interpreted as stellar models because the matching conditions to the Schwarzschild solution in some cases make it possible to draw conclusions about their evolution.

The earliest solution of this family was found by McVittie (1933), and it is still probably the most interesting. It is a superposition of the FLRW metrics with the Schwarzschild solution, and as such has been thought-provoking. In the same year, Dingle (1933) reduced the Einstein equations for a spherically symmetric metric of the form (4.1.1) to a single equation equivalent to (4.1.3). Unfortunately, the parametrization he chose made the equation unreadable. In the simple form (4.1.3), the equation for the spherically symmetric case was first derived by Kustaanheimo and Qvist (K–Q, 1948), and then rederived in nearly 20 other papers, including one by this author (Krasiński 1984b). The conformally flat case was completely solved by Stephani (1967a), and the whole family was derived by Barnes (1973, see also a general discussion by Barnes 1984). It consists of the Stephani (1967a) class, the K–Q class, and the plane and hyperbolically symmetric counterparts of the K–Q class. Barnes was the first to show that this family exhausts the perfect fluid solutions with $\sigma=\omega=0\neq\theta$. In the same paper, Barnes also considered static solutions with $\sigma=\omega=0$, but these are not taken into account in this book. All three type D classes were united by this author (Krasiński 1989) into a single two-parameter family, within which continuous deformations of the spherically symmetric case into the hyperbolically symmetric case via the plane symmetric case, and vice versa, can be considered.

Maiti (1984) proved, without solving the Einstein equations, that a perfect fluid solution with $\sigma=\omega=0\neq\theta$ must be either spherically or plane or hyperbolically sym-metric. Maiti concluded this from the equations of motion and the equations of evolution of kinematical tensors (Ellis 1971), tacitly assuming that the conformal curvature is nonzero (otherwise the statement would not be true).

In Wainwright's (1979 and 1981) classification scheme, the whole family is of extrinsic type D. Solutions with zero heat-flow have the expansion scalar dependent only on the comoving time, and so are of class D_3. The solutions with heat-flow and of Petrov type D are of Wainwright class D_2; the conformally flat ones with heat-flow are of class D_1. All the solutions are of intrinsic class V.

As before, the solutions will be presented in the order of decreasing generality, and, until further notice is given, Figure 4.1 should be consulted. The solutions by Shah and Vaidya (1968) in Box 4.1.5 and by Srivastava and Prasad (1991) are generalizations to nonzero electric charge; they will be described in Sections 4.12 and 4.2.2 respectively.

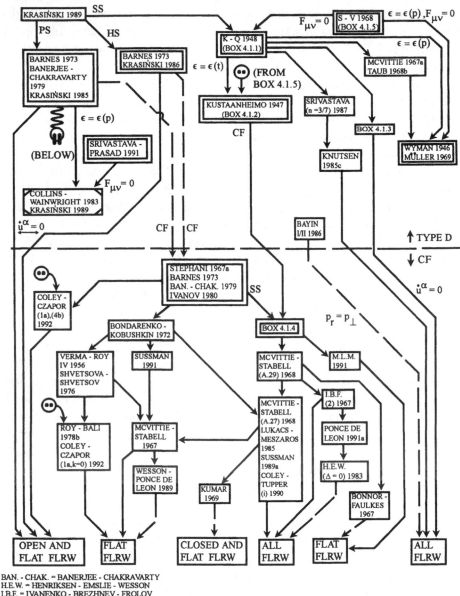

Figure 4.1. Perfect fluid solutions with $\omega = \sigma = 0$ and their charged generalizations.

4.2 The type D solutions

A coordinate representation covering all three type D models at once was found by this author (Krasiński 1989). In it, the functions w and V from (4.1.1)–(4.1.3) are related as follows:

$$V=(z+b)w(t,u), \tag{4.2.1}$$

where b is an arbitrary constant, and the variable u is defined by:

$$u=\tfrac{1}{2}(x^2+y^2+z^2-a)/(z+b), \tag{4.2.2}$$

a being another arbitrary constant. Although no symmetries other than rotations in the (x,y) plane are evident in this representation, these solutions have three-dimensional symmetry groups acting on two-dimensional orbits. The symmetry is spherical when $a<b^2$, plane when $a=b^2$ and hyperbolic when $a>b^2$.

It must be stressed that this representation does not provide any new solutions; the collection of solutions of (4.1.3) with (4.2.1)–(4.2.2) is exactly the same as Barnes' (1973) collection of type D solutions with $\theta\neq0$. However, the unified representation may be useful in looking for further generalizations (nonexistent so far). The equivalence of (4.2.1)–(4.2.2) to the Barnes solutions is demonstrated in the paper by the present author (Krasiński 1989) by explicit transformations. Since the transformations preserve the general form (4.1.1) of the metric, it is no surprise that they are closely related to the conformal symmetries of a three-dimensional flat space (for details see Krasiński 1989; see also Haantjes 1937 and 1940). The present author's paper (Krasiński 1989) contains explicit formulae for the symmetry transformations of (4.1.1) in the coordinates of (4.2.1)–(4.2.2). This representation was obtained by working from the metric ansatz (4.1.1) in comoving coordinates. In this approach, the invariant definition of the class was not visible. A few other papers by this author (Krasiński 1984b, 1985 and 1986) reported on the progress that resulted in the 1989 paper. As a historical curiosity, let us note that the ansatz (4.1.1) was proposed by Dirac (1981) as a way of generalizing the FLRW models. Unfortunately, Dirac considered the resulting Einstein equations only perturbatively.

4.2.1 The hyperbolically symmetric Barnes models

In the case of hyperbolic symmetry, there exists a simpler representation in which:

$$u=x/y, \quad V=yw(t,u). \tag{4.2.3}$$

The matter-density and pressure are given by:

$$\kappa\epsilon=3/F^2+2(u^2+1)fw^3+6uww,_u-3(u^2+1)w,_u^2-3w^2, \tag{4.2.4}$$

$$\kappa p=-3/F^2-2uww,_u+(u^2+1)w,_u^2+w^2-2F,_t/(F^2D)$$
$$+2(F/D)[u-(u^2+1)w,_u/w](ww,_{tu}-w,_tw,_u). \tag{4.2.5}$$

These models were found by Barnes (1973) and rediscovered by this author (Krasiński 1986). They reproduce the open and flat FLRW models in their full generality. The transformation to a standard FLRW form is nontrivial; see Krasiński 1989 for explicit formulae. Collins and Wainwright (1983) showed that these models have no subcase with a barotropic equation of state, $p=p(\epsilon)$, apart from the FLRW spacetimes. This author (Krasiński 1989) showed that the only ways in which the models (4.2.3)–(4.2.5) can acquire additional symmetry are by degenerating to either a static solution or a vacuum solution (possibly with Λ) or to a FLRW model. Barnes (1973) provided coordinate representations other than (4.2.3), but we prefer (4.2.3) because it allows one to find the hyperbolically symmetric counterpart to every spherically symmetric solution of this family immediately (see later).

4.2.2 The plane symmetric Barnes models

In this case, coordinates can be chosen so that:

$$u=z, \quad V=w(t,z) \tag{4.2.6}$$

in (4.1.1)–(4.1.3). The matter-density and pressure are:

$$\kappa\epsilon=3/F^2+2fV^3-3V_{,z}^2, \tag{4.2.7}$$

$$\kappa p=-3/F^2+V_{,z}^2-2F_{,t}/(F^2D)-2[FV_{,z}/(DV)](VV_{,tz}-V_{,t}V_{,z}). \tag{4.2.8}$$

These models were found by Barnes (1973), and then rediscovered by Banerjee and Chakravarty (1979) as one of the cases resulting from the ansatz (4.1.1) in comoving coordinates. In fact, the latter authors found a coordinate representation that covers the plane symmetric case and the spherically symmetric case, but not the hyperbolically symmetric case. Later, these models were rediscovered again by this author (Krasiński 1986). They reproduce the open and the flat FLRW models in their full generality in the limit $\dot{u}^\alpha=0$ (see Krasiński 1989 for explicit formulae). The barotropic equation of state $p=p(\epsilon)$ reduces the models either to FLRW or to a metric that was first identified by Collins and Wainwright (1983). In the latter case $F=$const and:

$$V=azv(X), \tag{4.2.9}$$

where a is an arbitrary constant, $X=azt$, and v obeys the following equation:

$$X^2v_{,XX}+2Xv_{,X}=Bv^2, \tag{4.2.10}$$

B being another arbitrary constant. Equation (4.2.10) results from (4.1.3) when $w=azv$ and $f(z)=B/(az^3)$.

In the limit of spatially homogeneous matter-density, the Collins–Wainwright solution reduces to the de Sitter solution, and so has no nontrivial FLRW limit. It is mentioned here for completeness because it is the plane symmetric counterpart of the Wyman (1946) solution that plays an important role in the spherically symmetric family. This author (Krasiński 1989) identified the same solution as the only nontrivial subcase of the plane symmetric Barnes subfamily that has a higher symmetry. Nontrivial means nonstatic, nonvacuum and inhomogeneous. The solution defined by (4.2.9)–(4.2.10) has a four-dimensional symmetry group acting on three-dimensional orbits; it is a tilted Bianchi V model with an additional symmetry (see Collins and Wainwright 1983, Krasiński 1989 and, especially, Collins 1985 and 1986 for more details). Since we will not be returning to this solution, let us note that it was generalized to the case of a charged perfect fluid obeying a barotropic equation of state by Srivastava and Prasad (1991). Taub (1972) discussed a family of self-similar plane symmetric spacetimes that includes the Collins–Wainwright (1983) solution as an example. However, Taub did not discuss any explicit solutions.

Barnes (1973) provided another coordinate representation of this case, but we prefer (4.2.6) for the same reason as before.

4.2.3 The Kustaanheimo–Qvist (spherically symmetric Barnes) models

In the case of spherical symmetry, coordinates can be chosen so that:

$$u = x^2 + y^2 + z^2 \equiv r^2, \quad V = w(t,u), \tag{4.2.11}$$

in (4.1.1)–(4.1.3), and so:

$$ds^2 = D^2 dt^2 - V^{-2}(t,u)[dr^2 + r^2(d\vartheta^2 + \sin^2\vartheta \, d\varphi^2)]. \tag{4.2.12}$$

The matter-density and pressure are:

$$\kappa\epsilon = 3/F^2 + 8uf V^3 + 12VV,_u - 12u V,_u^2, \tag{4.2.13}$$

$$\kappa p = -3/F^2 - 4VV,_u + 4u V,_u^2 - 2F,_t/(F^2 D)$$

$$+ 4(F/D)(1 - 2u V,_u/V)(VV,_{tu} - V,_t V,_u). \tag{4.2.14}$$

Now that the relation between the metric variables (x, y, z, V) in (4.1.1) and the variables (u,w) in (4.1.3) has been determined for every case, let us observe that different solutions of (4.1.3) do not necessarily define different geometries. The metric of the hypersurfaces t=const in (4.1.1) is defined up to conformal symmetries of a three-dimensional Euclidean space, so different solutions of (4.1.3) may in fact correspond to the same geometries. As an example, let us take the following transformation in (4.2.12):

$$(x, y, z) = a(x', y', z')/u', \tag{4.2.15}$$

where a is an arbitrary constant and $u' = x'^2 + y'^2 + z'^2$. This induces the transformation

$$u = a^2/u' \tag{4.2.16}$$

for the variable u. Then the form of (4.2.12) is preserved, but V is transformed to:

$$V' = u' V/a. \tag{4.2.17}$$

Hence, the transformation (4.2.16) accompanied by (4.2.17) leads to another coordinate representation of the same geometry, and so $V'(u')$ defined in this way should be another solution of (4.1.3). Indeed, it can be verified that it is so, the new $f'(u')$ being:

$$f'(u') = a^5 f(a^2/u')u'^{-5}. \tag{4.2.18}$$

It follows that the transformation (4.2.16)–(4.2.18) leads from one solution of (4.1.3) to another solution, but the metrics (4.2.12) corresponding to the two solutions differ only by a coordinate transformation, and so represent the same geometry.

The FLRW models (all of them) result from (4.1.1)–(4.1.3) with (4.2.11) when $f(u)=0$ and $V=(1+ku/4)/R(t)$, where $k=$const and $R(t)$ is an arbitrary function. The limit $f(u)=0$ without further conditions leads to the spherically symmetric Stephani (1967a) solution, see Section 4.10.

This subfamily of the Barnes solutions was first investigated by Dingle (1933). He derived (4.1.2) and the time-derivative of (4.1.3) but stopped short of integrating it to (4.1.3) because his parametrization was inconvenient for this purpose.

As a historical curiosity, let us note that Dingle in fact predicted the later discovery of "intrinsically homogeneous" spacetimes, in particular of the Stephani (1967a) solution (see our Section 4.9). This is the prediction: ". . . a line-element which reduces to that of a homogeneous three-dimensional subspace at each instant of time is not necessarily the line-element of a universe whose space is homogeneous".

After deriving the two equations, Dingle interpreted them in three cases only: I. approximately, the small parameter being the acceleration; II. with matter-density and pressure depending only on time; the FLRW models result then; III. in the static limit. Dingle used the approximate result to investigate the stability of both the static Einstein and the FLRW models. The result was that the Einstein and the collapsing FLRW models are unstable, while the expanding FRLW models are stable against such a perturbation (see the note on the Raychaudhuri 1952 paper given in the next section). Dingle was aware that the perturbation he considered was not general. He then showed that the static model would not be consistent with the Hubble law connecting redshifts to distances. The paper closes with a rather strong criticism of the postulate of spatial homogeneity of the Universe – see quotations in Appendix C.

Formulae equivalent to (4.1.2)–(4.1.3) arose in the paper by Wyman (1946, his eqs. (2.5) and (3.3)), but the author paid no attention to them because he was interested only in those solutions that obey the barotropic equation of state. Thus his "prediscovery" of this class of solutions can only be recognized *ex post*. The Wyman solution will be described in Section 4.4.

V. V. Narlikar (1947) presented another partial integration of the Einstein equations for this case. He presented an equation equivalent to (4.1.3), without considering (4.1.2) (which is rather strange because (4.1.3) is a consequence of (4.1.2)).

The Einstein equations for this case were first reduced to the simple form (4.1.1)–(4.1.3) by Kustaanheimo and Qvist (1948), and therefore this class is usually named after them. They were also the first to attempt an integration of (4.1.3). This problem, in full generality, remains unsolved to this day; there are only partial results, for specific forms of the arbitrary function $f(u)$. Kustaanheimo and Qvist identified one special form of $f(u)$ when (4.1.3) can be integrated explicitly. We shall return to this point in Section 4.6.

Examples published in the literature (and described later in this book) show that solutions of (4.1.3) with various explicit forms of $f(u)$ usually involve elliptic or other transcendental functions. The question as to when (4.1.3) can be integrated in terms of quadratures was investigated directly by Stephani (1983), who used the theory of invariance transformations of differential equations developed by Lie (1912). The results of Stephani's paper are, in short, as follows:

1. Equation (4.1.3) can be integrated to a first-order equation when $f(u)=u^n$ or $f(u)=e^u$ or $f(u)=(u+\alpha)^n (u+\beta)^{-n-5}$, where n, α and β are arbitrary constants (of course, this also applies to the cases when $f(u)$ can be transformed to one of these forms by substitutions, e.g by (4.2.15)–(4.2.18)). The corresponding first-order equations are given in the paper.
2. Equation (4.1.3) can be explicitly solved by quadratures when $f(u)=(ax^2+bx+c)^{-5/2}$ or $f(u)=u^{-15/7}$, where a, b and c are arbitrary constants. The integrals implicitly defining the solutions are presented in the paper, see also our Section 4.6.

Unfortunately, Stephani did not make it clear that the conditions given above are sufficient, but not necessary. An example exists in the literature (Srivastava 1987, see our Section 4.8) of a solution that obeys only the first condition of Stephani $(f(u)=(u+\alpha)^{-15/7}(u+\beta)^{-20/7})$, but is expressible in quadratures.

The paper by Srivastava (1987) is a broad overview of solutions of (4.1.1)–(4.1.3) in the spherically symmetric case, but the author did not investigate interrelations among them. It is rather strange that he treated (4.1.2) as an equation to be integrated, when it is in fact just a formula for calculating D given V.

4.3 Rediscoveries of the Kustaanheimo–Qvist (K–Q) class of metrics

The K–Q paper remained unknown for a long time, and the result was repeatedly rederived, so far in 17 papers, including one by this author. The list of rediscoveries is given in Box 4.1.1. We shall briefly comment on the papers in the list, and mention other papers where related results were obtained.

Raychaudhuri (1952) discussed the metric (4.1.1) as a perturbation of the FLRW models, without finding explicit solutions of the Einstein equations. He found that

DINGLE 1933
WYMAN 1946
V.V. NARLIKAR 1947
KUSTAANHEIMO - QVIST 1948
WAGH 1955a
WAGH 1958
NARIAI 1968
TAUB 1968
CAHILL - MCVITTIE 1970a
KRISHNA RAO 1973
BARNES 1973
WYMAN 1976
MANSOURI 1977
GLASS 1979
BANERJEE - CHAKRAVARTY 1979
MANSOURI 1980
KRASINSKI 1984b
DYER - MCVITTIE - OATTES 1987
SUSSMAN 1989a
HOGAN 1990
NOLAN 1993

NOTE: THE PAPERS BY DINGLE 1933,
WYMAN 1946 AND NARLIKAR 1947
ARE "PRE-DISCOVERIES", SEE TEXT
FOR AN EXPLANATION. THE STORY
BEGINS WITH KUSTAANHEIMO AND QVIST 1948.

Box 4.1.1. Rediscoveries of the K–Q spacetimes.

as long as the Universe is expanding (i.e. $(1/V)_{,t} > 0$), the spatial gradient of matter-density $\epsilon_{,r}$ will decrease with time, but the density contrast, $(\epsilon_{,r}/\epsilon)$, will increase. This conclusion is independent of any assumptions about pressure.

Wagh (1955a) rederived (4.1.1)–(4.1.3), and then presented the most general solution for $f(u)$=const. The solution will be mentioned in Section 4.6. In a companion paper (Wagh 1955b), the author considered examples of solutions for $f(u)$=0, but they were: two expressions for a FLRW metric, a metric whose limit $\dot{u}^\alpha$=0 is the Minkowski space, a metric that has no conformally flat limit and a static metric.

Later, the same author (Wagh 1958) rederived (4.1.1)–(4.1.3) by working from the equations given by Dingle (1933). The three cases that Wagh considered were, respectively, with no conformally flat limit (two of them) and static.

Taub (1968b) reobtained the K–Q class in a different parametrization. That paper contains contributions to a few other entries in Figure 4.1. and will be mentioned again later on. The other paper (Taub 1968a) is a short conference report announcing the 1968b paper.

Nariai (1968) rederived the class simply in order to place his older solution (Nariai 1967a, see our Section 4.6 and Box 4.1.3) within a broader context.

Cahill and McVittie (1970a) derived the K–Q class from a consideration similar to our Section 1.4. The paper will be mentioned again in Section 4.10 for its other derivation. In a companion paper (Cahill and McVittie 1970b), the authors showed that the K–Q metrics can be matched to the Schwarzschild solution and that among them there are some that describe oscillating and bouncing perfect fluid spheres. The latter are nonsingular and never collapse under their Schwarzschild horizons.

Krishna Rao (1973) rederived the K–Q class while proving the theorem that a shearfree spherically symmetric solution with spatially homogeneous matter-density can be nonsingular at the centre only if it is conformally flat.

Barnes (1973) reobtained the K–Q class as a subclass of the perfect fluid solutions with $\sigma = \omega = 0$.

Wyman (1976) rediscovered the K–Q class and made an extensive study of the space of solutions of (4.1.3). He produced a large family of "explicit" examples – represented parametrically in terms of the Weierstrass elliptic function. The examples could very possibly reproduce several of the solutions published by other authors by limiting transitions. Tracing such relationships would be rather laborious and not very illuminating, so we shall not do it here. We want, however, to give the readers a rather strong warning that, as far as solving eq. (4.1.3) is concerned, Wyman did a substantial part of the job, and all claims of finding a "new" solution should be checked against his paper.

Mansouri (1977) derived the K–Q class as a basis for the theorem that a spherically symmetric shearfree perfect fluid solution obeying $p = p(\epsilon)$ cannot be matched to the Schwarzschild solution except in the trivial case $p = 0$ (which is the FLRW dust). Note that this result automatically implies that the Wyman (1946) solution (see our Section 4.4) cannot be matched to the Schwarzschild solution. This conclusion was rederived by Lambas, Lamberti and Hamity (1987).

Glass (1979) rederived the class and presented some of its properties, among them the conditions for forming trapped surfaces and for the matter-density to be spatially homogeneous. The paper will be quoted in respect of some other results later on. Glass also showed how some special solutions published earlier arise as special cases of the K–Q class.

Banerjee and Chakravarty (1979) rederived the K–Q class by working from the metric ansatz (4.1.1). As already mentioned in Section 4.2.2, their coordinate representation covers both the plane and the spherically symmetric case.

Again Mansouri (1980) rederived the same class for proving the same theorem as before (Mansouri 1977) and comparing the result to the Newtonian case.

This author (Krasiński 1984b) rederived the K–Q class while working out the metric ansatz (4.1.1).

Dyer, McVittie and Oattes (1987) rederived the class (in their section 3) as a by-product of considerations about conformal Killing vectors in spherically symmetric spacetimes. The other metric (their eq. (32)) is just the Kottler solution (i.e. the Schwarzschild solution generalized for Λ) expressed in expanding coordinates; see Krasiński and Plebański (1980) and Krasiński (1989).

Sussman (1989a) rederived the class while discussing spherically symmetric shearfree perfect fluid solutions with an additional conformal symmetry.

Hogan (1990) rederived the class while deriving anew the McVittie (1933) solution. His eq. (4.25b) is the time-derivative of (4.1.3) expressed in another parametrization. We shall come back to Hogan's paper in Section 4.7. The two solutions given in the last section of the paper become static in the limit $(\delta^{\alpha}_{\ \beta} - u^{\alpha} u_{\beta}) p,_{\alpha} = 0$, and so do not admit a FLRW limit.

Nolan (1993) rediscovered the K–Q class while looking for solutions regular at the centre that can be matched to the McVittie (1933) solution (see our Sections 4.6 and 4.7).

4.4 Invariantly defined subcases of the K–Q class

Wyman (1946) showed that apart from the FLRW models, the only solutions of the K–Q class that obey the barotropic equation of state are those for which:

$$V = \mathcal{P}(v)/C, \quad v = t + C_2 u/2, \quad u = x^2 + y^2 + z^2,$$

$$F(t) = [2(\alpha C_2 C_3 t + \beta)]^{-1/2},$$

(4.4.1)

where C, C_2, C_3, α and β are arbitrary constants, and $\mathcal{P}(v)$ is the subcase of the Weierstrass elliptic function defined by:

$$\mathcal{P},_v^2 = \mathcal{P}^3/3 - 12^2 C_3.$$

(4.4.2)

Equation (4.4.2) arises by integrating (4.1.3) when $f(u) = \frac{1}{8} CC_2^2 = \text{const}$ (note that the Wyman solution is not the most general one with $f(u) = \text{const}$; see Section 4.6 for more remarks on this point). The matter-density and pressure are:

$$\kappa \epsilon = 6\alpha C_2 (\mathcal{P}\mathcal{P},_v/12^2 + C_3 v) + K,$$

$$\kappa p = -\alpha C_2 (\mathcal{P}\mathcal{P},_v/12^2 + 6C_3 v - 5C_3 \mathcal{P}/\mathcal{P},_v) - K,$$

(4.4.3)

where K is another arbitrary constant (a misprint in the last formula was corrected here). As stated in Section 4.1, the solution becomes conformally flat when $8f(u) = CC_2^2 = 0$, that is, when $C_2 = 0$ (because the limit $C = 0$ is singular). Then $\epsilon + p = 0$ and the de Sitter solution results. Hence, the Wyman solution does not in fact fulfil the criteria for being included in this book, but it is described here because of the important role it plays in the K–Q family. It demonstrates how restrictive the barotropic equation of state is. In general, the set of solutions of (4.1.3) is labelled by three arbitrary functions: $f(u)$ and the two functions of time that arise when (4.1.3) is integrated. The equation of state $p=p(\epsilon)$ splits this set into two very restricted subsets that have only the de Sitter solution in common: (1) The FLRW collection with $f=0$ and one arbitrary function of time; (2) The Wyman solution with $f=$const and both functions of time explicitly specified. This shows, moreover, that the FLRW models have no inhomogeneous parent solutions in the set of K–Q spacetimes obeying $p=p(\epsilon)$ (see Krasiński 1991 for a discussion of this point).

The second of Wyman's solutions (his case 2) results from (4.4.1)–(4.4.3) by the coordinate transformation $r=1/r'$.

The Wyman solution is indeed the unique inhomogeneous K–Q spacetime obeying $p=p(\epsilon)$, as stated by Wyman himself, although Wyman's presentation of the proof contains a gap that may give rise to doubts about the result. A complete proof was presented by Müller (1969), who pointed out the gap in Wyman's proof, but did not explain how to close it. Srivastava and Prasad (1983) presented another proof of uniqueness of the Wyman solution, but using the assumption of "regularity at the centre" of the metric. They thus created a new mystery: how Wyman and Müller succeeded without such an assumption. One more proof was published by Collins and Wainwright (1983), by a still different method, that apparently settled the question of uniqueness but did not solve the mystery.[36] Properties of the Wyman solution were described in much detail by Collins (1985 and 1986), and its global geometry was discussed by Sussman (1988a,b). The reasoning that leads to the Wyman solution was partly worked out by Taub (1968), but he only presented several necessary conditions to be obeyed by the solution. In fact, McVittie (1967) did very similar work, but apparently did not notice that it leads towards imposing the condition $p=p(\epsilon)$.

As follows from the papers by Mansouri (1977 and 1980, see the preceding section), the Wyman solution cannot be matched to the Schwarzschild solution, so it cannot be a stellar model; this fact has been observed in a few other papers. That it is no good as a cosmological model follows from the fact that it reduces to the de Sitter solution in the conformally flat limit. Further objections to its cosmological interpretation can be found in the papers by Srivastava and Prasad (1983) and Collins (1985). Hence, the main importance of the Wyman solution is that it reveals the effect of the barotropic equation of state on the K–Q class. In the opinion of this author, the Wyman solution proves that $p=p(\epsilon)$ is not a reasonable equation of state to impose on inhomogeneous models, in contrast to the spatially homogeneous models where it follows necessarily. For a single-component perfect fluid, $p=p(\epsilon)$ implies that the entropy per particle is a universal constant in the spacetime (see Krasiński 1974 for a proof; but this fact was shown by several other authors). This

---

[36] A paper by this author (A. K.) reconciling the proofs by Wyman and by Srivastava and Prasad was successfully prevented from being published by three teams of referees. It is thus concluded that there is no public demand for the mystery to be explained.

is natural in a spatially homogeneous fluid, but finds no explanation in inhomogeneous media where all hydrodynamical characteristics are allowed to vary with spatial position.

QVIST 1947 / KUSTAANHEIMO 1947
TAUB 1968b
BARNES 1973
EISENSTAEDT 1975a
EISENSTAEDT 1976
GLASS 1979

TO BOX 4.1.3
(EISENSTAEDT)

Box 4.1.2. Rediscoveries of the K–Q spacetimes with $\epsilon=\epsilon(t)$.

Another invariantly defined limit of the K–Q class is the collection of solutions with spatially homogeneous matter-density, $\epsilon=\epsilon(t)$ (see Box 4.1.2). This happens when:

$$f(u)=C/u^{5/2} \tag{4.4.4}$$

in (4.1.3), where $C=$const. The first to consider this case was Qvist (1947); his presentation was developed and explained by Kustaanheimo (1947). In the notation adapted to ours, they found the following:

1. The case when $\epsilon(t)=\frac{1}{3}\theta^2(t)$ has to be considered separately, and it leads to the subcase $k=0$ of the McVittie (1933) solution, see Section 4.6.
2. When $\epsilon(t)\neq\frac{1}{3}\theta^2(t)$, the solution is given by (4.1.1) with (4.1.2), $F(t)=3/\theta$, and:

$$V=\tfrac{1}{2}|\theta^2/3-\epsilon|^{1/2}r/\zeta, \quad r=(x^2+y^2+z^2)^{1/2}, \tag{4.4.5}$$

where ζ is defined by:

$$r^2\zeta_{,r}^2=-\tfrac{4}{3}\varepsilon\zeta^4+\zeta^2+\tfrac{4}{3}C|\theta^{2/3}-\epsilon|^{1/2}\zeta, \tag{4.4.6}$$

C being the constant from (4.4.4), and ε being the sign of $(\epsilon-\theta^2/3)$.

The special cases with zero pressure considered by Kustaanheimo (1947) are the following:

Case 1, when worked out in detail, turns out to have constant density.

Case 2 turns out to be just the FLRW models.

Case 3 is static.

In the limit $C=0$, the Qvist/Kustaanheimo solution becomes conformally flat, but not necessarily FLRW. This limit is the spherically symmetric subcase of the Stephani (1967a) solution and will be discussed in Section 4.10. The sign of $(\epsilon - \theta^2/3)$ coincides with the sign of the curvature index of the resulting FLRW limit, in particular the case with $\varepsilon=0$ (the special McVittie solution) has a flat FLRW limit. In fact, Qvist and Kustaanheimo considered only the case $\varepsilon=-1$ (a generalization of the open FLRW model), but the modification to include arbitrary ε is straightforward.

Taub (1968b), working in different variables, observed that the case equivalent to $f(u)u^{5/2}=C=$const leads to $\epsilon=\epsilon(t)$, but he derived the formulae for the metric only in the case $C=0$, that is, in the conformally flat limit.

The solution (4.4.5)–(4.4.6) was derived by Barnes (1973, his case IB) together with the whole Stephani–Barnes family. He expressed the solution of (4.4.6) in terms of the Weierstrass elliptic function.

Eisenstaedt (1975a) derived the solution in a form very similar to (4.4.5)–(4.4.6) and showed how it can be reduced or matched to the FLRW and Schwarzschild solutions. He discussed at more length the solutions of (4.4.6) in terms of elliptic integrals. As the main topic, he considered a model of the Universe with a spherical condensation; it consisted of a region with $C=0$ (i.e. a conformally flat one) matched to a region with $C\neq0$. In a subsequent paper (Eisenstaedt 1975b), the author showed that the solutions can be composed into a multi-layered model with different densities $\epsilon_i(t)$ in each layer. In another paper (Eisenstaedt 1976), the author presented a much more detailed derivation and description of the same solution, with all possible limiting cases. In that paper, the author studied the solution perturbatively, assuming that $\alpha \overset{\text{def}}{=} (\theta^2/3-\epsilon)$ is a small parameter, up to linear terms in α. From this study, Eisenstaedt concluded that the solution represents a point-mass (of positive or negative mass, the sign being the sign of $(-C)$ in (4.4.6)) superimposed on a FLRW background. Indeed, this interpretation is plausible, taking into account the fact that the solution contains as subcases both the Schwarzschild solution and the FLRW models (see Eisenstaedt 1975a) and that its limiting case includes the special McVittie (1933) solution which, as is evident, represents such a superposition.

In a follow-up paper (Eisenstaedt 1977), the author investigated geodesics in this solution and showed that particles moving on them may experience repulsive forces when $(-Cq)<0$, where q is the deceleration parameter of the background FLRW model. Eisenstaedt tried to use this observation as an explanation of clustering in the Universe. However, in this particular model the cosmic medium does not move on geodesics and its matter-density is permanently spatially homogeneous, so such an explanation does not sound plausible.

The solution was rederived once more by Glass (1979).

General analytic properties of the K–Q solutions with spatially homogeneous density (including the conformally flat ones) were described by Knutsen (1983b).

4.5 The McVittie (1967) class

The class of metrics introduced by McVittie (1967a) has to be described here
because the author himself, and a few other authors, derived explicit solutions using
this approach. However, the approach has the distinctive feature of obscuring rather
than clarifying the properties of solutions and relationships among them.

Guided by the similarity of the resulting equations to the equations of Newtonian
blast waves, McVittie proposed to consider the following metrics:

$$ds^2 = e^{\lambda(z)}dt^2 - R_0^2 S^2(t)e^{\eta(z)}[dr^2 + f^2(r)(d\vartheta^2 + \sin^2\vartheta\, d\varphi^2)], \qquad (4.5.1)$$

where R_0 is a constant, $S(t)$, $f(r)$, $\lambda(z)$ and $\eta(z)$ are functions to be determined from
the Einstein equations and z is a variable defined by:

$$e^z = Q(r)/S(t), \qquad (4.5.2)$$

$Q(r)$ being another function to be determined. The source is assumed to be a perfect
fluid, and the coordinates of (4.5.1) are assumed comoving. The Einstein equations
then imply:

$$y \overset{\text{def}}{=} e^{\lambda/2} = 1 - \eta,_z/2, \qquad (4.5.3)$$

$$Q,_{rr}/Q - Q,_r f,_r/(Qf) = aQ,_r^2/Q^2, \qquad (4.5.4)$$

$$f,_{rr}/f - f,_r^2/f^2 + 1/f^2 = bQ,_r^2/Q^2, \qquad (4.5.5)$$

$$y,_{zz} + (a-3+y)y,_z + [a+b-2-(a-3)y-y^2] = 0, \qquad (4.5.6)$$

where a and b are arbitrary constants. Particular solutions can be found by assuming various definite values of a and b. It can be verified (albeit with some effort) that
equations (4.5.1)–(4.5.6) do imply (4.1.1)–(4.1.3) with (4.2.11)–(4.2.12), and so the
McVittie class is a subset of the K–Q class.

Calculating the FLRW limit of (4.5.1)–(4.5.6) reveals the main disadvantage of
this approach: the coordinates of (4.5.1) are nonunique, and the FLRW limit can be
obtained in several ways. Moreover, it happens that explicit solutions that appear to
be different in the McVittie coordinates can in fact be transformed one into the
other (see an example in Section 4.6). The FLRW limit results when $\dot{u}^\alpha = 0$, that is,
when $y,_r = 0$, that is, when either $y = \text{const}$ or $Q = \text{const}$ in the set (4.5.3)–(4.5.6).

Equations (4.5.1)–(4.5.6) are necessary (but not sufficient) conditions for the
barotropic equation of state to hold, as was shown by Taub (1968). Hence, the
Wyman (1946) solution is contained in the McVittie class; equations equivalent to
(4.5.2)–(4.5.6) were derived in Wyman's paper in isotropic coordinates.

McVittie's 1967a paper contains partial integrations in a few special cases, and
one explicit solution that will be mentioned in Section 4.6. The other paper by
McVittie (1967b) is a very short account of the same material.

In a later paper (McVittie 1984), the author discussed various possible solutions
of (4.5.6) (without, however, deriving the corresponding solutions of (4.5.4) and

(4.5.5)), and showed that several solutions found in the literature are subcases of (4.5.1)–(4.5.6). That consideration was developed even further by Srivastava (1987).

In other papers, various authors discussed the possible solutions of (4.5.4)–(4.5.5). It must be stressed that all the complications are caused by the coordinates of (4.5.1). Equations equivalent to (4.5.4)–(4.5.5) can be explicitly integrated in full generality in the isotropic coordinates (verified by this author, unpublished).

Knutsen (1983a) investigated the McVittie class for the existence of oscillating and nonoscillating configurations, and of configurations with spatially homogeneous density. The discussion did not invoke or produce explicit solutions. Knutsen and Stabell (1979) performed a similar analysis on a few subcases of the McVittie metrics.

4.6 Explicit K–Q solutions corresponding to $f(u)=(au^2+bu+c)^{-5/2}$

As was mentioned in Section 4.2, eq. (4.1.3) can have simpler or more complicated solutions, depending on the form of the arbitrary function $f(u)$. As shown by Stephani (1983, see our Section 4.2), one of the cases when (4.1.3) can be solved in quadratures is:

$$f(u)=(au^2+bu+c)^{-5/2}\stackrel{\text{def}}{=}[h(u)]^{-5/2}, \qquad (4.6.1)$$

where a, b and c are arbitrary constants. Several authors found out independently that explicit metrics (4.1.1) can be derived with this form of $f(u)$. The independently published solutions corresponding to (4.6.1) form an interlocked network (with repetitions, of course) that is displayed in Box 4.1.3. It is interesting to note that nearly all the solutions are duplicates or subcases of the results already found by Kustaanheimo and Qvist (1948). In Box 4.1.3, they are classified by the algebraic properties of the trinomial $h(u)=[f(u)]^{-2/5}$.

The general formula for a solution of (4.1.3) with $f(u)$ given by (4.6.1) was first derived by Kustaanheimo and Qvist (1948), it is:

$$V(t,u)=v(t,u)\,(au^2+bu+c)^{1/2}, \qquad (4.6.2)$$

where $v(t,u)$ is defined by:

$$\int[\tfrac{2}{3}v^3+(b^2/4-ac)v^2+f_1(t)]^{-1/2}dv=\int(au^2+bu+c)^{-1}du+f_2(t), \qquad (4.6.3)$$

$f_1(t)$ and $f_2(t)$ being arbitrary functions of time. This result does not depend on the sign of:

$$\Delta\stackrel{\text{def}}{=}b^2/4-ac, \qquad (4.6.4)$$

although Kustaanheimo and Qvist seem to suggest that (4.6.3) results when $\Delta\geq0$. The authors then listed all the explicit solutions arising in the cases when the left-hand side of (4.6.3) produces an elementary function. We shall come back to that list soon.

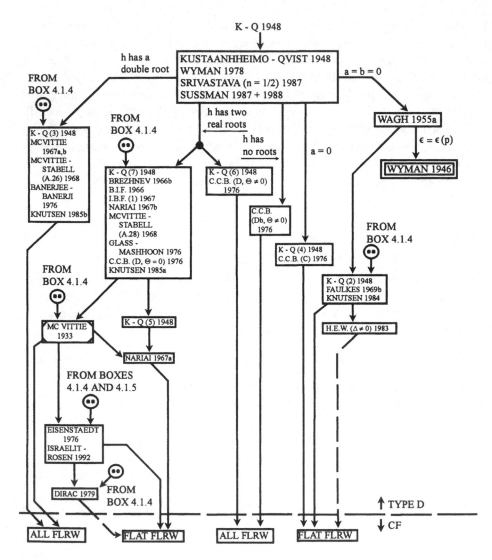

B.I.F. + I.B.F. = PERMUTATIONS OF BREZHNEV - IVANENKO - FROLOV
C.C.B. = CHAKRAVARTY - CHOUDHURY - BANERJEE
H.E.W. = HENRIKSEN - EMSLIE - WESSON

Box 4.1.3. The perfect fluid solutions generated from the K–Q equation
$$V_{,uu}/V^2 = [h(u)]^{-5/2}, \text{ where } h(u) = au^2 + bu + c.$$

Wyman (1978) basically repeated the same derivation and observed that (4.6.3) is solved in terms of the Weierstrass elliptic function. Srivastava (1987) derived the general solution again; this is his case $n=1/2$. It agrees with Wyman's result, provided that a misprint is corrected: the second parameter of the Weierstrass function should actually read $+(vm-wl)^4/192$. Srivastava apparently assumed $\Delta \geq 0$, that is, that $h(u)$ factorizes, but the result obtained is also valid when $\Delta < 0$. Sussman (1987) considered charged generalizations of the K–Q class, and we shall come back to his paper in Section 4.12. In the case of vanishing electromagnetic field, he presented the most exhaustive classification and discussion of all solutions of (4.6.3), elementary and otherwise, and identified those previously published in his scheme. The classification is a little obscured by the fact that Sussman used the coordinates in which the subspaces t=const of (4.1.1) have the metric $ds_3^2 = V^{-2}(t,r)[dr^2 + \bar{f}^2(r)(d\vartheta^2 + \sin^2\vartheta \, d\varphi^2)]$, where $\bar{f}(r)$ is either sin r or r or sinh r. The three cases can be transformed into one another by coordinate transformations, and in such an approach every solution will show up three times. In the discussion, however, Sussman used the variable $y = \int \bar{f}(r)dr$ which is insensitive to the choice of $\bar{f}(r)$. Two follow-up papers (Sussman 1988a,b) discuss the geometrical and physical properties of the solutions from the first paper. The Sussman papers will occasionally be invoked here, but the wealth of material contained in them cannot be fairly represented in a short review, and they are very strongly recommended as further reading on the subject. The earlier paper (Sussman 1986) is an introductory short description of the same material.

The integral on the left-hand side of (4.6.3) is elementary when the polynomial $[\frac{2}{3}v^3 + \Delta v^2 + f_1(t)]$ has a multiple root. This happens when $f_1(t)=0$ or $f_1(t)=-\Delta^3/3$. Kustaanheimo and Qvist (1948) listed all the corresponding explicit solutions, but omitted the cases when $\Delta<0$. We present the various cases reparametrized so that the FLRW limits can be easily calculated.

Case 1, when $f(u)=0$, is a conformally flat solution; see our Section 4.10. Formally, it results from (4.6.3) when one of the parameters a, b or c goes to infinity.

Case 2: $a=b=f_1(t)=0$, then:

$$V(t,u)=[R^{1/2}(t)+Cu]^{-2}, \tag{4.6.5}$$

where $C=(6c^{5/2})^{-1/2}$ and $R(t)$ is an arbitrary function. The (flat only) FLRW limit results when $C=0$.

Case 3: $\Delta=f_1(t)=0\neq a$, that is, $h(u)$ has a double root $u=u_0$:

$$V(t,u)=(u-u_0)^3[R^{1/2}(t)(u-u_0)-C]^{-2}, \tag{4.6.6}$$

where $C=(6a^{5/2})^{-1/2}$. All the FLRW models result from here when $C=0$, but the standard representation (1.3.1) results after the coordinate transformation $r=1/r'$, the curvature index is $k=-4u_0$.

Case 4: $b\neq0=a$, $f_1(t)=-\Delta^3/3$:

$$V=\tfrac{1}{8}B^2(1+BCu)^{1/2}\{1+3\tan^{-2}[\ln(S(t)(1+BCu)^{1/4})]\}, \tag{4.6.7}$$

where $S(t)$ is an arbitrary function, $C=c^{-5/4}$, $B=bc^{1/4}$. The (flat only) FLRW limit follows when $C=0$, with $R(t)=\{\tfrac{1}{8}B^2[1+\tan^{-2}(\ln S)]\}^{-1}$.

Case 5: $a=f_1(t)=0\neq b$:

$$V=R^{-1}(t)(1+BC^4u)[(1+BC^4u)^{1/2}-\tfrac{2}{3}C/(B^2R)]^{-2}, \qquad (4.6.8)$$

where $C=c^{-1/2}$, $B=bc$. The (flat only) FLRW limit results when $C=0$.

Case 6: $\Delta\neq0\neq a$, $f_1(t)=-\Delta^3/3$:

$$V=C[(u-u_1)(u-u_2)]^{1/2}\left[1+3\tan^{-2}\left\{\ln\left[S(t)\left(\frac{u-u_1}{u-u_2}\right)^{1/4}\right]\right\}\right], \qquad (4.6.9)$$

where u_1 and u_2 are the roots of $h(u)$ and $C=\tfrac{1}{8}a^{5/2}(u_1-u_2)^2$. All the FLRW models follow when $u_1=u_2$; then $R(t)=C^{-1}\{1+3\tan^{-2}[\ln S(t)]\}^{-1}$, and the coordinate transformation $r=1/r'$ reduces the metric to (1.3.1).

Case 7: $\Delta\neq0\neq a$, $f_1(t)\neq0$:

$$V=R^{-1}(t)[(u-u_1)^{-1/2}-CR^{-1}(u-u_2)^{-1/2}]^{-2}, \qquad (4.6.10)$$

where u_1 and u_2 are the roots of $h(u)$ and $C=[\tfrac{3}{2}a^{5/2}(u_1-u_2)^2]^{-1}$. All the FLRW models result when either $C=0$ (then $u_2=\tfrac{1}{4}k$ and the transformation $r=1/r'$ is needed) or $u_1=u_2$ (the same remark applies, and the scale factor is $R/(1-C/R)^2$). Note that case 5 follows from case 7 as the limit $u_2=0$ accompanied by the transformation $r=1/r'$ and the reparametrizations $u_1=-BC^4$, $C=\tfrac{2}{3}\tilde{C}/B^2$.

In the terminology of Sussman (1987), cases 2, 3, 5 and 7 are "McVittie-type", and cases 4 and 6 are "Wyman-type". However, neither of them can reproduce the Wyman solution, which is not expressible in elementary functions.

Now we shall relate the other solutions found in the literature to the above scheme. They are classified in Box 4.1.3 by the properties of $h(u)$, but the inscriptions on the arrows do not denote limits; most solutions are only subcases of the classes indicated, expressible in terms of elementary functions.

We begin with the case when $h(u)$ is of second degree and has a double root, that is, with Kustaanheimo and Qvist's (1948) case 3. McVittie (1967a – part 5 of the appendix, and 1967b) was the first to reobtain it, just as an example of how his approach (4.5.1)–(4.5.6) might be applied.

McVittie and Stabell (1968) discussed the solutions of McVittie's equations (4.5.3)–(4.5.6), but displayed explicit solutions only for $b=0$. The solutions will be labelled (A.26), (A.27), (A.28) and (A.29) as in the paper (the labelling is carried over from McVittie 1967a, and has already become a standard). The solution (A.26) – in both cases $a=1/2$ and $a=4/3$ – is equivalent to the K–Q case 3.

The same solution was derived again by Banerjee and Banerji (1976). They found the condition imposed on $R(t)$ by matching the solution to the Schwarzschild solution, and concluded that initial expansion of such a sphere can be reversed into collapse leading to a singularity, or initial collapse can be halted and reversed, all depending on the initial conditions.

Knutsen (1985b) found the same solution in a very complicated disguise by working from the McVittie (1967) approach. He calculated several analytic properties of it.

We shall now discuss the cases when $h(u)$ is of second degree and has two distinct roots. These are K–Q cases 6 and 7. Case 7 was first rediscovered by Brezhnev (1966b). Since his result was subject to some physical discussion, we shall quote his formula for V:

$$V = R^{-1}\left\{(1 + \tfrac{1}{4}kr^2)^{-1/2} + \frac{m}{2R}[a(1 + \tfrac{1}{4}kr^2) + r^2]^{-1/2}\right\}^{-2}, \qquad (4.6.11)$$

where k, m and a are constants. This is equivalent to (4.6.10) through simple reparametrizations. Brezhnev found the following interesting properties of the solution:

1. In the limit $R = 1$, $k = 0$ and $a \to 0$, the mass-density distribution defined by (4.6.11) tends to $6m\delta(r)$, where $\delta(r)$ is the Dirac function, so in this limit the solution describes a mass-point at $r = 0$. This agrees with the commonly accepted interpretation of the McVittie (1933) solution that follows from (4.6.11) when $a = 0$; see below.
2. In the Newtonian limit with $R = 1$, $k = 0$, the mass distribution defined by (4.6.11) becomes $\kappa\epsilon = 6ma^{-3}(1 + r^2/a^2)^{-5/2}$. This agrees with Shuster's empirical formula (see Parenago 1954) for mass distribution in globular clusters. Brezhnev thus concluded that (4.6.11) should be interpreted as a model of a globular cluster in an expanding Universe.

The same solution was introduced by Brezhnev, Ivanenko and Frolov (1966), and discussed in another way. These authors argued that in the Newtonian limit, eq. (4.6.11) describes the gravitational field of a spherical body with the gravitational constant changing according to $G(t) = G_0/R(t)$, $G_0 = $const. This interpretation is dubious (neither Einstein's nor Newton's theory is compatible with variable G); the conclusion should rather be that (4.6.11) describes the gravitational field of a body with variable mass $M(t) = M_0/R(t)$, the variation being due to the mass flowing past the observer measuring it. Brezhnev *et al.* then state that planetary orbits in a Universe described by (4.6.11) should expand by the law $R_{,t}/R = H$ (the Hubble constant), and so the Earth's orbit should expand by 0.5 mm/year. This statement tacitly assumes that the planets are freely swept along by the cosmological expansion, which is incorrect. Compare the more careful discussion by Gautreau (1984) of the same problem, based on the L–T model (see our Section 3.3).

The same solution was derived again by Ivanenko, Brezhnev and Frolov (1967, the first of the two solutions considered in the paper).

Again the same solution was found by Nariai (1967b), who showed that it can be matched to the Schwarzschild solution, and it then describes a collapsing sphere that bounces outside the Schwarzschild horizon. Nariai, Tomita and Hayakawa (1968) analysed numerically in detail the redshift of light emitted from such an object and received by a distant observer, with various values of the parameters. Knutsen (1982) found several analytic properties of the solution using the Nariai (1967b) paper as a reference.

McVittie and Stabell's (1968) case (A.28) is just another parametrization of (4.6.10)–(4.6.11) (the two cases, $\delta = +1$ and $\delta = -1$, that follow from the authors' discussion are isometric to each other).

Glass and Mashhoon (1976) rediscovered the solution in still another parametrization which is more convenient for considering limits, so we shall quote their formula for V:

$$V = R^{-1}(t)[(\gamma u + \delta)^{-1/2} + \tfrac{1}{2}\lambda_0 R^{-1}(\alpha u + \beta)^{-1/2}]^{-2}, \qquad (4.6.12)$$

where α, β, γ, δ and λ_0 are arbitrary constants. All the FLRW models follow when either $\lambda_0 = 0$ or $\alpha\delta - \beta\gamma = 0$. Glass and Mashhoon interpreted this solution as a model of a globular cluster in the last stages of collapse with a central collapsed core.

Chakravarty, Choudhury and Banerjee (1976) presented a set of handpicked cases when (4.1.3) can be integrated in elementary functions. They identified a few older solutions as subcases in their scheme, then presented several solutions that they expected to be new. However, most of their solutions are in fact subcases of the Kustaanheimo–Qvist collection. All the subcases of their case D with $\theta = 0$ are within K–Q's case 7, and, taken together, exhaust case 7. The other cases will be placed in our scheme where appropriate.

The same solution was rediscovered yet again by Knutsen (1985a) by working from the McVittie (1967) approach. Knutsen described several analytic properties of the solution in this paper, and in another one (Knutsen 1987a). Note that the solution derived and discussed by Knutsen in the 1985a paper is just a coordinate transform of the solution discussed by Knutsen in the 1982 paper, yet he apparently overlooked their equivalence. This shows how misleading and obscuring the McVittie approach is (and how critical and suspicious we should all be about claims of a new solution being discovered).

McVittie (1933) derived the solution that follows from (4.6.11) as the limit $a = 0$ and from (4.6.12) as the limit $\alpha = 0$, $\beta = \delta = 1$ (it will follow from (4.6.10) after a reparametrization). Since it is important for historical reasons, and was the basis of some physical considerations, we shall reproduce the full metric here (in a slightly changed notation):

$$ds^2 = \left[\frac{1 - \mu(t,r)}{1 + \mu(t,r)}\right]^2 dt^2 - \frac{[1 + \mu(t,r)]^4}{(1 + \tfrac{1}{4}kr^2)^2} R^2(t)[dr^2 + r^2(d\vartheta^2 + \sin^2\vartheta\, d\varphi^2)], \quad (4.6.13)$$

where:

$$\mu(t,r) = \frac{m}{2rR}(1 + \tfrac{1}{4}kr^2)^{1/2}, \qquad (4.6.14)$$

m and k being arbitrary constants. In fact, McVittie wrote k in the form $(1/\bar{R}^2)$, $\bar{R}$ being a constant, which (incorrectly) suggests that $k > 0$ necessarily, but the relaxation of this limitation is immediate. When $k = 0$ and $R = 1$, the McVittie solution reproduces the Schwarzschild solution in isotropic coordinates; when $m=0$, it reproduces all the FLRW models. Hence, it is a superposition of the Schwarzschild and FLRW solutions, and may be understood as a representation of the gravitational field of a point-mass placed in a FLRW Universe. Physical interpretation of this solution was considered in several papers; see next section. McVittie himself used this solution for a preliminary discussion of influence of the cosmological expansion on planetary orbits. However, his discussion was coordinate-dependent. He found that, with respect to the comoving coordinates, the planetary orbits are shrinking (and this is intuitively clear: the cosmic medium is expanding and flowing past the planets).

As already mentioned in Section 4.4, the subcase $k=0$ of the McVittie (1933) solution has matter-density depending only on time in the comoving coordinates, and so is a subcase of the Qvist (1947)/Kustaanheimo (1947) class from Box 4.1.2. This subcase was rediscovered by Eisenstaedt (1976) while investigating the solutions with $\epsilon=\epsilon(t)$, and then rediscovered again by Israelit and Rosen (1992). The last two authors analysed the equations of motion of a test particle in this spacetime in the Newtonian limit and showed that the gravitational force acting on the particle can be decomposed into three components: the attraction of the point-mass at $r=0$, the attraction of the cosmic medium between the centre and the particle's position, and the force generated by the pressure of the cosmic medium (also attractive!).

Dirac (1979a,b) rediscovered an even more special subcase of the McVittie (1933) solution, in which $R(t)=t^{2/3}$, $k=0$. Hence, Dirac's solution becomes the flat FLRW dust in the spatially homogeneous limit. Dirac derived and discussed this solution in connection with his Large Numbers Hypothesis, and interpreted various cosmological effects in the model from the point of view of that hypothesis.

Let us observe that Kustaanheimo and Qvist's (1948) case 5 follows from (4.6.11) as the subcase $k=0$, but with $a\neq0$. Hence, it cannot in general be a subcase of the McVittie (1933) solution in which $a=0$.

The solution found by Nariai (1967a) is a subcase of the K–Q case 5 resulting from (4.6.11) when $k=0$ and $m=2/a$. In the parametrization of (4.6.11) it will reproduce the (flat only) FLRW model when $a\to\infty$. Nariai's paper contains some discussion of properties of this solution. It can also be obtained from the McVittie (1933) solution through the transformation $r=1/r'$ as the subcase $k=(m/2)^2$.

This completes the discussion of K–Q's case 7. Case 6 is another one in which $h(u)$ is of second degree and has two roots. It was rediscovered by Chakravarty, Choudhury and Banerjee (1976, their case D with $\theta=2p/3$). Actually, Chakravarty *et al.* extended K–Q's case 6 to the situation when the roots of $h(u)$ are complex; this is the CCB case Db with $\theta=2p/3$. This solution was new.

K–Q's case 4 arises when $h(u)$ degenerates to a linear expression in u (when $a=0$). It was rediscovered by Chakravarty, Choudhury and Banerjee (1976) as their case C, eq. (6).

When $a=b=0$, eq. (4.1.3) integrates to:

$$V_{,u}^2 = \tfrac{1}{6}c^{-5/2}V^3 + f_1(t), \tag{4.6.15}$$

and the general solution of the above can be expressed through elliptic functions. This was done by Wagh (1955a). The resulting expression is not very illuminating and does not allow for the limiting transition to the conformally flat case. The subcase of (4.6.15) with $\epsilon = \epsilon(p)$, that is, when $V(t,u)$ depends on t and u through the variable $(t + C_2 u/2)$ while $f_1(t) = $ const, is the Wyman (1946) solution. The subcase of (4.6.15) with $f_1(t) = 0$ leads to elementary functions; it is the K–Q case 2 (eq. (4.6.5)), and was rediscovered by Faulkes (1969b) and by Knutsen (1984) (some physical properties of this subcase were discussed by Banerjee, 1971). The further subcase of (4.6.5) when $R(t) \propto t^{4/5}$ was rediscovered by Henriksen, Emslie and Wesson (1983, their case $\Delta \neq 0$), and discussed by Alexander, Green and Emslie (1989).

In connection with the paper by Faulkes (1969b) we must mention the paper by Meszaros (1985), who showed in effect that the Faulkes (1969b) solution may be understood as the limit of vanishing background curvature of the Banerjee–Banerji (1976) solution (i.e. of the K–Q case 3).

This completes the discussion of Box 4.1.3. Geometric properties of solutions generated by (4.6.1) were described by Sussman (1987, 1988a,b). The properties he discussed are possibilities that arise with various forms of $V(t,u)$ and do not correlate simply with the separately published explicit solutions. Since Sussman considered charged perfect fluids, we shall describe (briefly) his results in Section 4.12.

4.7 Physical and geometrical considerations based on the McVittie (1933) solution

Järnefelt (1940b) made a detailed discussion of timelike geodesics (orbits) in the subcase of the McVittie (1933) solution where the background curvature is zero and the scale factor $R(t) = e^{2kt}$, $k = $ const $\neq 0$. This is a disguised representation of the Kottler solution (i.e. the Schwarzschild solution generalized for Λ; see Krasiński and Plebański 1980), so it is of no relevance for this book.

In another paper, the same author (Järnefelt 1940a) discussed the problem of influence of cosmological expansion on planetary orbits, using the McVittie solution. He first concentrated on the flat background case, $k = 0$, and substituted for m and $\dot{R}/R$ the values implied by the astronomical observations (from that time). Unfortunately, Järnefelt's aim was to show that there is no such influence, and so he concentrated on showing that the change of the orbital radius from the Newtonian value is very small compared with the cosmological change of distances caused by expansion. No definite number is given in the paper, Järnefelt says only that: "The planetary orbit will thus, during the time given [the age of the Universe] not depart significantly from the Newtonian orbit. However, as emphasized on p. 7, in the same time the expansion of the space by the factor 10^4 occurred. It was thus shown, under the special assumptions that were made, that the planetary orbit does not participate in the expansion."[37]

---

[37] Translated from the German by A.K.

This should mean that the expansion of orbits (the paper does not say which orbit is investigated) relative to their initial radii is much smaller than the cosmological expansion. However, the influence clearly exists. Then Järnefelt showed that the effect is of the same order of magnitude when the background is of positive curvature. Finally, he showed that there will be a place in the Universe where a particle can remain permanently at rest because the cosmological expansion will balance the attraction of the star. He even derived the formula for the distance of equilibrium (the distance is given as a value of the curvature coordinate r), but did not calculate its numerical value. The formula is $r^3 = m/H^2$, where m is the (time-independent) mass of the source and $H = \dot{R}/R$ is the Hubble constant. The results of this paper were all derived by estimating the orders of magnitude of the various quantities from the equations of timelike geodesics.

In still another paper, the author (Järnefelt 1942) supported the same conclusion with perturbative calculations. The result was proved up to linear terms in the small parameter $T = k_0/(1 + \frac{3}{2}k_0\tau)$, where $k_0 \cong 0.6 \cdot 10^{-22}$ km^{-1} (a constant) and τ is the time coordinate in the curvature coordinates. The transformation to the curvature coordinates cannot, in general, be integrated in elementary functions, and so it is expressed through power series in $1/r$ where r is the (curvature) radial coordinate. Hence the paper in fact makes use of a power series in two variables. As can be seen, the small parameter T itself depends on time, so it becomes unclear what the statement that the coefficient of such a power series is constant in time means. The paper does not seem to have paid sufficient attention to the problem of what the standard unit of length should be, against which the possible expansion of orbits would be gauged. Still, Järnefelt misled Einstein and Straus (1946) who were apparently convinced that Järnefelt had already proved the lack of influence of cosmic expansion on the orbits, and they believed they in turn had only provided a clearer proof of the same result (see our Section 3.3).

McVittie (1966) reworked the solution into the coordinates defined in our Section 4.5, and then discussed the subcase $k=0$ as a model of stellar collapse. The conclusion was that a configuration of initial surface temperature 5000 K will collapse to the Schwarzschild radius in $2.83 \cdot 10^4$ years, reaching the temperature $4.71 \cdot 10^{10}$ K. The applicability of the model can be questioned, however, because it has a spatially homogeneous density distribution – an unrealistic assumption under such extreme conditions.

Noerdlinger and Petrosian (1971) considered the problem of whether clusters and superclusters of galaxies participate in the cosmological expansion. Their discussion was mostly Newtonian, and the McVittie (1933) solution was used only to estimate the relativistic correction to the Newtonian effect. Thus, their study was analogous to the one by Gautreau (1984) based on the L–T model; see our Section 3.3. Noerdlinger and Petrosian assumed that the cosmological background is a spatially homogeneous radiation, with the equation of state $p = \epsilon/3$, and found that the average distance of a test particle from the centre of the cluster, $\langle R \rangle$, will change according to:

$$\langle R \rangle_{,t}/\langle R \rangle = -[\varepsilon/(1+4\varepsilon)]\epsilon_{,t}/\epsilon, \qquad (4.7.1)$$

where:

$$\varepsilon \overset{\text{def}}{=} 8\pi\langle R \rangle^3\epsilon/(3M), \qquad (4.7.2)$$

M being the mass of the cluster and ϵ the cosmological background density. The parameter ε thus measures the ratio of the cosmological density to the mean mass density in the cluster. If $\varepsilon \ll 1$, then $\langle R \rangle_{,t}/\langle R \rangle \cong +4\varepsilon H$, where $H \overset{\text{def}}{=} -\frac{1}{4}\epsilon_{,t}/\epsilon$ is the Hubble parameter. Hence, dense clusters will expand much slower than the background. For $\varepsilon \gg 1$, $\langle R \rangle_{,t}/\langle R \rangle \cong H(1-1/\varepsilon)$, that is, the "cluster" (which is in fact a void in this case) will expand nearly as fast as the background medium (in this case, eq. (4.7.1) does not apply, and a different method must be used). Noerdlinger and Petrosian did not calculate any definite numbers. Then they took the McVittie (1933) solution with $k=0$ in the linear approximation of small m (see (4.6.13) and (4.6.14)) and with the background obeying $p=\epsilon/3$ in the same approximation. It turns out that at this order of approximation, there is no relativistic correction to (4.7.1)–(4.7.2).

Sussman (1988b) described the global properties of the McVittie (1933) solution, and showed that the common interpretation (a point particle in the expanding Universe) is not consistent with the global geometry. The set $r=0$ is a null boundary and its intersection with any $t=$const hypersurface is at an infinite geodesic distance from all other points in the hypersurface. When $k=+1$, the spacetime evolves from a source at infinity to a finite-density singularity, or the other way round. When $k \leq 0$, the spacetime has no regular centre, that is, it has the $S^2 \times R^2$ topology, similar to the K–S spacetime. The solution with $k=+1$ can be matched to the Schwarzschild spacetime so that the fluid region has a regular centre, but then, if $r=r_0$ is the hypersurface of matching, the Schwarzschild region must extend from $r=0$ to $r=r_0$, and the McVittie region from $r=r_0$ to $r=\pi$. The paper contains many more details.

The McVittie (1933) solution is spherically symmetric, and, literally as given, describes the Schwarzschild particle superimposed on the closed FLRW background. Hogan (1990) showed how McVittie's construction can be performed with any background curvature, with different expressions being used for the background FLRW metric. Hogan presented one hyperbolically symmetric and two plane symmetric analogues of the McVittie solution. Of the plane symmetric analogues, the first degenerates into a vacuum with Λ in the limit $p_{,\mu}(\delta^\mu{}_\alpha - u^\mu u_\alpha)=0$, and the other one does not allow such a limit at all. The hyperbolically symmetric analogue does not allow the limit either.

Nolan (1993) showed that the spherical limit of the Stephani (1967a) solution can be matched to the McVittie solution, and so can be interpreted as the latter's source. The McVittie solution, when extended to all values of r, has a singularity at $r=0$, while the spherical Stephani solution does not have it.

4.8 Other solutions of the K–Q class

Srivastava (1987) in his review paper derived two general classes of solutions. One (his case $n=1/2$) turned out to coincide with the class described in Section 4.6 and displayed in Box 4.1.3. The other is Srivastava's case $n=3/7$. In it, the function $f(u)$ has the form:

$$f(u)=6(vu+w)^{-15/7}(lu+m)^{-20/7}, \tag{4.8.1}$$

where v, w, l and m are arbitrary constants. The solution of (4.1.3) is:

$$V=(vu+w)^{3/7}(lu+m)^{4/7}\mathscr{P}(\eta+f_2(t),\,0,\,f_1(t))$$
$$-\frac{1}{49}(vm-wl)^2(vu+w)^{1/7}(lu+m)^{6/7}, \tag{4.8.2}$$

where $f_1(t)$ and $f_2(t)$ are arbitrary functions,

$$\eta=7(vm-wl)^{-1}[(vu+w)/(lu+m)]^{1/7}, \tag{4.8.3}$$

and $\mathscr{P}(\cdot,\cdot,\cdot)$ is the Weierstrass elliptic function. This can become conformally flat only after the following rather refined reparametrization. The Weierstrass elliptic function with the second parameter equal to zero can become independent of u, which happens when $\mathscr{P}=(f_1/4)^{1/3}$. Then we redefine $f_1=[4F_1(t)/m]^3$, $v=wl/m$, $w=Wm$, $l=Lm$, and take the limit $m\to\infty$. The result is $f(u)=0$ and $V=W^{3/7}(1+Lu)F_1(t)$, that is, the limit leads directly to the FLRW models. We leave the question open as to whether a nontrivial conformally flat generalization of the FLRW models can be generated from this Srivastava solution by another reparametrization.

A subcase of the solution (4.8.1)–(4.8.3) was found by Knutsen (1985c). He worked from the McVittie (1967a) approach and produced a complicated family of metrics that can be explicitly assembled only in the cases $\alpha=-1/6$ and $\alpha=1$. The two cases turn out to be equivalent (under a coordinate transformation), and they result from the solution given above when $f_1(t)=0$. Then the Weierstrass elliptic function degenerates into the elementary function:

$$\mathscr{P}(\eta+f_2,\,0,\,0)=(\eta+f_2)^{-2}, \tag{4.8.4}$$

and (4.8.1)–(4.8.3) reduces to the Knutsen (1985c) solution in a different parametrization. The calculations are lengthy and difficult, but not very rewarding, so they are omitted here. Some analytic properties of the solution were discussed in a subsequent paper (Knutsen 1987b). The solution reproduces all the FLRW models when $f(u)=0$.

To close the discussion of the type D Barnes models, we shall mention the two solutions by Bayin (1986). These do not quite belong to the family because they have anisotropic pressure, but they have $\sigma=\omega=0=\dot{u}^\alpha$, and in the limit of isotropic pressure they reduce directly to the FLRW models.

They are of type D, and have the metric of the form:

$$ds^2 = dt^2 - R^2(t)[dr^2/S(r) + r^2(d\vartheta^2 + \sin^2\vartheta\, d\varphi^2)], \qquad (4.8.5)$$

where $S(r) = 1 + s_0(r)r^2 - C_0/r^{n-2}$ for the first solution and $S(r) = 1 + s_0(r)r^2 - C_0\, r^2 e^{br}$ for the second one, s_0, C_0, n and b being constants. The function $R(t)$ is required to obey $RR_{,tt} - \frac{1}{4}m_{,t}^2 = s_0\, m$. The FLRW limit, with the equation of state limited by the equation imposed on R, results when either $C_0 = 0$ or $n = 0$ in the first case and when $b = 0$ in the second case. Bayin invoked Letelier's (1980) two-fluid formalism (see our Section 6.6), but did not really use it.

The main result of Bayin's paper is that if the first solution is matched to the FLRW background, then the density contrast will grow with time. Bayin concluded from this that the solution could be used to describe the formation of voids. It is regrettable, though, that it totally lacks an invariant definition.

This closes the discussion of the type D perfect fluid Barnes models. The remaining solutions displayed in Figure 4.1 are conformally flat perfect fluid solutions, and generalizations of perfect fluid solutions for electromagnetic field. These are described in subsequent sections.

4.9 The conformally flat solutions

The most general conformally flat perfect fluid solution with nonzero expansion was found by Stephani (1967a). In comoving coordinates, this is:

$$ds^2 = D^2 dt^2 - V^{-2}(t,x,y,z)(dx^2 + dy^2 + dz^2), \qquad (4.9.1)$$

where:

$$D = F(t)V_{,t}/V, \qquad (4.9.2)$$

$$V = R^{-1}\left\{1 + \tfrac{1}{4}k(t)[(x - x_0(t))^2 + (y - y_0(t))^2 + (z - z_0(t))^2]\right\}, \qquad (4.9.3)$$

$F(t)$, $R(t)$, $k(t)$, $x_0(t)$, $y_0(t)$ and $z_0(t)$ are arbitrary functions of time. The matter-density and pressure are given by:

$$\kappa\epsilon = 3C^2(t),$$

$$\kappa p = -3C^2(t) + 2CC_{,t}\,V/V_{,t}, \qquad (4.9.4)$$

where $C(t)$ is a function of time, connected with the others by:

$$k(t) = [C^2(t) - 1/F^2(t)]R^2(t). \qquad (4.9.5)$$

As in the Barnes (1973) models, the scalar of expansion is:

$$\theta(t) = 3/F(t). \qquad (4.9.6)$$

Note that the solution results only from the assumptions of conformal flatness and a perfect fluid source with $\theta \neq 0$; the properties $\sigma = \omega = 0$ then follow automatically.

The FLRW models result from here when k, x_0, y_0 and z_0 are all constant. Then

(4.9.5) becomes the Friedmann equation (remember that in the proper time coordinates $1/F = \theta/3 = R_{,t}/R$). This solution has in general no symmetry, and so is not a subcase of the Barnes (1973) and K–Q models. We stress that it is the most general of all conformally flat perfect fluid solutions with nonzero expansion, and not just the most general one in the Stephani–Barnes conformally flat subfamily; see theorem 32.15 in Kramer *et al.* (1980), and Stephani (1967b). It was first found by Stephani (1967a) as one of the solutions of embedding class 1, then it was identified by this author (Krasiński 1981) as an example of spacetimes with intrinsic symmetries in the sense of Collins (1979) (see also a discussion of embedding properties of the Stephani model in the paper by Kramer, Neugebauer and Stephani 1972, and a very brief description of the solution in the appendix to the paper by Papini and Weiss, 1986). The hypersurfaces t=const in this spacetime are homogeneous and isotropic (they are the same as the slices t=const of the FLRW models), but they are put together in such a way that distances between them are different along different t-coordinate lines. As a result, the spacetime does not inherit the symmetry (a similar thing happens in the hypersurfaces t=const of the Szekeres spacetimes with the surfaces r=const; see our Section 2.1).

The Stephani solution has a property that is dual to a certain property of the Szekeres–Szafron spacetimes. In the latter, the function generalizing the FLRW curvature index is dependent on spatial location. In the Stephani solution, the function $k(t)$ becomes the curvature index in the FLRW limit, but it depends on time, and in particular may change its sign during evolution. It would be interesting to discuss a definite example of the evolution of such a model. Unfortunately, the equations of state $p=p(\epsilon)$ and $p=0$ both reduce (4.9.1)–(4.9.6) to a FLRW model, and so the Stephani solution could be discussed only from a general geometrical point of view.[38] This author (Krasiński 1983) observed that when C=const, the Stephani solution reduces to the de Sitter solution. In this limit, the spacetime can be represented as a four-dimensional hyperboloid embedded in a flat (pseudoeuclidean) five-dimensional space.[39] The slices t=const are intersections of this hyperboloid with the planes, all containing the same straight line tangent to the equator of the hyperboloid. Such foliation of the de Sitter manifold has all the qualitative properties of the t=const foliation of the Stephani manifold (see Krasiński 1983 for details).

Several other explicit subcases of the Stephani solution (all of them spherically symmetric) were discussed by Dąbrowski (1993). Two of the subcases are non-vacuum and inhomogeneous. Dąbrowski presented the embedding diagrams and Penrose diagrams for each case, with detailed descriptions. They give a deeper insight into the geometry of the Stephani manifold.

Other descriptions of the geometry of the Stephani solution were published by this author (Krasiński 1984a) and by Cook (1974 and 1975). Cook rediscovered the spherically symmetric limit of the Stephani solution (see next section), and

---

[38] See "Notes added in proof", Note 4.
[39] As proved by Stephani (1967a), such embedding exists for the general case of (4.9.1)–(4.9.6), but is difficult to use explicitly because of the many arbitrary functions of t.

then discussed its geometrical properties, coming to similar conclusions as this author. The main conclusion was that when $k(t)<0$, pressure must be negative in part of the space. The earlier paper (Cook 1974) is a short abstract of the same work.

The paper by Lorenz-Petzold (1986) must be mentioned here. It was supposed to show whether the Stephani Universe is a viable model of the real Universe. Lorenz-Petzold wished to imply that it is not, but the argument cannot be taken seriously. Lorenz-Petzold argues that the evolution of the model is unpredictable because it contains arbitrary functions of time. This is so because in all previous discussions of the Stephani solution, no definite equation of state was assumed. Without an equation of state imposed on the solution of Einstein's equations even the FLRW models contain an arbitrary function of time (the scale factor) and evolve unpredictably. The other arguments in the paper are: (1) One can choose the arbitrary functions so that the energy conditions are violated (this is besides the point – it can also happen in the FLRW models). (2) The equations of null geodesics are complicated, and therefore impossible to solve. The paper is a misunderstanding and is mentioned here for completeness only.

The parametrization of (4.9.3) does not conveniently cover the case when V degenerates into a linear function of x, y and z; it can be changed by transformations of coordinates and renaming of the functions. However, the resulting limit can be transformed back to the form (4.9.3) by means of conformal symmetries of the Euclidean metric $(dx^2+dy^2+dz^2)$ (see Krasiński 1989 for a short account of such transformations).

The Stephani solution was rediscovered by Barnes (1973) as a member of the $\sigma=\omega=0$ family of perfect fluid solutions, by Banerjee and Chakravarty (1979) while following the metric ansatz (4.1.1) and by Ivanov (1980) while investigating the Bianchi identities in a conformally flat perfect fluid spacetime. See also a short description of the model by Bona (1985) from the point of view of intrinsic symmetries, and somewhat more extended descriptions by Bona and Coll (1985 and 1988).

Note that the function $F(t)$ in (4.9.2) can be manipulated by transformations of the t-coordinate. Let us choose a new coordinate $T(t)$ and let us write (4.9.2) as:

$$D=F(T)V,_T/V\equiv F(T)V,_R R,_T/V. \qquad (4.9.7)$$

We can then define T as we wish, for example by $F(T)R,_T=1$, that is, $T=\int F(R)dR$, and the result is equivalent to $F=1/R,_t$ in (4.9.1)–(4.9.3). Hence, the subcases of the Stephani (1967a) solution that differ only by the function $F(t)$ are not geometrically different.

Obozov (1979) claimed to have proved that a perfect fluid solution with $\sigma=\omega=0\neq\theta$ can only be conformally flat, but the Barnes (1973) solutions are counterexamples. Obozov deduced his proof from the integrability conditions $u_{\alpha;\beta\gamma}-u_{\alpha;\gamma\beta}=-R^\rho{}_{\alpha\beta\gamma}u_\rho$ for each Petrov type in turn. The error is where he discarded the possibility that $C_{0123}=0=C_{0231}=C_{0312}$. This shows that the Barnes solutions have their Weyl tensor less general than the type D would maximally allow.

On first encounter with the Stephani (1967a) solution, several people react by saying "it must be wrong because it contradicts the theorem by Geroch (1967)". It is commonly believed that the Geroch theorem prohibits the change of topology of spatial slices of a manifold, while in the Stephani manifold the topology may change from S^3 to E^3 or vice versa. It must therefore be recalled that a theorem applies only in those situations in which its assumptions are fulfilled, and the Geroch theorem states the following:

Theorem 2. Let M be a compact geometry whose boundary is the disjoint union of two compact spacelike 3-manifolds, S and S'. Suppose M is isochronous, and has no closed timelike curve. Then S and S' are diffeomorphic, and further M is topologically $S \times [0,1]$. (Geroch 1967).

An E^3 slice of the Stephani manifold is obviously noncompact, and a segment of the Stephani manifold contained between an E^3 slice and an S^3 slice is not compact either, so the assumptions of the Geroch theorem are not fulfilled in this case and there is no contradiction (to appreciate this last argument, the readers should consult the figures given in the paper by Krasiński 1983 or 1984a). One can choose a segment of the Stephani manifold that is contained between two S^3 slices – then Geroch's assumptions will be fulfilled, and such a set will be an example for the theorem.

In connection with the Stephani model we have to mention the papers by Martinez and Sanz (1985) and by Argueso and Sanz (1985). In both, the authors considered solutions with intrinsic axial symmetry in the t=const hypersurfaces. They would have obtained subcases of the Barnes and Stephani models, but they explicitly excluded all solutions where intrinsic symmetries were actually ordinary symmetries. In the first paper, the solutions that remained are either static or have no limit $\sigma \to 0$, and so have no FLRW limit. In the second paper, the solutions are either a vacuum or a vacuum with Λ, or have $\theta=0$, or else have $\epsilon < 0$. The solutions with $\epsilon < 0$ are defined only through their metric; no physical quantities are defined for them. The two solutions with $\epsilon = p$ have $\epsilon > 0$ and are nonstationary, but they cannot have the limit $\sigma = 0$ and so have no FLRW limit.

4.10 The spherically symmetric Stephani solution

The spherically symmetric limit of the Stephani (1967a) solution is at the same time the conformally flat limit of the Kustaanheimo–Qvist (1948) solutions. Being so strongly distinguished geometrically, it was prone to being rediscovered, and is now the most spectacular evidence that the dictum "Einstein equations are difficult, so keep solving them" no longer applies. The total number of rediscoveries is, as at the moment of writing, 23, and this does not include separately discovered subcases. The solution is obtained from (4.9.3) when all the functions $x_0(t)$, $y_0(t)$ and $z_0(t)$ are constant; then they can be transformed away. The list of papers presenting the rediscoveries is given in Box 4.1.4; the text below also includes those papers in which

WYMAN 1946
KUSTAANHEIMO - QVIST (1) 1948
RAYCHAUDHURI 1955
GUPTA 1959
BREZHNEV 1966a
BONDI 1967
THOMPSON - WHITROW 1967
NARIAI - TOMITA 1968
STROBEL 1968
TAUB 1968b
VAIDYA 1968b
CAHILL - MCVITTIE 1970a
BANERJEE 1972
MISRA - SRIVASTAVA 1973
COOK 1974 + 1975
EISENSTAEDT 1975a
GLASS 1979
NDUKA 1981
PANDEY - GUPTA - SHARMA 1983
PONCE DE LEON 1986
SUSSMAN 1989a
SRIVASTAVA 1992
NOLAN 1993

NOTE: THE PAPER BY WYMAN IS A "PRE-DISCOVERY",
SEE TEXT

Box 4.1.4. Rediscoveries of the spherically symmetric subcase of the Stephani
Universe.

properties of the solution were discussed. The papers containing rediscoveries are
marked by the letter R.

The first to derive the solution was Wyman (1946). The formulae are present in
his paper (the paragraph after his eq. (3.2)), but Wyman dismissed the solution after
observing that with $p=p(\epsilon)$ it reduces to the FLRW models.

Kustaanheimo and Qvist (R, 1948) listed the solution among their explicitly solv-
able cases as case 1.

Raychaudhuri (R, 1955) used the solution to discuss the development of inhomogeneities in the FLRW background. Having found that in comoving coordinates the
matter-density is spatially homogeneous, the author calculated the mass-density
along a surface of constant proper time of the perfect fluid source. The result was

that galaxies could not have formed in such a model during the lifetime of the Universe. The author's rough, order-of-magnitude estimate was based on the observational data then believed to be correct. However, this solution is inadequate for describing any kind of condensation: even if matter-density in it is calculated along a noncomoving hypersurface (e.g. the past light cone), the variability of the mass-density with distance will be a rigidly determined global function of r, which does not allow for independent evolution of localized condensations.

Gupta (R, 1959) found that the solution can describe oscillatory motions of the medium. In a subsequent paper (Gupta 1962), the author calculated the equations of timelike and null geodesics and found that circular orbits do not exist in this model. Light rays should undergo deflection.

Brezhnev (R, 1966a) spotted the solution as an example of a spherically symmetric spacetime and showed that it can be matched to the Schwarzschild solution.

Bondi (R, 1967) derived the solution as an example of a spherically symmetric spacetime with spatially homogeneous matter-density, and showed that it can describe a bouncing sphere. In a subsequent paper (Bondi 1969), the solution was discussed in more detail as a stellar model. Bondi investigated analytic properties of the metric, including special and limiting cases. Among other things, he showed that if sufficient conditions for the positivity of pressure are obeyed, then the spatial gradient of pressure will automatically be negative. The evolution of the sphere was discussed for a few definite equations of state assumed at the centre.

Thompson and Whitrow (R, 1967) showed that if the configuration described by the solution continues to contract (i.e. does not bounce), then it will collapse to a point singularity of infinite density in a finite comoving time. An external observer, however, would see the collapse continuing for an infinite time. In a follow-up paper (Thompson and Whitrow 1968), the authors discussed the solution in more detail. In particular, they formulated conditions under which the configuration collapses asymptotically to a nonsingular state. Thompson and Whitrow made the evolution definite by assuming an equation of state at the centre.

Nariai and Tomita (R, 1968) showed that the solution can describe a bouncing sphere. In another paper (Tomita and Nariai 1968), the authors showed that, with appropriately chosen parameters, the solution can describe an oscillating sphere.

Strobel (R, 1968) derived the solution in passing, while investigating solutions with heat-flow. The paper made important contributions to that subject; see Section 4.14.

Taub (R, 1968b) derived the solution in his exhaustive discussion of spherically symmetric shearfree solutions, and briefly discussed its analytic properties.

Vaidya (R, 1968b) derived the solution in curvature coordinates which are non-comoving. However, his result can be explicitly transformed to isotropic comoving coordinates, and then it turns out to be identical with the limit $x_0 = y_0 = z_0 = 0$ of (4.9.1)–(4.9.5). The earlier conference abstract (Vaidya 1968a) is a short announcement of the same result. Vaidya formulated the conditions for bounce, collapse and staticity of the configuration.

Cahill and McVittie (R, 1970a) derived the solution alongside the K–Q class (see Section 4.3).

Banerjee (R, 1972) discussed in detail the evolution of the case corresponding, in our notation, to $k=(4/R^2)^{1/\alpha-1}$, where α is a constant, $1<\alpha<2$. With this range of α, Banerjee's subcase has no nontrivial FLRW limit.

Misra and Srivastava (R, 1973) showed that there exist no charged spherically symmetric perfect fluid solutions with $\sigma=0$ and $\epsilon=\epsilon(t)$ that could be matched to a vacuum solution. (This is how their statement is worded, but in fact they demanded the solution to be regular at the centre, and it is this requirement that resulted in the nonexistence of such solutions. Without this requirement, a point charge at the centre could exist. It is not clear where and how the matching condition influences the result.)

The paper by Cook (1974) is a short abstract of the results published later (Cook, R, 1975). Cook discussed the global geometric properties of the solution; these were mentioned in Section 4.9.

Eisenstaedt (R, 1975a) rediscovered the solution while discussing the solutions with $\epsilon=\epsilon(t)$.

Glass (R, 1979) found it as a special subcase of the K–Q class and showed how it can describe an explosion or collapse of a fluid sphere.

Nduka (R, 1981) rediscovered the solution while looking for charged fluid solutions in the Lyttleton–Bondi (L–B, 1959) theory. The metric he derived is exactly the one discussed in this section, with no trace of charge in it. Nduka's paper will be described at more length in Section 4.12, together with the L–B theory.

Pandey, Gupta and Sharma (R, 1983) derived the solution from the assumptions of spherical symmetry, perfect fluid source, conformal flatness and embedding class 1 (the last assumption is in fact redundant, as follows from Kramer et al., 1980, theorem 32.15). Their result is represented in the curvature (i.e. noncomoving) coordinates, and the authors were aware that it was the same as Vaidya's (1968b) result.

Ponce de Leon (R, 1986) derived the solution from the assumptions of spherical symmetry, perfect fluid source, spatially homogeneous density in comoving coordinates, and the vanishing mass and radius of the orbit of $O(3)$ at the centre. Then he discussed the evolution of four subcases defined by explicit (but *ad hoc*) choices of the arbitrary functions. The examples describe: (1) a sphere exploding out of a singularity through the horizon and tending asymptotically to the interior Schwarzschild solution; (2) a sphere collapsing asymptotically towards the interior Schwarzschild solution; (3) a sphere exploding from a singularity out through the horizon and collapsing back to the singularity; (4) a sphere collapsing from an initial configuration outside the horizon and bouncing.

Knutsen (1983c) discussed various analytic properties of this solution, crediting it to Gupta (1959).

Sussman (R, 1989a) rediscovered the solution while looking for solutions with additional conformal symmetry (see later).

Srivastava (R, 1992, his section 4.2) derived the solution as one example in a broad overview of spherically symmetric shearfree charged perfect fluid solutions.

Nolan (R, 1993) showed that this solution can be matched to the McVittie (1933) solution as its source.

Karmarkar (1948) investigated when a spherically symmetric spacetime is of class 1, then considered the perfect fluid field equations for the resulting metric. Unfortunately, he did not press on to solve the field equations and stopped just short of discovering the solution discussed here (otherwise, he would have been a co-discoverer, along with Wyman and Kustaanheimo and Qvist).

Kuchowicz (1973) very nearly reobtained the solution by solving Einstein's equations in a spherically symmetric conformally flat spacetime with a perfect fluid source. He tacitly assumed the shear to be zero (by taking the coordinates to be simultaneously comoving and isotropic). However, he did not make use of the equation $g_{00} = F(t)(\ln g_{rr})_{,t}$, even though he derived it. Hence his result is not quite a solution.

This closes the list of rediscoveries of the spherically symmetric subcase of the Stephani (1967a) solution. In addition to these, several authors have derived subcases corresponding to different special choices of the arbitrary functions in (4.9.1)–(4.9.6). These are listed in the next section.

4.11 Other subcases of the Stephani (1967a) solution

Several papers listed in this section do contain physical discussion of the solutions. Since, however, the solutions lack invariant definitions, and the properties may be specific to the respective subcases, we shall mention the properties only very briefly.

Coley and Czapor (1992) investigated plane symmetric perfect fluid spacetimes that admit an inheriting conformal Killing vector ξ (by definition, $\pounds_\xi g_{\alpha\beta} = 2\psi g_{\alpha\beta}$, $\pounds_\xi u^\alpha = -\psi u^\alpha$). They found several classes of solutions. Case 1(a) has two subcases, $k=0$ and $k\neq0$. The subcase $k=0$ will be mentioned further, the subcase $k\neq0$ is conformally flat and plane symmetric, and so is a common subcase of the plane symmetric Barnes (1973) solutions and of the Stephani (1967a) solutions. It can reproduce the open and flat FLRW models. Case 1(b) has no FLRW limit; case 2(a) is equivalent to case 1(a) with $k=0$; case 2(b) is just a FLRW model itself; case 3(a) has no FLRW limit; in case 3(b) no solutions are presented and it is suggested that possibly none exist. Of the shearfree solutions, to which we shall refer as case 4, case 4(a) is a FLRW model when $v_{,x}=0$, and is equivalent to case 1(a) with $k\neq0$ when $v_{,x}\neq0$. Case 4(b) is another plane symmetric subcase of the Stephani solution that can reproduce the open and flat FLRW limits, and case 4(c) has no FLRW limit.

Bondarenko and Kobushkin (1972) derived a subcase of the Stephani solution corresponding to $x_0=y_0=0$, and $k(t)$, $R(t)$, $z_0(t)$ and $F(t)$ all being expressed in terms of a single function of time. In a slightly modified notation, adapted to cover all the cases in one formula, their metric in isotropic coordinates is:

$$ds^2 = H^{-2}(t,r,\vartheta)\left\{ dt^2 - (1+\tfrac{1}{4}kr^2)^{-2}[dr^2 + r^2(d\vartheta^2 + \sin^2\vartheta\, d\varphi^2)]\right\}, \quad (4.11.1)$$

where:

$$H=(1+\tfrac{1}{4}kr^2)^{-1}[K(r^2+k)+C_0r\cos\vartheta+g(t)(1+\tfrac{1}{4}kr^2)], \qquad (4.11.2)$$

K and C_0 are arbitrary constants, $k=\pm1$, 0, and $g(t)$ is an arbitrary function. This solution is axially symmetric, and it reproduces all the FLRW models when $K=C_0=0$.

Sussman (1991) derived a solution which is a coordinate transform of the subcase $C_0=k$ of the Bondarenko–Kobushkin (1972) solution. It is not equivalent to the latter because $C_0=k$ cannot be achieved by coordinate transformations if it does not hold at the beginning. This subcase can reproduce only the flat FLRW limit. Sussman discussed thermodynamical properties of the solution.

Verma and Roy (1956) presented four metrics derived from purely *ad hoc* assumptions. Their case (i) belongs to the Szafron (1977) subfamily with $\beta'=0$, but has no FLRW limit. Case (ii) has constant matter-density and so has no FLRW limit. Case (iii) are just the FLRW models themselves. Case (iv) results from the Bondarenko–Kobushkin (1972) class when $k=0$. It can reproduce only the flat FLRW model. It is not a subcase of the Sussman (1991) solution because here $k=0\neq C_0$. The solution was in fact new when published, but it lacks an invariant definition.

Exactly the same solution as the Verma–Roy case (iv) was found by Shvetsova and Shvetsov (1976).

Roy and Bali (1978b) obtained two solutions. The second of them has no FLRW limit, while the first one can be transformed to such a form in which it is the subcase $k=K=0$ of the Bondarenko–Kobushkin (1972) solution. It is thus the subcase $K=0$ of the Verma–Roy (1956) solution (iv), and it coincides with the Coley–Czapor (1992) case 1(a) with $k=0$.

The solutions listed above were either plane symmetric (Coley and Czapor 1992, Roy and Bali 1978b) or only axially symmetric (all the others). Consequently, they were not subcases of the spherically symmetric Stephani solution. The Bondarenko–Kobushkin (1972) solution does become spherically symmetric when $C_0=0$, and this subcase was, sure enough, published separately (see below). The solutions discussed below are all spherically symmetric and are subcases of the spherically symmetric limit of the Stephani (1967a) solution. For convenience, the metric of that limit is:

$$ds^2=D^2dt^2-V^{-2}(t,r)\,[dr^2+r^2(d\vartheta^2+\sin^2\vartheta\,d\varphi^2)], \qquad (4.11.3)$$

where:

$$V=[1+\tfrac{1}{4}k(t)r^2]/R(t), \qquad (4.11.4)$$

$$D=F(t)V_{,t}/V. \qquad (4.11.5)$$

Roy and Bali (1978a) derived a certain subcase of this metric in noncomoving coordinates. However, their description of the solution contains basic errors, and so the solution is not placed in the diagrams. The errors are as follows: (1) Roy and Bali claim that this spherically symmetric perfect fluid solution has nonzero rotation (!)

and nonzero shear (although shear is zero here). (2) The shear tensor as given by Roy and Bali has nonzero trace. On closer inspection it turns out that the authors used incorrect formulae for the velocity invariants.

We shall first discuss those subcases of (4.11.3)–(4.11.5) that can reproduce all the FLRW models. McVittie and Stabell's (1968) case (A.29) is a subcase with $R=S/(1+\beta S^{a-1})$ and $k=(k_M+4\alpha S^{a-1})/(1+\beta S^{a-1})$, where $S(t)$ is an arbitrary function and α, β, a and k_M are arbitrary constants. It reproduces the FLRW models when $\alpha=\beta=0$, and is only a subcase of (4.11.3) because it implies an algebraic relation between $k(t)$ and $R(t)$. Case (A.27) from the same paper results from (A.29) when $a=2$.

Lukacs and Meszaros (1985) discussed extensively various methods of modifying the cosmological principle so that it can allow for inhomogeneous models. They invoked several solutions published earlier, and as one of the modifications they proposed conformal flatness of the spacetime. The example of a solution they proposed, when transformed to isotropic coordinates, has V of the form:

$$V=[A(t)-kW][1+\tfrac{1}{4}(W+kA)r^2/(A-kW)],\qquad(4.11.6)$$

where $k=\pm 1, 0$, W is an arbitrary constant and $A(t)$ is an arbitrary function. This is simply a different parametrization of the McVittie–Stabell (1968) case (A.27).

The very same solution was derived by Sussman (1989a) as a spherically symmetric shearfree perfect fluid solution with a radial conformal vector. Sussman's parametrization makes it easier to see that the solution is the spherically symmetric subcase ($C_0=0$) of the Bondarenko–Kobushkin (1972) solution. The solution was introduced again and briefly discussed in another paper by the same author (Sussman 1990).

Coley and Tupper (1990a) considered perfect fluid solutions that admit an inheriting conformal Killing vector (see the comment above on the paper by Coley and Czapor 1992). Their case (i) is identical to the McVittie–Stabell (1968) case (A.27) as represented by Sussman (1989a). Their case (ii) is called a "generalized Gutman–Bespalko–Wesson" spacetime, but no explicit solution is presented. Their case (iii) is the interior Schwarzschild solution, case (iv) are plane and hyperbolically symmetric counterparts of case (ii); case (v) are stationary solutions; and case (vi) are the plane symmetric members of case (ii). No explicit solutions were presented except in case (i).

In a companion paper, Coley and Tupper (1990b) discussed spherically symmetric perfect fluid spacetimes that admit an inheriting conformal Killing vector, but found no explicit examples other than those mentioned above.

A note about the so-called "Gutman–Bespalko" spacetimes must be made here. They came to be known under this name after the book by Kramer *et al.* (1980, p. 173), where the paper by Gutman and Bespalko (1967) is mentioned as an example of a study of a spherically symmetric solution with $\sigma\theta\dot{u}^\alpha\neq 0$. In the paper by Gutman and Bespalko, four specific solutions are considered, of which the first one is static, the second and the third have no FLRW limit, and the fourth is no solution at all (due to a misprint or other error). For some reason, though, the metric:

$$ds^2 = [H(r) + G(t)]^{-2}[dt^2 - dr^2 - S^2(t)(d\vartheta^2 + \sin^2\vartheta \, d\varphi^2)], \qquad (4.11.7)$$

where $G(t)$, $H(r)$ and $S(t)$ are functions to be determined from the Einstein equations, is sometimes called a "generalized Gutman–Bespalko(–Wesson) spacetime" (because the second Gutman–Bespalko solution is a very simple subcase of (4.11.7)). This author (A.K., unpublished) verified the Einstein equations for the metric (4.11.7) using the program Ortocartan. The results are as follows:

1. The Einstein tensor for the metric (4.11.7) is diagonal. Therefore, if the source in the Einstein equations is a perfect fluid, then the coordinates of (4.11.7) are necessarily comoving.
2. When $H = $const, (4.11.7) is simply a Kompaneets–Chernov (1964) spacetime; see Section 2.11.
3. When $H_{,r} \neq 0$, the limit $\dot{u}^\alpha = 0$ requires $H_{,r} \to 0$ and the limit $\sigma = 0$ requires $G_{,t} = 0$, the resulting metric is a very special subcase of the K–S metric.

Hence, (4.11.7) does not contain any new generalization of the FLRW or K–S spacetimes, and is not an interesting class for cosmology. The perfect fluid solutions for the metric (4.11.7) were found and investigated by Kitamura (1994).

This completes the discussion of those conformally flat spherically symmetric solutions that reproduce all the FLRW models. We will now discuss those that reproduce only subcases of the FLRW models.

The solution presented by McVittie and Stabell (1967) results from (4.11.6) when $k = 0$ and can reproduce the general flat FLRW model. It is also the limit $C_0 = k = 0$ of the Sussman (1991) solution and a limit of the Verma–Roy (1956) solution (iv).

Wesson and Ponce de Leon (1989) presented and discussed a solution that results from the one by McVittie and Stabell (1967) as the limit $R(t) = t^{2/3}$, where $R(t)$ is the scale factor. It reproduces the flat FLRW dust in the spatially homogeneous limit. Sussman (1989b) showed that this case belongs to the spherically symmetric Stephani class and discussed its global properties. Wesson (1990) advocated this solution as a useful model for considering astrophysical processes.

Kumar (1969) presented five solutions, all of which are subcases of the McVittie–Stabell (1968) case (A.27). We shall not discuss them in detail because the profit from the discussion would not justify the space it would require. Kumar's case 1 can reproduce the flat and the closed FLRW models, each with a specific equation of state. His case 2 can reproduce only a flat FLRW model with a specific equation of state. Case 3 is contained in case 1; case 4 is equivalent to case 2; and case 5 can reproduce only a flat FLRW limit again.

Maharaj, Leach and Maartens (1991) presented two solutions. The first of them has no FLRW limit, the second one results from (4.11.3)–(4.11.5) when k and R are connected by:

$$b(k/R)_{,t} = a(1/R)_{,t}/t^2, \qquad (4.11.8)$$

where a and b are arbitrary functions. It reproduces the flat FLRW limit when

$ab=0 \neq a^2+b^2$. Since it involves two arbitrary functions of t, it is not a subcase of the McVittie–Stabell (1968) solutions (A.29) or (A.27), nor does it contain them because the latter contain all the FLRW models.

Bonnor and Faulkes (1967) derived and discussed a special subcase of (4.11.3)–(4.11.5) in which the functions $R(t)$ and $k(t)$ are expressed in terms of a single function of time that obeys in addition a certain differential equation. Their solution admits only a flat FLRW limit with a specific equation of state, and is a subcase of the McVittie–Stabell (1968) case (A.29).

Ivanenko, Brezhnev and Frolov (1967, second solution in the paper), working by inspired guesswork, found a solution that results from the McVittie–Stabell (1968) solution (A.29) when $a=3$. It can reproduce all the FLRW models.

Ponce de Leon (1991a) derived a family of self-similar solutions. In the case $\Delta \neq 0$, he presented no solutions. In the case $\Delta=0$, the solutions with $a \neq 0$, when transformed to isotropic coordinates, follow from the McVittie–Stabell (1968) case (A.29) as the subcase $\beta=0=k_M$, $a=3$ and $S(t)=t^{1-C}$, $C=$const. They are thus subcases of the Ivanenko–Brezhnev–Frolov (1967) second solution. They can reproduce all the FLRW models with a definite equation of state, but only after reparametrization tricks. They do not obey the Maharaj–Leach–Maartens (1991) condition on k and R. The solution with $\Delta=0=a$ is simply a flat FLRW model with a specific equation of state.

The solution by Henriksen, Emslie and Wesson (1983) with $\Delta=0$ is identical to that by Ponce de Leon (1991a) in the case $a>0$. It can reproduce all the FLRW models (with a specific equation of state each time) after reparametrizations. Various properties of the solution were discussed further by Alexander, Green and Emslie (1989).

Ray (1978, not shown in Figure 4.1) discussed four metrics. The first of them has no FLRW limit, the other three are conformally flat and spherically symmetric, so they must be subcases of the Stephani Universe. However, not enough information is given to classify them in Figure 4.1.

Narlikar and Singh (1950) considered Einstein's equations with a perfect fluid source for a spherically symmetric conformally flat spacetime. Thus the authors were on their way to finding the spherically symmetric subcase of the Stephani solution, but they worked in the curvature coordinates in which the field equations are not easily integrable. They left the equations unsolved and only showed that the interior Schwarzschild solution fulfils them.

This completes the overview of perfect fluid solutions in the Stephani–Barnes family. In the following sections we shall consider generalizations of them for electromagnetic field and heat-flow.

4.12 Electromagnetic generalizations of the K–Q spacetimes

In this section, we shall consider spherically symmetric solutions of the Einstein–Maxwell equations in which the source is a shearfree, nonrotating charged

fluid. The metric form in the isotropic coordinates is in this case the same as for the
K–Q spacetime, that is:

$$ds^2 = D^2(t,r)\, dt^2 - V^{-2}(t,r)\, [dr^2 + r^2(d\vartheta^2 + \sin^2\vartheta\, d\varphi^2)], \qquad (4.12.1)$$

where:

$$D(t,r) = F(t)V,_t/V, \qquad (4.12.2)$$

but this time, the function V must obey:

$$V,_{uu} = f(u)V^2 + g(u)V^3, \qquad (4.12.3)$$

where $u = r^2$, and f and g are arbitrary functions. The single nonvanishing component of the electromagnetic field is:

$$F_{01} = -F_{10} = -F(t)E(r)V,_t, \qquad (4.12.4)$$

where:

$$E^2(r) = 2r^2 g(u). \qquad (4.12.5)$$

Hence, when $g(u) = 0 = E(r)$, the electromagnetic field vanishes and the K–Q class
(eq. (4.1.3)) is recovered. In the case when $Er^2 = E_0 = $ const, the charge density and
the current defined by (4.12.4) vanish, but the electromagnetic field will be nonzero
if $E_0 \neq 0$. This is the case when there is a charge at (or around) the centre, and the
fluid described by (4.12.1)–(4.12.5) moves in its exterior electric field. The various
papers discussing (4.12.1)–(4.12.5) are displayed in Box 4.1.5.

Equations (4.12.1)–(4.12.5) were first derived by Shah and Vaidya (1968), and later
rediscovered by Faulkes (1969a), Mashhoon and Partovi (1979) and in two papers by
Nduka (1979 and 1981). In the first three papers, other results were obtained as well,
and we shall come back to these later on. Nduka (1979) used (4.12.1)–(4.12.5) to
obtain an approximate solution for the case when $f(u)$ and $g(u)$ are both constant and
$gV^3 \ll fV^2$. The meaning of this approximation was not explained. In the next paper
(Nduka 1981), the author considered the Lyttleton–Bondi (1959) theory, which,
based on the steady-state idea, assumes that either the absolute value of the proton
charge is slightly greater than the absolute value of the electron charge, or else more
protons are created than electrons in any unit of time. The net electric charge would
then create electrostatic repulsion that would power the expansion of the Universe.
However, what Nduka derived in the 1981 paper is again the ordinary
Einstein–Maxwell model described by (4.12.1)–(4.12.5). The Lyttleton–Bondi
corrections cancel out in the energy-momentum tensor in this case and leave no trace
in the metric. The explicit solution discussed as an example is the spherical Stephani
(1967a) model, see our Section 4.10.

The case when:

$$f(u) = [h(u)]^{-5/2}, \quad g(u) = d[h(u)]^{-3}, \quad h(u) \stackrel{\text{def}}{=} au^2 + bu + c, \qquad (4.12.6)$$

where a, b, c and d are arbitrary constants, allows explicit solutions of (4.12.3) to be
derived, and has been identified by many authors independently. In full generality,

$$V_{,uu} = f(u) V^2 + g(u) V^3,$$

where $f(u) = \overline{f(u)} = [h(u)]^{-5/2}$, $g(u) = \overline{g(u)} = [h(u)]^{-3}$ and $h(u) = au^2 + bu + c$.

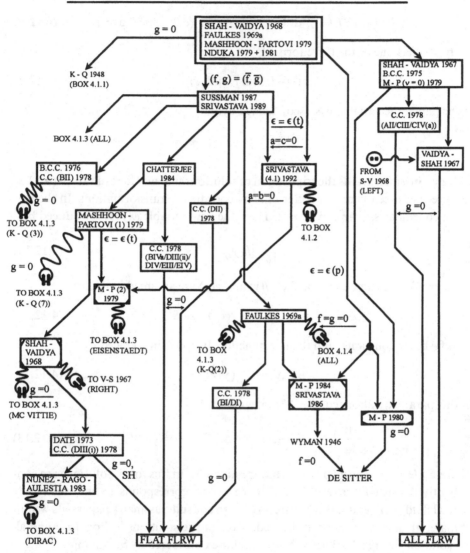

B.C.C. = BANERJEE - CHAKRAVARTY - CHOUDHURY
C.C. = CHAKRAVARTY - CHATTERJEE
M - P = MASHHOON - PARTOVI

Box 4.1.5. The charged fluid solutions generated from the Shah-Vaidya equation
$V_{,uu} = f(u)V^2 + g(u)V^3$, where $f(u) = \overline{f(u)} = [h(u)]^{-5/2}$, $g(u) = \overline{g(u)} = d[h(u)]^{-3}$ and
$h(u) = au^2 + bu + c$.

it was considered by Sussman (1987), who investigated the general solution and classified all separately published explicit subcases. He considered the metric in the form (notation adapted to ours):

$$ds^2=[3V_{,t}/(\theta V)]^2 dt^2 - V^{-2}[dr^2 + \bar{f}^2(r)(d\vartheta^2 + \sin^2\vartheta\, d\varphi^2)], \qquad (4.12.7)$$

where $\bar{f}(r)$ has one of the three forms:

$$\bar{f}(r) = \{r, \sin r, \sinh r\}. \qquad (4.12.8)$$

In the paper, the variable y is used, defined by:

$$y(r) = \int_0^r \bar{f}(r')dr'. \qquad (4.12.9)$$

Using y, Sussman avoids the triplication of solutions that would result from (4.12.8) because the three forms of $\bar{f}$ differ only by coordinate transformations. In the isotropic coordinates, $\bar{f}=r$ and $y=r^2/2$. Then, the new variable $\omega(y)$ is introduced by:

$$\frac{d\omega}{dy}=[h(y)]^{-1}, \qquad (4.12.10)$$

where $h(y)$ is the same as in (4.12.6), $f(y)$ and $g(y)$ are assumed in the form:

$$f(y)=-3\mu h^{-5/2}, \quad g(y)=2e^2 h^{-3}, \qquad (4.12.11)$$

μ and e being constants; and the new function $W(t,y)$ is introduced by:

$$W(t,y)=h^{-1/2}(y)V(t,y). \qquad (4.12.12)$$

In this parametrization, eq. (4.12.3) integrates to:

$$\left(\frac{\partial W}{\partial \omega}\right)^2 = e^2 W^4 - 2\mu W^3 + \Delta W^2 + L(t) \overset{\text{def}}{=} Q(W), \qquad (4.12.13)$$

where $\Delta = b^2/4 - ac$ and $L(t)$ is an arbitrary function. In this form, solutions are conveniently discussed and classified. The case $e=0$ corresponds to the electrically neutral fluid. For details, readers are advised to consult Sussman's paper – its wealth of material cannot easily be abstracted. Most of the solutions in Box 4.1.5 and all the solutions in Box 4.1.3 are subcases of those considered by Sussman.

In two follow-up papers (Sussman 1988a,b), the author investigated properties of the solutions. Again, the results are not easy to report in brief, especially since they do not necessarily correlate with the separately published solutions. The most thought-provoking part of the papers is the discussion of the possible types of singularities in the 1988a paper. They are:

1. finite-density singularities, at which $|\epsilon|<\infty$, but $\theta_{,\alpha}u^\alpha$ and $\dot{u}^\alpha{}_{;\alpha}$ diverge;
2. asymptotically delayed Big Bang, at which curvature scalars diverge, but its locus is infinitely far in the future or in the past;

3. finite-volume singularities, at which θ, ϵ and p diverge, but the proper volume remains finite;
4. a localized singularity which is a null segment of the singularity, hit by a set of observers of measure zero.

While discussing the collapse to a black hole, Sussman considered the solutions matched to the Schwarzschild or the Reissner–Nordström spacetime. Then, it can happen that an exterior observer will see the matter disappear under the Schwarzschild horizon, but a comoving observer may exist for an infinite time and follow the evolution up to an asymptotically delayed Big Bang. In this case, the interior spacetime pinches off from the exterior one and closes upon itself.

Global geometry of the spacetimes was discussed in the 1988b paper. The t=const hypersurfaces may avoid or hit the various singularities, and may be homeomorphic to either S^3 or R^3 or to $S^2 \times R^1$. Some of the spacetimes have a built-in asymptotic de Sitter-like phase of evolution, and so may be termed asymptotically inflationary. In them, the hypersurfaces t=const may undergo a change of topology, just as in the Stephani (1967a) Universe. There exist uniform density solutions in which matter is neutral, but the electric field is nonzero. It is then interpreted as being due to charges of opposite sign located at opposite boundaries of the spacetime which are possibly singular. For the remaining information (a great amount of it), readers are advised to look up the papers.

The Sussman (1987) class of solutions was also identified by Srivastava (1989) as one in which explicit solutions may be found. His "new solution" (eq. (59)) has no limit $F_{\mu\nu}$=0, and so will not be classified here.

As we did before for the neutral solutions, we shall not quote the physical discussion of the particular solutions in detail (unless they have some remarkable properties), but will simply register the interrelations between them.

Banerjee, Chakrabarty and Dutta Choudhury (1976) found the solution corresponding to Sussman's $\Delta=L=0$. When e=0, it reproduces the K–Q case 3 from Box 4.1.3.

Chakravarty and Chatterjee (1978) identified a combination of Sussman's parameters with which a large number of particular explicit solutions may be derived. The solutions are spread all over Box 4.1.5, and some were new in 1978. To consider limits and coincidences, the solutions must usually be reparametrized. Case BII is identical to the Banerjee, Chakrabarty and Dutta Choudhury (1976) solution.

Chatterjee (1984) presented a formal general formula for a solution of (4.12.3) in the case corresponding to Sussman's (1987) e^2=1 and $\Delta=\mu^2$. Because of e^2=1, reparametrizations are necessary when generating some other solutions from this one, and, because of $\Delta=\mu^2$, only a subset of Sussman's cases is covered.

Mashhoon and Partovi (1979) derived and discussed three solutions. The first of them is a subcase of the Chatterjee (1984) class corresponding to Sussman's (1987) L=0, and in the case $g(u)$=0 it reproduces the K–Q case 7 from Box 4.1.3. Mashhoon and Partovi interpreted it as a charged fluid falling into a charged black hole. The second solution is the subcase of the first one corresponding to spatially homogeneous matter-density. In this case, the matter is neutral, but is permeated by electric field. Mashhoon and Partovi interpreted the solution as a neutral fluid falling into a charged black hole. In the limit of zero charge, the second solution reproduces the subcase $k=0$ of the McVittie (1933) solution. The third solution will be mentioned later on.

Shah and Vaidya (1968), in the same paper in which they introduced the whole class, derived a generalization of the McVittie (1933) solution to the case when the point-mass has nonzero electric charge. Since the solution has a clear physical interpretation, we shall present it here (in somewhat changed notation):

$$\mathrm{d}s^2=\left\{\frac{1-\mu^2(t,r)+\varepsilon^2(t,r)}{[1+\mu(t,r)]^2-\varepsilon^2(t,r)}\right\}^2\mathrm{d}t^2-R^2(t)(1+\tfrac{1}{4}kr^2)^{-2} \qquad (4.12.14)$$

$$\times\left\{[1+\mu(t,r)]^2-\varepsilon^2(t,r)\right\}^2[\mathrm{d}r^2+r^2(\mathrm{d}\vartheta^2+\sin^2\vartheta\,\mathrm{d}\varphi^2)],$$

where, as in the McVittie model:

$$\mu(t,r)=\frac{m}{2rR(t)}(1+\tfrac{1}{4}kr^2)^{1/2}, \qquad (4.12.15)$$

and

$$\varepsilon(t,r)=\frac{e}{2rR(t)}(1+\tfrac{1}{4}kr^2)^{1/2}, \qquad (4.12.16)$$

k, m and e being constants; these are the background curvature index, mass and charge of the point particle respectively. This is the second solution in the paper; the first one is the subcase $k=0$ of the above. When $e=0$, the McVittie (1933) solution (4.6.13)–(4.6.14) is reproduced. The solution (4.12.14)–(4.12.16) results from Mashhoon and Partovi's (1979) first solution when the M–P parameters have the values $\alpha=\delta=1$, $\beta=0$, $\gamma=k/4$, $\lambda_0=m$, $\eta_0=e$.

Date (1973) published the derivation of the subcase $k=0$ of the above, even though he knew about the paper by Shah and Vaidya (1968). He showed how other well-known solutions arise as limits of this subcase, which can be reparametrized to coincide with the Chakravarty–Chatterjee (1978) case DIII(i).

Nuñez, Rago and Aulestia (1983) found the subcase of (4.12.14)–(4.12.16) which results when $k=0$, $R=t^{2/3}$, that is, a mass-charge-point on the background of a flat FLRW dust. This is the limit $R=t^{2/3}$ of the solution discussed by Date (1973). The limit $e=0$ of Nuñez et al.'s solution is the case found by Dirac (1979a,b, see Box 4.1.3).

Chakravarty and Chatterjee's (1978) cases BIVa, DIII(ii), DIV, EIII and EIV are within the Chatterjee (1984) class. We will not present them here because it would require much space, while the solutions do not have any clear physical definition or interpretation. In the limit $g(u)=0$ (which, in most cases, can be calculated only after reparametrizations) they reproduce the flat FLRW model directly.

The Chakravarty–Chatterjee (1978) case DII is not within the Chatterjee (1984) class, but it is within the Sussman (1987) class with $\Delta=0$, and it reproduces the flat FLRW model when $g(u)=0$.

Faulkes (1969a) found the first integral of (4.12.3) with (4.12.6) that arises when $a=b=0$. It results from (4.12.13) when $\Delta=0$ and $\omega=y/c$. In the limit $e=0=L$ it reproduces the K–Q case 2 from Box 4.1.3. Chakravarty and Chatterjee's (1978) cases BI and DI are contained within the Faulkes class, and they reproduce the flat FLRW model when $g(u)=0$. This Faulkes class reproduces the spherical Stephani (1967a) solution in the limit $c\rightarrow\infty$, that is, $f=g=0$.

The paper by Srivastava (1992) is a broad overview of solutions of the Einstein–Maxwell equations for a spherically symmetric shearfree charged perfect fluid. Srivastava sorted the solutions by the method used to obtain them, and integrated (4.12.3) to a first order equation in the case when $\epsilon=\epsilon(t)$. This is a subcase of (4.12.6) with $a=c=0$, that is, it is within the Sussman (1987) class. The label [4.1] in Box 4.1.5 refers to section 4.1 in the paper. In the limit $d=0$, Srivastava's case [4.1] reduces to the class considered by Qvist (1947) and Kustaanheimo (1947), and it also reproduces the Mashhoon–Partovi (1979) second solution.

Mashhoon and Partovi, in two papers, found all those solutions of the Shah–Vaidya (1968) class that obey the barotropic equation of state. In the 1980 paper, the authors derived the equation (4.12.3) and stated that with $p=p(\epsilon)$ and $g(u)=0$ the FLRW models uniquely follow. This denies the existence of the Wyman (1946) solution, as pointed out by Srivastava and Prasad (1983) and by Collins and Wainwright (1983). Mashhoon and Partovi corrected the oversight in the 1984 paper in which they obtained the charged generalization of the Wyman (1946) solution. The solution from the 1980 paper is (in notation adapted to ours):

$$ds^2=3(8\pi p_0)^{-1}[t+w(r)]^{-2}dt^2$$
$$-(24\pi p_1)^{-1}(t+w)^2[dr^2+\sinh^2r\,(d\vartheta^2+\sin^2\vartheta\,d\varphi^2)], \qquad (4.12.17)$$

where p_0 and p_1 are constants, and $w(r)$ is given by:

$$\frac{dw}{dr}=-\tfrac{3}{4}[\sinh(2r)-2(r+r_0)]/\sinh^2r, \qquad (4.12.18)$$

where r_0 is another constant. The pressure, matter-density and charge within the sphere of radius r are:

$$p=-p_0+3p_1/(t+w)^2,$$
$$\epsilon=-p+2(p+p_0)[-1+(3/p_1)^{1/2}(p+p_0)^{1/2}], \qquad (4.12.19)$$
$$E(r)=\pm\left(\frac{3}{128\pi p_1}\right)^{1/2}[\sinh(2r)-2(r+r_0)].$$

Contrary to the statement by Srivastava (1986), this solution does have the limit $E=0$. In order to find it, the following reparametrization and transformation must be made:

$$p_1=a^2, \quad r_0=a^2\bar{r}_0, \quad r=a\bar{r}, \tag{4.12.20}$$

and then the limit $a\to0$ leads to $w=$const, $p=-p_0=$const, $\epsilon=-p$ and $E=0$; the resulting metric is the de Sitter solution. Mashhoon and Partovi (1980) discussed several properties of the solution.

In the 1984 paper, the same authors derived the other solution that they missed in 1980. This obeys the barotropic equation of state, too, and generalizes the Wyman (1946) solution. It is (in adapted notation):

$$ds^2=(C-4\xi t)^{-2}(V_{,u}/V)^2dt^2-V^{-2}[dr^2+r^2(d\vartheta^2+\sin^2\vartheta\,d\varphi^2)], \tag{4.12.21}$$

where C and ξ are constants, $u=t+r^2$ and V is determined by:

$$V_{,u}^2=\xi+\eta V^3+|\sigma|V^4, \quad \eta=1+|\sigma|(\lambda-1), \tag{4.12.22}$$

η, σ and λ being other constants. The Wyman solution results when $\sigma=0$. Mashhoon and Partovi argue that their two solutions provide a uniqueness argument for the FLRW models: they are the only nonstatic isotropic shearfree solutions of the Einstein equations obeying a physically reasonable equation of state. The "physical reasonability" requirement excludes the solutions (4.12.17)–(4.12.22).

The Mashhoon–Partovi (1984) solution is within the Faulkes (1969a) class. It was discovered independently and discussed by Srivastava (1986).

All the solutions obeying a barotropic equation of state, for all G_3/S_2 symmetries, were found and discussed by Srivastava and Prasad (1991). Among other results, these authors found the plane and hyperbolically symmetric counterparts of the Mashhoon–Partovi (1980) solution. They claimed that those solutions have no limit $F_{\mu\nu}=0$. We shall take that claim for granted and not display the solutions in Box 4.1.5, although, as shown above, the claim is incorrect in the spherically symmetric case.

Shah and Vaidya (1967) showed that the metric:

$$ds^2=(F+G)^{-2}dt^2-(F+G)^2[dr^2/(1-kr^2)+r^2(d\vartheta^2+\sin^2\vartheta\,d\varphi^2)], \tag{4.12.23}$$

where $F(r)$ and $G(t)$ are arbitrary functions, obeys the Einstein–Maxwell equations for a charged perfect fluid. After it is transformed to the isotropic coordinates, it obeys (4.12.3) with:

$$V=(1+\tfrac{1}{4}kr^2)/(F+G), \quad u=r^2,$$

$$f(u)=-\tfrac{1}{2}kF_{,u}/(1+\tfrac{1}{4}ku)^2-F_{,uu}/(1+\tfrac{1}{4}ku), \tag{4.12.24}$$

$$g(u)=2F_{,u}^2(1+\tfrac{1}{4}ku)^2.$$

Hence, it is within the Shah–Vaidya (1968) class, but not within the Sussman (1987) class. It does contain the Mashhoon–Partovi (1980) solution as the limit $k=-1$, $G=$const$\cdot t$, $F=$const$\cdot w$, and it reproduces all the FLRW models when $F_{\mu\nu}=0$; then $g=0$ and $F=$const.

The very same solution was rediscovered by Banerjee, Chakravorty and Dutta Choudhury (1975). The solution with $\nu=0$ in Mashhoon and Partovi's (1979) appendix D is (4.12.23)–(4.12.24) again.

The subcase of (4.12.23)–(4.12.24) corresponding to $F=m(1-kr^2)^{1/2}/r$ was derived by Vaidya and Shah (1967). It can be recognized as the limit $m^2=e^2$ of the solution (4.12.14)–(4.12.15), and it reproduces all the FLRW models when $m=0$. For this particular solution, Banerjee and Dutta Choudhury (1977) took into account the conditions for matching it to the Reissner–Nordström solution. They calculated and discussed the redshift/blueshift from the surface of a collapsing/expanding charged perfect fluid sphere. The matching conditions were worked out earlier by Banerjee, Chakravorty and Dutta Choudhury (1975).

The Chakravarty–Chatterjee (1978) cases AII, CIII and CIVa are subcases of the Shah–Vaidya (1967) class, and they can reproduce all the FLRW limits.

4.13 Other solutions with charged fluid, related to the Stephani–Barnes family

In a few papers, A_2- and plane symmetric[40] solutions with charged fluid source were considered which reduce directly to the FLRW models when the electromagnetic field is zero. They are related to those of Box 4.1.5 because $\sigma=\omega=0$ here also. They are shown in Figure 4.2.

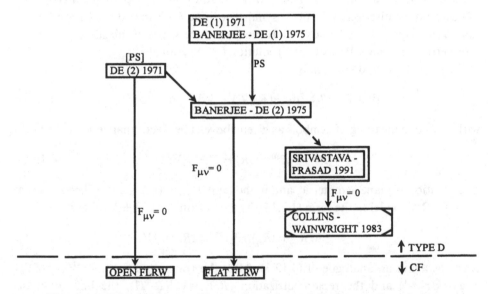

Figure 4.2. Other solutions with charge and $\omega=\sigma=0$ (A_2-symmetric and plane symmetric).

---

[40] The symbol A_2 denotes an Abelian two-dimensional group acting on spacelike orbits. Invariance with respect to A_2 is sometimes called "cylindrical symmetry", but this latter name is appropriate only when one of the Killing vectors has closed trajectories and an axis can be defined. These questions were not studied in most papers.

De (1971) found a few solutions just by guessing simple subcases of the Einstein–Maxwell equations. His first solution is (in his original notation):

$$ds^2 = (R+T)^{-2}dt^2 - (R+T)^2 e^{2kr}(dr^2 + dz^2) - (R+T)^2 d\varphi^2, \qquad (4.13.1)$$

where $R(r)$ and $T(t)$ are arbitrary functions and k is an arbitrary constant; the coordinates of (4.13.1) are comoving. The only nonvanishing components of the electromagnetic field are:

$$F^{01} = -F^{10} = \pm R_{,r} e^{-2kr}/(R+T)^2. \qquad (4.13.2)$$

The solution is A_2-symmetric, and this is why it does not fit into Box 4.1.5, but has zero rotation and shear. In the limit $k=0$, it becomes the plane symmetric counterpart of the Shah–Vaidya (1967) class from Box 4.1.5. Indeed, the function $V = (R+T)^{-1}$ then obeys:

$$V_{,rr} = -R_{,rr}V^2 + 2R_{,r}^2 V^3, \qquad (4.13.3)$$

which is a plane symmetric analogue of (4.12.3), but with very special forms of $f(r)$ and $g(r)$. The electromagnetic field will vanish when $R_{,r} = 0$, but then only the flat FLRW model results.

Banerjee and De (1975, their first solution, eq. (17d)) rediscovered (4.13.1)–(4.13.2) in other coordinates, with the roles of z and φ interchanged. The same authors rediscovered the plane symmetric limit, $k=0$, of (4.13.1) (second solution in the paper, eqs. (20) and (24)). It can reproduce a special subcase of the plane symmetric Srivastava–Prasad (1991) solution (see Section 4.12).

De's (1971) second solution is:

$$ds^2 = (R+T)^{-2}dt^2 - (R+T)^2 r^{-2}(dr^2 + dz^2 + d\varphi^2), \qquad (4.13.4)$$

with the same meaning of symbols as given above. The electromagnetic field is:

$$F^{01} = -F^{10} = \pm R_{,r} r^2/(R+T)^2. \qquad (4.13.5)$$

The solution is plane symmetric, and in the limit $F_{\mu\nu} = 0$ (i.e. R const) becomes the open FLRW model in the form (1.3.15). The function $V = r/(R+T)$ obeys:

$$V_{,rr} = -[(rR_{,rr} + 2R_{,r})/r^2]V^2 + (2R_{,r}^2/r^2)V^3, \qquad (4.13.6)$$

which is again an analogue of (4.12.3). After the transformation $r = e^{-kr'}$, $t = t'/k$, $(z,\varphi) = k(z',\varphi')$ and the reparametrization $(R,T) = k^{-1}(R',T')$, the limit $k=0$ of (4.13.4) will reproduce the limit $k=0$ of (4.13.1), that is, Banerjee and De's (1975) second solution.

De's third case (1971, the "new line element") contains solutions that do not belong in this book: the "general solution" and the "special case 2" have no limit $\dot{u}^\alpha = 0$; the "special case 1" is an approximate solution and has no limit $F_{\mu\nu} = 0$; the "special case 3" is the open FLRW model itself.

4.14 Generalizations of the Stephani–Barnes models with heat-flow

Just as with the corresponding perfect fluid solutions, the solutions that are the subject of this section divide into the Petrov type D solutions (all of them spherically symmetric) and the conformally flat ones. These solutions are displayed in Figure 4.3.

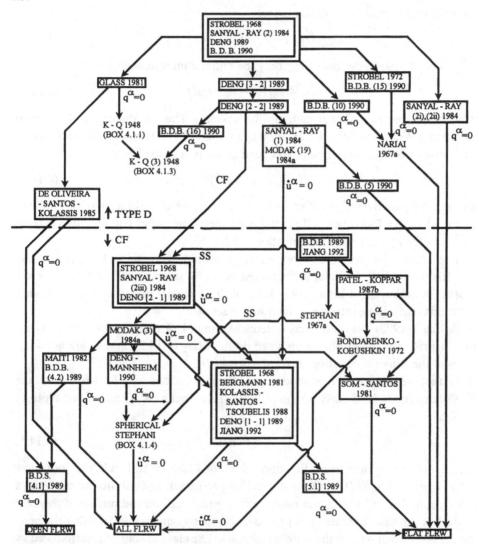

B.D.B. = BANERJEE - DUTTA CHOUDHURY - BHUI
B.D.S. = BONNOR - DE OLIVEIRA - SANTOS
REFERENCES WITHOUT FRAMES ARE PERFECT FLUID SOLUTIONS
FROM OTHER DIAGRAMS

Figure 4.3. Generalizations to Figure 4.1 with heat-flow.

The most general spherically symmetric class of solutions (and the earliest one historically) was found by Strobel (1968). It was obtained under the assumptions of spherical symmetry and zero shear. In notation adapted to our usual one, the metric is:

$$ds^2 = D^2 dt^2 - V^{-2}[dr^2 + (d\vartheta^2 + \sin^2\vartheta \, d\varphi^2)], \tag{4.14.1}$$

where D and V must obey the equation:

$$D,_{uu}/D + 2D,_u V,_u/(DV) - V,_{uu}/V = 0, \tag{4.14.2}$$

$u \overset{\text{def}}{=} r^2$, and the heat-flow vector has only the radial component equal to:

$$q = (4r/\kappa)V^2[-V,_t/(DV)],_r \tag{4.14.3}$$

(this formula is not in fact given in Strobel's paper). The coordinates of (4.14.1) are comoving. When $q=0$, eq. (4.1.2) follows, and then (4.14.2) turns out to be the time-derivative of (4.1.3). Equation (4.14.2) can be written in the form:

$$\tfrac{1}{2}(DV),_{uu}/(DV) = V,_{uu}/V \overset{\text{def}}{=} \bar{f}(t,u), \tag{4.14.4}$$

and Strobel gave the full list of special cases of $\bar{f}$ for which a solution of (4.14.4) can be found in handbooks on differential equations. The paper contains two explicit solutions for $\bar{f}=0$; these will be mentioned later on.

The formulation described above was rediscovered by Sanyal and Ray (1984, case 2 in the paper).[41] It was later rediscovered again by Deng (1989) and by Banerjee, Dutta Choudhury and Bhui (1990). Each of those papers presented special explicit solutions that will be introduced later in this section.

Bokhari (1992) was on the way to rediscovering (4.14.2), but got it wrong. Then, not surprisingly, the results he obtained do not obey (4.14.2), and so are not solutions of the Einstein equations (except when $k=0$).

Glass (1981) invented a method of generating solutions with heat-flow from the K–Q class. In the K–Q class, $q=0$ in (4.14.3). If one takes a K–Q metric and replaces its V by:

$$V_1 = V/h(t), \tag{4.14.5}$$

where $h(t)$ is an arbitrary function, then (4.14.2) is still fulfilled, but now $0 \neq q = (4\pi)^{-1}V^2[h,_t/(hD)],_r$. Glass showed how to match this solution to the Vaidya (1953) solution and stated that pressure may equal zero on the surface of the fluid body, but this last statement was proved incorrect by Santos (1985): either the pressure is nonzero or $q=0$ at the matching surface. The class of solutions introduced by Glass contains the whole K–Q family and is contained in the Strobel (1968) class.

De Oliveira, Santos and Kolassis (1985) studied a particular example of the Glass (1981) family of solutions, in which:

---

[41]   The other paper by Sanyal and Ray, immediately following this one, will not be reviewed here because the solutions in it are either static or equivalent to those of Minkowski or de Sitter.

$$V_1(t,u)=V_0(u)/R(t), \quad D=D(u). \tag{4.14.6}$$

In this paper, and in another one (de Oliveira, Kolassis and Santos 1988), various analytic properties of this model were studied. In the limit $q=0$ it reproduces the FLRW solutions only. Bonnor (1987) used this solution to show that the arrow of time provided by radiation (which always travels away from the source) is opposite to the gravitational arrow of time provided by $P=C^{\alpha\beta\gamma\delta}C_{\alpha\beta\gamma\delta}/R^{\mu\nu}R_{\mu\nu}$, where $C_{\alpha\beta\gamma\delta}$ and $R_{\mu\nu}$ are the Weyl and the Ricci tensors respectively. This shows that the definition of the gravitational arrow of time through P is not satisfactory (see our Section 2.5, after eq. (2.5.40), for a more detailed description of that definition).

Deng (1989) invented a more powerful method of generating solutions of (4.14.2). He observed that (4.14.2) may be understood as an equation for D given V or vice versa. Hence, the following procedure may be applied:

1. Choose a simple function $D=D_1$ and find the most general solution $V=V_1$ of (4.14.2).
2. Take $V=V_1$ and find the most general $D=D_2$ obeying (4.14.2).
3. Take $D=D_2$ and find the most general $V=V_2$,

and so on. Note that eq. (4.14.2), after it is multiplied by (DV), becomes linear in D and in V, so the procedure should work rather simply.

One such sequence of solutions is presented in Deng's paper. It begins with:

$$D_1=1, \quad V_1=a(t)u+b(t), \tag{4.14.7}$$

where $a(t)$ and $b(t)$ are arbitrary functions. $V=V_1$ is the most general solution of (4.14.2) with $D=D_1$. For later reference, we shall label this case Deng [1–1] (1989).

In the next step, V_1 from (4.14.7) is substituted into (4.14.2), and the solution for D is:

$$D=D_2=[c(t)u+d(t)]/[a(t)u+b(t)], \quad V=V_1, \tag{4.14.8}$$

where each new letter denotes an arbitrary function of time. We shall label this case Deng [2–1] (1989).

Next, D_2 is used in (4.14.2) to find:

$$D=D_2, \tag{4.14.9}$$

$$V=V_2=g(t)(au+b)-[h(t)/(3a)]\left[\frac{(cu+d)^2}{(au+b)^2}+\frac{c}{a}\cdot\frac{cu+d}{au+b}+\frac{c^2}{a^2}\right].$$

This will be the Deng [2–2] solution. In the next step, with $V=V_2$:

$$D=D_3=l(t)\cdot\frac{cu+d}{au+b}+m(t)\cdot\frac{cu+d}{au+b}\int_{u_0}^{u}\left(\frac{ay+b}{cy+d}\right)^2 [V_2(t,y)]^{-2}dy. \tag{4.14.10}$$

The integrand in (4.14.10) is a rational function, so the integral could be calculated explicitly, but (4.14.10) is a more compact notation. This will be the Deng [3–2] solution, and it is the last element of the sequence explicitly given in the paper. The solutions [1–1] and [2–1] were known earlier, and we shall mention them later on. The cases [2–2] and [3–2] were new. The case [2–2] is contained in [3–2] as the limit $m=0$, $l=1$, the case [2–1] then follows when $h=0$, $g=1$, and so on. Already the solution [1–1] contains all the FLRW models as the limit $a/b=k=$const.

The particular solution of Banerjee, Dutta Choudhury and Bhui (1990) given by their eq. (16) is a subcase of Deng's case [2–2] (resulting after a reparametrization), and it reproduces the K–Q case 3 in the limit $q=0$.

Sanyal and Ray's (1984) case 1 is another subcase of Deng's solution [2–2]; reparametrizations are also necessary here. The same solution, in a different parametrization, was found by Modak (1984a, his eq. (19); Modak's eq. (16) is a simple subcase of (19) and will not be considered separately). The conformally flat limits of the Deng [2–2] (1989) and Sanyal–Ray (1) (1984) solutions are still nontrivial generalizations of the FLRW models, and will be described later on.

The solution given by Banerjee, Dutta Choudhury and Bhui's (1990) eq. (5) is a subcase of Sanyal and Ray's (1984) case 1, and in the limit $q=0$ it reproduces the flat FLRW model.

Banerjee, Dutta Choudhury and Bhui's (1990) solution 10 is not within the Deng (1989) family (its V is not a rational function in u), and in the limit $q=0$ it reproduces the Nariai (1967a) solution.

Strobel (1972) published an explicit example of a solution which is not in Deng's class for the same reason. The solution given by Banerjee, Dutta Choudhury and Bhui's (1990) eq. (15) is equivalent to this one. In the limit $q=0$ it reduces to the Nariai (1967a) solution again.

Sanyal and Ray's (1984) cases 2i and 2ii are other explicit examples of Petrov type D spacetimes of the Strobel (1968) class, and in the limit $q=0$ they reduce to the flat FLRW model.

This completes the overview of spherically symmetric Petrov type D solutions with heat-flow and with $\sigma=\omega=0$.

Of the conformally flat solutions, the most general one was found by Banerjee, Dutta Choudhury and Bhui (1989). It has the metric (in notation adapted to ours):

$$ds^2=D^2(t,x,y,z)dt^2-V^{-2}(t,x,y,z)(dx^2+dy^2+dz^2), \qquad (4.14.11)$$

where:

$$V=B(t)(x^2+y^2+z^2)+B_1(t)x+B_2(t)y+B_3(t)z+B_4(t), \quad D=W/V,$$

$$W=A(t)(x^2+y^2+z^2)+A_1(t)x+A_2(t)y+A_3(t)z+A_4(t), \qquad (4.14.12)$$

all the functions of time being arbitrary. The heat-flux vector is:

$$\kappa q^i=2V^2(V_{,t}/W)_{,i}, \qquad (4.14.13)$$

where $i=1$, 2, 3; $q^0=0$. The coordinates of (4.14.11) are comoving. In the limit $q^i=0$, the general Stephani (1967a) solution results from this one.

Exactly the same result was rediscovered by Jiang (1992). Both papers contain additional solutions that will be mentioned later on. Neither paper presented a proof that (4.14.11)–(4.14.13) is the most general conformally flat solution with $\sigma=\omega=0\neq\theta$ and $q^\alpha\neq0$, and so (as yet) the solution has no invariant definition.

Sussman (1993) considered a larger class of solutions with a heat-conducting and viscous fluid source, generalizingthe Stephani–Barnes family. However, as shown by Sussman himself, his type D solutions do not admit a perfect fluid limit, and his conformally flat solution coincides with that of Banerjee, Dutta Choudhury and Bhui (1989).

The spherically symmetric limit of (4.14.11)–(4.14.13) was first derived by Strobel (1968, his eq. (14)).[42] It results when $A_i=B_i=0$, $i=1, 2, 3$. The same solution was later rediscovered by Sanyal and Ray (1984, their case 2iii); it coincides with Deng's (1989) solution [2–1]. The limit $q^\alpha=0$ of this solution is the spherical Stephani (1967a) solution of Box 4.1.4.

The subcase of the last solution corresponding to $a=c$ in (4.14.8) was derived by Modak (1984a, his eq. (3)). Its limit $q^\alpha=0$ is still the spherical Stephani (1967a) solution.

The further subcase of (4.14.8) when $a=c=$const was found by Maiti (1982). In the limit $q^\alpha=0$, this subcase reproduces all the FLRW models. It was rediscovered by Banerjee, Dutta Choudhury and Bhui (1989, their eq. (4.2)), who showed that it can be matched to the Vaidya (1953) solution.

The paper by Bonnor, de Oliveira and Santos (1989) is a broad overview of spherically symmetric solutions with a heat-conducting fluid source. It is aimed at a description of collapsing bodies with internal energy transport that emit null radiation. Its subject are the general problems of matching such solutions to the Vaidya solution, and their thermodynamical properties. Two explicit examples of solutions are studied in detail. The first of them (in their sections 4.1 and 4.2) is a common subcase of the class considered by de Oliveira, Santos and Kolassis (1985) and of Maiti's (1982) solution. It can reproduce only the open FLRW model, and can do so only before the matching conditions are imposed (after they are imposed, the parameter β that should be zero in the limit becomes such a function of r_Σ – the radius of the body – that can never vanish). The second explicit solution will be mentioned later on.

Deng and Mannheim (1990) presented another subcase of Modak's (1984a) eq. (3). It results from (4.14.8) when a, b, c and d are all simple power functions of t. Deng and Mannheim interpreted it as a solution with heat-conduction and viscosity. However, the heat-conduction and viscosity terms refuse to vanish when the metric is reduced to a subcase of the spherical Stephani (1967a) solution. This should mean that the two dissipative terms cancel each other in this limit, that is, that the source was artificially extended for a component that does not enter the metric. Deng and Mannheim interpreted the solution in detail and showed how it

---

[42] Strobel's eq. (13) defines a static solution that will not be considered here.

avoids the "Problems" of horizon, entropy and flatness, known from inflationary models. Because of the special time-dependence, the solution can reproduce only a subcase of the spherical Stephani (1967a) model, and the FLRW models only with a specific equation of state.

The limit $\dot{u}^\alpha=0$ of the two solutions by Modak (1984a), and also of the conformally flat Strobel (1968) solution, has the following simple form:

$$ds^2 = dt^2 - R^2(t)[1 + \tfrac{1}{4}k(t)r^2]^{-2}[dr^2 + r^2(d\vartheta^2 + \sin^2\vartheta \, d\varphi^2)]. \qquad (4.14.14)$$

It was first found by Strobel (1968, his eq. (22)), and then rediscovered by Bergmann (1981), Kolassis, Santos and Tsoubelis (1988), Deng (1989, the case [1–1]) and Jiang (1992, his "almost Robertson–Walker metric"). Kolassis, Santos and Tsoubelis considered the metric (4.14.14) matched to the Vaidya (1953) solution as a model of a collapsing and radiating body. They showed that it will never form a black hole because, shortly before reaching the horizon, it will explode in consequence of the instability $p_{,r}>0$. Further details of the model were elaborated by Chan, Lemos, Santos and Pacheco (1989). Grammenos and Kolassis (1992) reinterpreted the solution as an anisotropic fluid with null radiation and heat-flow such that the null radiation and pressure anisotropy exactly cancel each other.

A subcase of the Strobel (1968)–Bergmann (1981) solution is the second explicit example considered by Bonnor, de Oliveira and Santos (1989, section 5.1 in the paper). It can reproduce only the flat FLRW model.

Patel and Koppar (1987b) found a solution that is within the Banerjee, Dutta Choudhury and Bhui (1989) class, but is not a subcase of any other in this section because it is not spherically symmetric. In the limit $q^\alpha=0$, it reproduces the Bondarenko–Kobushkin (1972) solution from our Section 4.11.

The solution by Som and Santos (1981) is a common subcase of the solution by Patel and Koppar (1987b) and of the first solution by Modak (1984a). In the limit $q^\alpha=0$, it reproduces the flat FLRW model.

This closes the overview of the Stephani–Barnes family and its generalizations.

5

Solutions with null radiation

5.1 General remarks

In this chapter, we shall study a family of solutions which were constructed with the explicit aim of superposing the FLRW models with various important vacuum or electrovacuum solutions. They become the FLRW models in the homogeneous perfect fluid limit, and they reduce to the Kerr or related solutions in the stationary (electro-) vacuum limit. They were guessed rather than derived by integration of the Einstein equations. The null radiation in them was not introduced as an additional physical component of matter, but appeared *ex post* as a device to interpret those components of the Einstein tensor that do not belong to the perfect fluid or electromagnetic field. As a result, the various components of the source (the fluid, the electromagnetic field and the null radiation) are coupled through the parameters and functions that they all contain, and cannot be set to zero separately. Usually, setting the null radiation component to zero results in trivializing the other components automatically (for example, it may result in reducing the solution to a FLRW model or to a vacuum). This is a disadvantage, of course, but otherwise the solutions constitute a very interesting experiment in combining different models that has already reached remarkable sophistication and provided new insights into the properties of known solutions. The papers from this family do not contain sufficient information to assign the solutions to the Wainwright (1979 and 1981) classes. The solutions are displayed in Figure 5.1.

Before discussing them, we must first recall a few properties of the null radiation and introduce the (electro-) vacuum solutions that will be superposed onto the FLRW background.

5.2 The energy-momentum tensor of null radiation

In our Section 1.2 (eq. (1.2.26) and following) we already described how the energy-momentum tensor (1.2.18) of null radiation may arise as a special case of null electromagnetic field. Simplified derivations of this result were presented by Vaidya (1951) and by Narlikar and Vaidya (1947)[43]; in the spherically symmetric case the implications of the Maxwell equations for (1.2.18) were considered by Griffiths (1974). Interpretation of the energy-momentum tensor (1.2.18) in terms of a stream

---

[43] Narlikar and Vaidya suggested that the Vaidya solution had already been derived by Mineur (1933). I looked up Mineur's paper, but could not recognize the Vaidya solution in it.

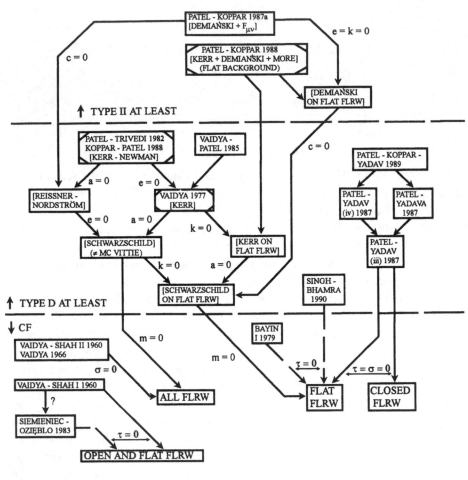

Figure 5.1. Solutions with null radiation component in the source.

of neutrinos obeying the Weyl equation was discussed by Griffiths (1973) and Griffiths and Newing (1974). These papers are suggested as further reading for those readers who are interested in deeper physical insight into the null radiation. The limitations imposed on (1.2.18) by other field theories are usually ignored in the papers considering the Einstein equations.

It must be stressed that the energy-momentum tensor (1.2.18) is not appropriate for describing the cosmic microwave background radiation. The cosmic radiation reaches every observer from all directions, and it can only be assigned energy-density at every point. It is usually described as a perfect fluid with the equation of state $p=\epsilon/3$. The null radiation (1.2.18) has a well-defined direction of flow at every point; for an observer with four-velocity u^α, the direction vector is $(k^\alpha - k^\mu u_\mu u^\alpha)$. Therefore, eq. (1.2.18) describes a directed stream of radiation.

A prototype, and the historically earliest example of all solutions with null radiation as a source, is the Vaidya solution. It has the simplest form in null coordinates in which:

$$ds^2 = [1 - 2m(u)/r]du^2 + 2du\,dr - r^2(d\vartheta^2 + \sin^2\vartheta\,d\varphi^2), \qquad (5.2.1)$$

where $m(u)$ is an arbitrary function (see Kramer et al. 1980, section 13.4, for generalizations of the above to include the plane and hyperbolically symmetric counterparts and the cosmological constant). The energy-momentum tensor of this metric has only one nonvanishing component:

$$\kappa T_{00} = -2m_{,u}/r^2, \qquad (5.2.2)$$

where $u = x^0$. In the limit $m = $ const, the metric (5.2.1) thus becomes the Schwarzschild solution. The function $m(u)$ is interpreted as the varying mass of the source of (5.2.1) that is decreasing because it is being converted into radiation. Even with the normalizing condition $k^\alpha_{;\beta}k^\beta = 0$, the decomposition of (5.2.2) into the radiation density τ and the wave vector k^α is nonunique (see our eq. (1.2.18) and comments following it), but the direction of k^α is unique, and one can take:

$$k_\alpha = \delta^0_{\ \alpha}, \quad \tau = -2m_{,u}/r^2. \qquad (5.2.3)$$

This solution was first found by Vaidya (1943 and 1951; in the null coordinates it was first presented in the 1953 paper), and its properties were discussed in more detail by Vaidya (1951) and by Lindquist, Schwartz and Misner (1965). It is frequently used as an exterior solution for solutions describing fluids with heat-flow (see examples in Section 4.14).

5.3 The Kerr, Kerr–Newman and Demiański (1972) solutions

We shall now very briefly describe those solutions whose superpositions with the FLRW backgrounds have been found.

The Kerr–Newman solution is a stationary axisymmetric electrovacuum solution with the following metric (Newman et al. 1965):

$$ds^2 = [1 - (2mr - e^2)/\Sigma]\,dt^2 + 2a\,[\sin^2\vartheta\,(2mr - e^2)/\Sigma]dt\,d\varphi$$

$$-\sin^2\vartheta[r^2 + a^2 + a^2\sin^2\vartheta(2mr - e^2)/\Sigma]d\varphi^2 - (\Sigma/\Delta)dr^2 - \Sigma d\vartheta^2, \qquad (5.3.1)$$

where:

$$\Sigma \overset{\text{def}}{=} r^2 + a^2\cos^2\vartheta, \quad \Delta \overset{\text{def}}{=} r^2 - 2mr + a^2 + e^2, \qquad (5.3.2)$$

m, a and e being arbitrary constants. The solution (5.3.1)–(5.3.2) describes the exterior gravitational and electromagnetic fields of a rotating charged body of mass m, angular momentum per unit mass a and charge e. It is only one of an infinity of possible exterior fields of rotating charged bodies. However, for reasons that will not be considered here (they are too far removed from cosmology), the Kerr–Newman solution is believed to be the unique future-asymptotic exterior field of a rotating charged body that collapsed to form a black hole. In the limit $a=0$, it becomes the Reissner–Nordström solution (the spherically symmetric electro-gravitational field of a charged mass point or a charged black hole), and in the limit $a=e=0$ it becomes the Schwarzschild solution. The metric (5.3.1)–(5.3.2) is of Petrov type D.

The limit $e=0$ of (5.3.1)–(5.3.2) is the (vacuum) Kerr solution, first found by Kerr (1963) and transformed to the simpler form given above by Boyer and Lindquist (1967). This one is believed to be the unique future-asymptotic field of a rotating neutral body that formed a black hole. The literature on the Kerr solution is enormous, and the introduction given above is only meant to be a quick reference for the metric.

For considering the superpositions presented later on, it is more convenient to use the Kerr metric transformed to the null coordinates. Take (5.3.1)–(5.3.2) and perform the transformation:

$$dt = -du - [(r^2+a^2)/\Delta]dr, \quad dr = dt' - du,$$

$$d\varphi = d\beta - (a/\Delta)dr, \quad \vartheta = \alpha \qquad (5.3.3)$$

(only the differential of r is to be replaced, not r itself). The result, dropping the prime on t, is:

$$ds^2 = 2dt(du + a\sin^2\alpha\,d\beta) - \Sigma(d\alpha^2 + \sin^2\alpha\,d\beta^2)$$

$$-[1 + (2mr - e^2)/\Sigma](du + a\sin^2\alpha\,d\beta)^2. \qquad (5.3.4)$$

The Kerr and Kerr–Newman metrics can both be derived by performing complex coordinate transformations on the Schwarzschild and Reissner–Nordström solutions respectively (this is how (5.3.1)–(5.3.2) was first found). There is no systematic theory behind such transformations, only the observation that the method leads to new solutions in some cases. Demiański (1972) showed that another complex transformation performed on the Schwarzschild solution leads to another vacuum solution. It is shown here in coordinates related to the original ones by $dr = dt - du$, $\alpha = \vartheta$, $\beta = -\varphi$:

$$ds^2 = 2\omega\,dt - (r^2 + Z^2)(d\alpha^2 + \sin^2\alpha\,d\beta^2)$$

$$-[1 + (2mr + 2bZ)/(r^2 + Z^2)]\omega^2 + 4b\cos\alpha\,\omega\,d\beta, \qquad (5.3.5)$$

where the function $Z(\alpha)$ and the differential form ω are:

$$Z = a \cos \alpha + b + c\{1 + \cos \alpha \ln [\tan(\alpha/2)]\},$$

$$\omega = du + \left(a \sin^2\alpha - 2b \cos \alpha + c\{\sin^2\alpha \ln[\tan(\alpha/2)] - \cos \alpha\}\right) d\beta, \qquad (5.3.6)$$

and a, b and c are arbitrary constants. This solution is in general of Petrov type II and it generalizes two well-known solutions. With $b = c = 0$, it is the Kerr metric; see (5.3.4) in the case $e = 0$. With $a = c = 0$, it is the NUT metric (see Kramer *et al.* 1980, p. 133). The parameter c, of unknown interpretation, defines a new structure; in Figure 5.1 it is referred to as the Demiański parameter.[44] For easy reference, we shall use the name "special Demiański solution" to denote the subcase $a = b = 0$ of (5.3.5)–(5.3.6).

There is a deeper connection between the Kerr and the special Demiański solutions, first discovered by Vaidya (1976; in fact Vaidya rediscovered the subcase $a = b = 0$ of the Demiański solution by the method shown below, and called it "the associated Kerr metric"). Let us take the Kerr metric in the form (5.3.4) (with $e = 0$) and substitute in it:

$$z = \cos \alpha, \quad Y(z) = a \cos \alpha = az, \quad Y,_z = a, \qquad (5.3.7)$$

so that z becomes a new coordinate and $Y(z)$ becomes a function to be determined from the vacuum Einstein equations. Then (5.3.4) changes to:

$$ds^2 = 2 \, dt \, (du + Y,_z \sin^2\alpha \, d\beta) - (r^2 + Y^2) \, (d\alpha^2 + \sin^2\alpha \, d\beta^2)$$

$$- [1 + 2mr/(r^2 + Y^2)] \, (du + Y,_z \sin^2\alpha \, d\beta)^2, \qquad (5.3.8)$$

where now $\alpha = \text{arc cos } z$. If (5.3.8) is substituted into the vacuum Einstein equations, they reduce to:

$$(1 - z^2) \, Y,_{zz} - 2z Y,_z + 2Y = 0. \qquad (5.3.9)$$

This is the Legendre equation with the index $n = 1$, and $Y = aP_1(z) = az$ is one of its solutions – the first Legendre polynomial. With $Y = az$, eq. (5.3.8) becomes the Kerr metric. But (5.3.9) has a second solution, $Y = cQ_1(z)$, where $Q_1(z)$ is the first associated Legendre function:

$$Q_1(z) = \tfrac{1}{2}z \ln [(1 + z)/(1 - z)] - 1 \qquad (5.3.10)$$

(see Ryżyk and Gradsztejn 1964, sections 6.84–6.85). With $cQ_1(z)$ in place of $Y(z)$, the metric (5.3.8) becomes the special Demiański solution, that is, the subcase $a = b = 0$ of (5.3.5)–(5.3.6).

The construction can go on further. Since (5.3.9) is the necessary and sufficient condition for (5.3.8) to be a vacuum solution of the Einstein equations, any linear combination of $P_1(z)$ and $Q_1(z)$:

$$Y = A \, P_1(z) + B \, Q_1(z), \qquad (5.3.11)$$

[44] Kramer *et al.* (1980, p. 342) use the name "Demiański solution" for the subcase $a = c = 0$ of (5.3.5)–(5.3.6) generalized to nonzero cosmological constant. That solution was derived in the same paper by Demiański (1972).

where A and B are constants, will generate a vacuum metric through (5.3.8) that will be the subcase $b=0$ of (5.3.5)–(5.3.6). A generalization of (5.3.8), where Y is expressed through higher Legendre polynomials and associated functions, is also known, but it is no longer a vacuum solution (Vaidya, Patel and Bhatt 1976). Instead, the source becomes the null radiation. The solution is:

$$ds^2 = 2\,dt[du + (Y_{,z}/c)\sin^2\alpha\,d\beta] - (X^2 + Y^2)(d\alpha^2 + \sin^2\alpha\,d\beta^2)$$

$$- \{1 - 2X[(1-c)t - m]/(X^2 + Y^2)\}[du + (Y_{,z}/c)\sin^2\alpha\,d\beta]^2, \qquad (5.3.12)$$

where:

$$z = \cos\alpha, \quad X = cu - t, \qquad (5.3.13)$$

c and m are constants, and Y obeys:

$$(1 - z^2)Y_{,zz} - 2zY_{,z} + 2cY = 0. \qquad (5.3.14)$$

The energy-momentum tensor generated by the metric (5.3.12) is:

$$T_{\alpha\beta} = -[2c(c-1)/(X^2 + Y^2)]k_\alpha k_\beta, \qquad (5.3.15)$$

where $k_\alpha = \delta^1{}_\alpha$, $x^1 = u$, $k^\alpha k_\alpha = 0$. With $c = n(n+1)/2$ and n a natural number, eq. (5.3.14) becomes the general Legendre equation. With $n=1$, the energy-momentum tensor (5.3.15) vanishes, and the combined Kerr/Demiański solution (5.3.8) with (5.3.11) is obtained. Equations (5.3.12)–(5.3.15) thus constitute a radiating generalization of the Kerr and the special Demiański solutions.

5.4 The FLRW metrics in null coordinates

Take the metric of the closed FLRW model in the form:

$$ds^2 = dt^2 - [R^2(t)/R_0^2][dr^2 + R_0^2\sin^2(r/R_0)(d\psi^2 + \sin^2\psi\,d\beta^2)], \qquad (5.4.1)$$

where R_0 is a constant. This is equivalent to (1.3.8) with $k = 1/R_0^2$. (The extensions to the cases $k<0$ and $k=0$ can be obtained by the reparametrizations $(R_0, R) = i(\tilde{R}_0, \tilde{R})$ and $\{R = R_0 S(t), R_0 \to \infty\}$ respectively.) Now transform:

$$dx = (R/R_0)dt, \quad du = (R_0^2/R^2)dx - dr. \qquad (5.4.2)$$

The metric (5.4.1) then becomes:

$$ds^2 = 2\,du\,dx - (R^2/R_0^2)du^2 - R^2\sin^2(r/R_0)(d\psi^2 + \sin^2\psi\,d\beta^2). \qquad (5.4.3)$$

In this form, the FLRW metric is most convenient for considering the superpositions. It was introduced by Patel and Koppar (1987a) and used in several papers of this family.

A still different form was introduced by Vaidya (1977). Transforming (5.4.1) to the quasi-Cartesian coordinates $x = R_0\sin(r/R_0)\sin\psi\cos\beta$, etc. for y and z, and going over to the conformal time $\tau = \int R^{-1}dt$, we obtain:

$$ds^2 = R^2(\tau)\{d\tau^2 - dx^2 - dy^2 - dz^2$$

$$- (xdx + ydy + zdz)^2/[R_0{}^2 - (x^2 + y^2 + z^2)]\}. \tag{5.4.4}$$

We introduce now a *different r* and *different α and β* by:

$$x = R_0 \sin(r/R_0) \sin \alpha \cos \beta - a \cos(r/R_0) \sin \alpha \sin \beta,$$

$$y = R_0 \sin(r/R_0) \sin \alpha \sin \beta + a \cos(r/R_0) \sin \alpha \cos \beta, \tag{5.4.5}$$

$$z = R_0 \sin(r/R_0) \cos \alpha,$$

and the null coordinate $u = \tau - r$. After such a transformation, the metric (5.4.4) changes to:

$$ds^2 = R^2(\tau)\Big\{2d\tau(du + a \sin^2\alpha \, d\beta)$$

$$- M^2[(1 - a^2\sin^2\alpha/R_0{}^2)^{-1}d\alpha^2 + \sin^2\alpha \, d\beta^2] - (du + a \sin^2\alpha \, d\beta)^2\Big\}, \tag{5.4.6}$$

where:

$$M^2 = (R_0{}^2 - a^2) \sin^2(r/R_0) + a^2\cos^2\alpha. \tag{5.4.7}$$

This form is convenient for considering some other superpositions.

5.5 Superpositions of the special Demiański (1972) solution with the FLRW backgrounds

The foregoing sections were just brief introductions to various auxiliary subjects. In this section, we shall consider the two most sophisticated composites of the FLRW models with the solutions of Section 5.3. In Figure 5.1, the comments in square brackets identify the vacuum solutions superposed on the FLRW background.

Patel and Koppar (1987a) found a superposition of the special Demiański (1972) solution with the FLRW models. The source includes null radiation and electromagnetic field. The metric is:

$$ds^2 = 2dx(du + Y_{,z}\sin^2\alpha \, d\beta) - (R^2/R_0{}^2)(X^2 + Y^2)[d\alpha^2/(1 + Y^2/R_0{}^2) + \sin^2\alpha \, d\beta^2]$$

$$- 2L(du + Y_{,z}\sin^2\alpha \, d\beta)^2, \tag{5.5.1}$$

where R_0, m, c and e are arbitrary constants, and:

$$dx = (R/R_0)dt, \quad z = \cos \alpha, \quad dr = (R_0{}^2/R^2)dx - du,$$

$$X = R_0\sin(r/R_0), \quad Y = c\{1 + \cos \alpha \ln [\tan(\alpha/2)]\}, \tag{5.5.2}$$

$$2L = (R^2/R_0{}^2) + [2mXX_{,r} - e^2(1 - 2X^2/R_0{}^2)]/(X^2 + Y^2). \tag{5.5.3}$$

The expressions for the source quantities are rather complicated, but they will be given for reference. The pressure, matter-density and radiation density are:

$$\kappa p = 2L(R_0^2/R^4 - R_{,xx}/R) - (R^2/R_0^2)(R_{,xx}/R - 3R_{,x}^2/R^2 - 2R_0^2/R^4), \qquad (5.5.4)$$

$$\kappa(\epsilon + p) = -4(R_0^2/R^4)(L - R^2/R_0^2) - 2RR_{,xx}/R_0^2$$
$$- 2Y^2/[R_0^2(X^2 + Y^2)] + (R_{,x}^2/R^2)(4L - 2R^2/R_0^2)$$
$$+ 2(R_{,x}/R^3)(X^2 + Y^2)^{-1}[mR_0^2(1 - 2X^2/R_0^2) + 4e^2 XX_{,r}], \qquad (5.5.5)$$

$$2\kappa\tau = -\kappa(\epsilon + p)/2 + 4\,[\kappa(\epsilon + p)]^{-1}(R_{,xx}/R - R_0^2/R^4)$$
$$\times \left\{ -\frac{R_{,x}}{R(X^2 + Y^2)}[m(1 - 2X^2/R_0^2) + 4e^2 XX_{,r}/R_0^2] \right.$$
$$\left. + 2L^2 R_{,xx}/R - 2(R_0^2/R^4)(L - R_0^2/R^2) \right\}, \qquad (5.5.6)$$

(Patel and Koppar included the cosmological constant, but it does not show up in the metric and can be set to zero by redefining p and ϵ.) The electromagnetic four-potential is:

$$A_\mu = A\{1, 0, c\,[\sin^2\alpha \ln(\tan(\alpha/2)) - \cos\alpha], 0\}, \qquad (5.5.7)$$

where:

$$A = eXX_{,r}/(X^2 + Y^2). \qquad (5.5.8)$$

For defining the velocity field of the fluid and the wave vector of the radiation, the following tetrad of forms is introduced:

$$\Theta^1 = du + c\,\{\sin^2\alpha \ln[\tan(\alpha/2)] - \cos\alpha\}\,d\beta,$$
$$\Theta^2 = (g_{\alpha\alpha})^{1/2}d\alpha, \qquad \Theta^3 = (g_{\beta\beta})^{1/2}d\beta, \qquad (5.5.9)$$
$$\Theta^4 = dx - L\Theta^1.$$

The tetrad components of the velocity field and of the radiation wave vector are now:

$$u_a = \{(2\lambda)^{-1}, 0, 0, \lambda\}, \quad k_a = \{\lambda^{-1}, 0, 0, 0\}, \qquad (5.5.10)$$

where:

$$\lambda^2 \stackrel{\text{def}}{=} 2[\kappa(\epsilon + p)]^{-1}(R_0^2/R^4 - R_{,xx}/R). \qquad (5.5.11)$$

Patel and Koppar did not identify the Petrov type of the solution, but it must be at least II because the Demiański (1972) solution is of type II.

With $m = e = c = 0$, the solution reduces to the FLRW metric (5.4.3). With $R = R_0$ and $R_0 \to \infty$, it becomes an electrovacuum generalization of the special Demiański (1972) metric, and in the further limit $e = 0$ it reproduces the special Demiański solution. With $c=0$, (5.5.1)–(5.5.11) represents a superposition of the Reissner–Nordström solution with a FLRW background, and with $R=R_0 \to \infty$ in addition it becomes the Reissner–Nordström metric. It follows that with $c=e=0$, (5.5.1)–(5.5.11) represents a superposition of the Schwarzschild and FLRW solutions. However, the superposition is different from the McVittie (1933) solution because its energy-momentum tensor contains a contribution from null radiation. In the limit of zero radiation, the solution discussed here becomes either pure FLRW or vacuum.

In the limit $R=R_0$, the solution (5.5.1)–(5.5.11) becomes stationary. Its cosmological background then reduces to the Einstein static Universe, which emerges in the further limit $e=c=m=0$.

The same authors (Patel and Koppar 1988) superposed another structure on the FLRW background: the higher Legendre polynomials and functions mentioned in Section 5.3 that generalize the Kerr and Demiański solutions. The FLRW background is this time flat, and the metric is:

$$ds^2 = R^2(t) \left\{ 2dt[du + (Y_{,z}/c) \sin^2\alpha \, d\beta] \right.$$
$$\left. + 2L[du + (Y_{,z}/c) \sin^2\alpha \, d\beta]^2 - (X^2 + Y^2)(d\alpha^2 + \sin^2\alpha \, d\beta^2) \right\}, \qquad (5.5.12)$$

where:

$$z = \cos\alpha, \quad X = cu - t, \quad 2c = n(n+1),$$

$$Y = KP_n(\cos\alpha) + NQ_n(\cos\alpha), \qquad (5.5.13)$$

$$2L = c + R^{-2}(t)\{1 - c + 2X[(1-c)t - m]/(X^2 + Y^2)\},$$

n is a natural number, K and N are arbitrary constants, P_n are the Legendre polynomials, Q_n are the associated Legendre functions, and $R(t)$ is an arbitrary function. The expressions for the source quantities are again rather complicated:

$$\kappa p = - R^{-2}[c(2R_{,tt}/R - R_{,t}^2/R^2) + R^{-2}(2L_0 - c)(R_{,tt}/R - 2R_{,t}^2/R^2)], \qquad (5.5.14)$$

$$\kappa(\epsilon + p) = -2R^{-2}[(c-1)/(X^2 + Y^2) + c(R_{,tt}/R - 2R_{,t}^2/R^2)]$$
$$- 2R^{-4}\{(1-c)/(X^2 + Y^2) - (2L_0 - c)R_{,t}^2/R^2$$
$$- 2(R_{,t}/R)[(c-1)t + m]/(X^2 + Y^2)\}, \qquad (5.5.15)$$

$$2\kappa\tau = - \kappa(\epsilon + p)/2 + 2[\kappa(\epsilon + p)]^{-1}\left(\frac{R_{,tt}}{R} - 2\frac{R_{,t}^2}{R^2}\right)$$
$$\times \left\{ \frac{4c}{R^6(X^2 + Y^2)}\{c - 1 - [c(c-1)u + m]R_{,t}/R\} + R^{-4}[c + (2L_0 - c)/R^2]^2 \right\}, \qquad (5.5.16)$$

where:

$$2L_0 = 2c - 1 - 2\{(c-1)Y^2 + [c(c-1)u + m]X\}/(X^2 + Y^2). \qquad (5.5.17)$$

In the tetrad constructed analogously to (5.5.9), the expressions for the velocity field of the fluid and for the wave vector of the radiation are the same as (5.5.10), but this time:

$$\lambda^2 = -2[\kappa(\epsilon+p)]^{-1}R^{-2}(R_{,tt}/R - 2R_{,t}^2/R^2). \tag{5.5.18}$$

This must again be of Petrov type II or more general. In the case $n=1$, $K=0$, the solution (5.5.12)–(5.5.18) coincides with the limit $e=0$, $R_0\to\infty$ of (5.5.1)–(5.5.11), that is, of the Patel–Koppar (1987a) solution (it can be calculated after the reparametrization $R=R_0S(t)$); this common subcase is the special Demiański solution superposed on the flat FLRW background. In the case $n=1$, $N=0$, (5.5.12)–(5.5.18) becomes the Kerr solution superposed on the flat FLRW background, and this subcase will show up in the next section. With $n=1$, $K=N=0=m$, the flat FLRW model results.

In the case $R=1$, the solution (5.5.12)–(5.5.18) reduces to the Vaidya–Patel–Bhatt (1976) solution given by (5.3.12)–(5.3.15). When $n=1$ in addition, it reduces to the subcase $b=0$ of the Demiański solution (5.3.5)–(5.3.6).

This completes the overview of superpositions involving the Demiański (1972) solution. The other solutions in this chapter are superpositions of Petrov type D vacuum solutions with the FLRW backgrounds.

5.6 Superpositions of the Kerr solution with the FLRW models and other conformally nonflat composites

The most elaborate solution in this class was found by Koppar and Patel (1988). It is a superposition of the Kerr–Newman solution (5.3.1)–(5.3.2) with an arbitrary FLRW background. It was obtained by gradually building up the metric tensor through the following steps:

1. Take the background metric $g^*_{\mu\nu}$, in this case the Einstein Universe in null coordinates.
2. Construct a Kerr–Schild metric:

$$\hat{g}_{\mu\nu} = g^*_{\mu\nu} + 2Hl_\mu l_\nu, \tag{5.6.1}$$

where H is a scalar to be determined from the field equations and l_μ is a null geodesic shearfree vector field.
3. Take a metric conformal to $\hat{g}_{\mu\nu}$:

$$g_{\mu\nu} = R^2(t)\hat{g}_{\mu\nu}, \tag{5.6.2}$$

where $R(t)$ is the FLRW scale factor.
4. Assume the electromagnetic potential in the form:

$$A_\mu = \phi l_\mu, \tag{5.6.3}$$

where ϕ is another scalar.

5. Insert everything into the Einstein–Maxwell equations with the source being a mixture of perfect fluid, electromagnetic field and null radiation. The vector field l_μ, in addition to the role it plays in (5.6.1) and (5.6.3), is also the wave vector of the null radiation.

The construction shows clearly that the solutions arise by interpreting the source in a preassumed metric, not by integrating the equations for the metric with a pre-assumed source. Still, the sophistication of the result is remarkable. The metric is:

$$ds^2 = R^2(t)\left[dt^2 - dr^2 + 2a\,\sin^2\alpha\,d\beta\,dr - (|\Sigma|^2/N^2)d\alpha^2\right.$$

$$\left. - (|\Sigma|^2 + a^2\sin^2\alpha)\sin^2\alpha\,d\beta^2\right] - H_0(-dt - dr + a\sin^2\alpha\,d\beta)^2, \quad (5.6.4)$$

where $R(t)$ is an arbitrary function, a and R_0 are arbitrary constants and:

$$\Sigma \stackrel{\text{def}}{=} (R_0^2 - a^2)^{1/2}\sin(r/R_0) + ia\cos\alpha, \quad (5.6.5)$$

$$N^2 \stackrel{\text{def}}{=} 1 - (a^2/R_0^2)\sin^2\alpha, \quad (5.6.6)$$

$$H_0 \stackrel{\text{def}}{=} -\frac{2m(R_0^2 - a^2)}{R_0|\Sigma|^2}\sin(r/R_0)\cos(r/R_0) + e^2/R_0^2$$

$$-\frac{e^2}{|\Sigma|^4}\left[a^2 N^2\cos^2\alpha + \frac{(R_0^2 - a^2)^2}{R_0^2}\sin^2(r/R_0)\cos^2(r/R_0)\right], \quad (5.6.7)$$

m and e being other arbitrary constants. The source quantities are:

$$-\kappa p = (2RR_{,tt} + R_{,t}^2 + 1/R_0^2)/R^2 + (H_0/R^4)(RR_{,tt} - R_{,t}^2 - 1/R_0^2), \quad (5.6.8)$$

$$-\kappa\epsilon = -3(R_{,t}^2 + 1/R_0^2)/R^2 + (H_0/R^4)(RR_{,tt} - R_{,t}^2 + 3/R_0^2)$$

$$-(2\sqrt{2}/R^4)R_{,t}(H_0 l^\mu)_{|\mu}, \quad (5.6.9)$$

$$-\kappa\tau R^2 = 4H_0/R_0^2 + 2\sqrt{2}\,R_{,t}(H_0 n^\mu)_{|\mu}$$

$$+2A[-2H_0/R_0^2 + H_0 RR_{,tt} + \sqrt{2}R_{,t}(H_0 l^\mu)_{|\mu}], \quad (5.6.10)$$

where:

$$A \stackrel{\text{def}}{=} 1 + \tfrac{1}{2}(H_0/R^4)[-2H_0/R_0^2 + H_0 RR_{,tt}$$

$$+\sqrt{2}R_{,t}(H_0 l^\mu)_{|\mu}]/(RR_{,tt} - R_{,t}^2 - 1/R_0^2), \quad (5.6.11)$$

$$(H_0 l^\mu)_{|\mu} = -(H_0 n^\mu)_{|\mu} \stackrel{\text{def}}{=} \tfrac{1}{2}m(Z\bar{Z} - 2/R_0^2) + e^2(Z + \bar{Z})/R_0^2, \quad (5.6.12)$$

$$Z \stackrel{\text{def}}{=} -(\sqrt{2}/|\Sigma|^2)\left\{[(R_0^2 - a^2)/R_0^2]\sin(r/R_0)\cos(r/R_0) - iaN\cos\alpha\right\}. \quad (5.6.13)$$

The electromagnetic tensor is:

$$F_{\mu\nu} = \Phi(l_{\mu,\nu} - l_{\nu,\mu}) + \Phi_{,\nu}l_\mu - \Phi_{,\mu}l_\nu, \quad (5.6.14)$$

where l_μ is the wave vector of the null radiation:

$$l^\mu = (\sqrt{2}R^2)^{-1}(1, -1, 0, 0), \quad (x^0, x^1, x^2, x^3) = (t, r, \alpha, \beta), \tag{5.6.15}$$

and Φ is defined by:

$$\Phi = e(Z + \bar{Z})/(2\sqrt{4\pi}). \tag{5.6.16}$$

The velocity field of the fluid is:

$$u_\mu = R(\alpha^* U_\mu + \sqrt{2}\beta^* l_\mu), \tag{5.6.17}$$

and the symbols in it are defined by the following equations:

$$U^\mu = R^{-2}\delta_0^\mu, \tag{5.6.18}$$

$$\alpha^{*2} = (1 + H)/[(1 + H)^2 - (2/X)(N^* - H_0 R^{-4})^{1/2}, \tag{5.6.19}$$

$$\alpha^{*2}(1 + H) + 2\alpha^* \beta^* = 1, \tag{5.6.20}$$

$$H = H_0/R^2, \quad X = (2/R^2)(RR_{,tt} - R_{,t}^2 - 1/R_0^2), \tag{5.6.21}$$

$$N^* = -[-N_0 + 2\sqrt{2}(H_0 n^\mu)_{|\mu}R_{,t} + 4H_0/R_0^2]/R^4, \tag{5.6.22}$$

$$N_0 = -H_{0|\mu\nu}g^{\mu\nu} - 2\Phi_{0\mu}n^\mu, \tag{5.6.23}$$

$$\Phi_{0\mu} = (H_0 l^\nu)_{|\nu\mu} - (H_0 l_{\mu|\rho\sigma} + 2H_{0,\sigma}l_{\mu|\rho})g^{\rho\sigma} - H_0 l^\sigma R_{\sigma\mu}, \tag{5.6.24}$$

the vertical stroke denoting the covariant derivative, and $R_{\sigma\mu}$ being the Ricci tensor of the metric (5.6.4).

The solution given by (5.6.4)–(5.6.24) is a superposition of the Kerr–Newman metric (5.3.1)–(5.3.2) (which results from the above in the limit $R = 1$, $R_0 \to \infty$) and the FLRW models (which result when $m = a = e = 0$). Since it contains the Kerr–Newman solution, its Petrov type must be at least D; Koppar and Patel gave no information about it.

As given, the FLRW background is closed (of positive curvature), but the open and the flat backgrounds can be obtained by the reparametrization $R_0 = iR_0$ and in the limit $R_0 \to \infty$ respectively.

The metric (5.6.4)–(5.6.24) is identical to that found by Patel and Trivedi (1982). However, Patel and Trivedi interpreted the source as an anisotropic fluid in electromagnetic field, while Koppar and Patel (1988) showed that it can be interpreted as a perfect fluid in electromagnetic field mixed with null radiation.

The limit $a = 0$ of the Koppar–Patel (1988) solution is the Reissner–Nordström solution superposed on a FLRW background; this is a common subcase with the Patel–Koppar (1987a) solution (resulting from the latter when $c = 0$). The limit $e = 0$ of (5.6.4)–(5.6.24) is the Kerr solution superposed on a FLRW background. This case was first found by Vaidya (1977), and that paper inspired all the activity that followed. Vaidya interpreted the source as a fluid with anisotropic pressure, but it follows from the present discussion that it can be interpreted as a mixture of a perfect fluid and null radiation. The subcase $a = 0$ of the Vaidya (1977) solution is a superposition of the Schwarzschild and FLRW solutions, but it still includes null radiation in the source. That subcase is simple enough to allow the limit $\tau=0$ to be calculated. It turns out that the limit implies either $m=0$ (the FLRW models) or $R,_t=0$ (a static solution, being a superposition of the Schwarzschild metric and the Einstein Universe). Hence, the limit $a=0$ of the Vaidya (1977) solution does not allow for a nontrivial (i.e. nonstationary and inhomogeneous) perfect fluid source, and so it cannot coincide with the McVittie (1933) solution. This limit is, at the same time, the limit $c=e=0$ of the Patel–Koppar (1987a) solution. The limit $R_0 \to \infty$ of the Vaidya (1977) solution is the superposition of the Kerr solution and the flat FLRW model, this one is contained in the Patel–Koppar (1988) solution (5.5.12)–(5.5.13) as the limit $n=1$ (i.e. $c=1$), $N=0$ and $K=a$.

Vaidya and Patel (1985) found another generalization of the Vaidya (1977) solution; its source is a mixture of perfect fluid and radiation. The metric is:

$$ds^2 = 2dx(du + g \sin \alpha\, d\beta) - M^2(d\alpha^2 + \sin^2\alpha\, d\beta^2)$$

$$- 2L(du + g \sin \alpha\, d\beta)^2, \tag{5.6.25}$$

where $g(\alpha)$ is an arbitrary function,

$$M^2 = (R/R_0)^2 F^2(y)[X^2(r) + Y^2(y)], \tag{5.6.26}$$

the variables r and $y(\alpha)$ are defined by:

$$dr = (R/R_0)^2 dx - du, \quad \frac{dy}{d\alpha} = g, \tag{5.6.27}$$

$R(x)$ is an arbitrary function, R_0 is an arbitrary constant and the functions $X(r)$, $Y(y)$, $F(y)$ and L are defined as follows:

$$X,_r^2 = C - X^2/R_0^2, \quad Y,_y^2 = C + Y^2/R_0^2, \tag{5.6.28}$$

$$F^2 = -(g,_\alpha + g \cot \alpha)/(2YY,_y),$$

$$2L = (R/R_0)^2 - S,_y/D,_y + (N+S)/D, \tag{5.6.29}$$

$$S = (l + Ay)YY,_y, \quad D = X^2 + Y^2, \quad N = (m + ACr)XX,_r, \tag{5.6.30}$$

l, m, A and C being other arbitrary constants. The formulae for the physical quantities are not given in the paper, and Vaidya and Patel do not discuss physical properties of the solutions. Because of the scant information it is rather difficult to compare this solution to the others, and we shall not attempt it beyond what Vaidya and Patel stated themselves: that in the limit:

$$A = l = 0, \quad g = a \sin \alpha/[\sin^2\alpha + (a/R_0 + \cos \alpha)^2], \quad a = \text{const}, \tag{5.6.31}$$

the solution (5.6.25)–(5.6.30) reproduces the Vaidya (1977) solution, and the following transformation is then required to make the metric identical to Vaidya's:

$$\alpha = \vartheta + \text{arc} \sin[(a/R_0) \sin \vartheta], \tag{5.6.32}$$

where ϑ is Vaidya's (1977) coordinate. However, this author (A.K.) was unable to obtain the same result: the limit (5.6.31) of (5.6.25)–(5.6.30) does not agree with the transform (5.6.32) of the Vaidya (1977) solution. I assume that a nonobvious misprint is involved somewhere.

Patel, Koppar and Yadav (1989) found a solution that generalizes solutions of Bianchi types II, VIII and IX simultaneously. Its source is a mixture of perfect fluid and null radiation, permeated by electromagnetic field. The solution is:

$$ds^2 = 2dx(du - 4m^2 d\beta) - B^2 K^2 (d\alpha^2 + \sin^2\alpha \, d\beta^2)$$
$$- \{A^2 + Fe^{-2\lambda_1 y} - 2\pi (q/\lambda_1)^2 e^{-4\lambda_1 y}\}(du - 4m^2 d\beta)^2, \qquad (5.6.33)$$

where λ_1 and q are constants, the functions $m(\alpha)$ and $K(\alpha)$ are defined by:

$$4mm,_\alpha /(K^2 \sin \alpha) = \lambda_1,$$
$$\qquad\qquad\qquad\qquad\qquad\qquad\qquad (5.6.34)$$
$$K,_{\alpha\alpha} - K,_\alpha^2 /K + \cot \alpha \, K,_\alpha - K = \lambda K^3,$$

λ is another constant, the variable y is defined by:

$$y,_\alpha = 4m^2 /\sin \alpha, \qquad (5.6.35)$$

the functions $F(r)$ and $E(r)$ are defined by:

$$F,_{rr} + 4\lambda_1^2 F = 0, \quad E,_r^2 + 4\lambda_1^2 E^2 = q^2, \qquad (5.6.36)$$

and the variable r is:

$$r = u - \int B^{-2} dx. \qquad (5.6.37)$$

Equation (5.6.35) is reproduced here as it is given in the paper; however, the limiting transitions to the other solutions consistently indicate that the right-hand side of (5.6.35) should be preceded by a minus. The function $E(r)$ defines the electromagnetic four-potential A_μ through:

$$A_\mu = \Phi(1, 0, -4m^2, 0), \quad \Phi = E(r)e^{-2\lambda_1 y}. \qquad (5.6.38)$$

The electromagnetic field defined by this four-potential is sourceless ($j^\alpha = 0$). It will vanish when $E = 0$. The background Bianchi metrics are:

$$ds^2 = dt^2 - A^2 (d\psi + 4m^2 d\beta)^2 - B^2 K^2 (d\alpha^2 + \sin^2\alpha \, d\beta^2), \qquad (5.6.39)$$

where the coordinates t and ψ are defined by:

$$dt = dx/A, \quad d\psi = dx/A^2 - du. \qquad (5.6.40)$$

In the Bianchi limit, the functions A and B must obey:

$$B,_{xx}/B + (B,_x /B)^2 - 4\lambda_1^2 /B^4 - \lambda/(AB)^2 - A,_{xx}/A - (A,_x /A)^2 = 0. \qquad (5.6.41)$$

The Bianchi type is II when $\lambda = 0$ and $K = 1/\sin \alpha$, VIII when $\lambda = 1$ and $K = \tan \alpha$, and IX when $\lambda = -1$, $K = 1$.

The pressure, matter-density and radiation density are given by:

$$\kappa p = - A^2[A,_{xx}/A + A,_x^2/A^2 + 2A,_x B,_x/(AB) + B,_{xx}/B + \lambda_1^2/B^4]$$
$$+ e^{-2\lambda_1 y}(B,_{xx}/B - \lambda_1^2/B^4)(2\pi e^{-2\lambda_1 y}q^2/\lambda_1^2 - F), \qquad (5.6.42)$$

$$\kappa(\epsilon + p) = -2A^2[A,_{xx}/A + A,_x^2/A^2 + 3\lambda_1^2/B^4 + \lambda/(AB)^2 - B,_x^2/B^2]$$
$$- 4\pi q^2 e^{-4\lambda_1 y}[1/B^4 + B,_x^2/(\lambda_1 B)^2]$$
$$- 2B^{-2}e^{-2\lambda_1 y}(F,_r B,_x/B + F\lambda_1^2/B^2 - FB,_x^2), \qquad (5.6.43)$$

$$\kappa(\epsilon + p)[\kappa\tau + \kappa(\epsilon + p)/4]/(B,_{xx}/B - \lambda_1^2/B^4)$$
$$= (B,_{xx}/B - \lambda_1^2/B^4)[A^2 + Fe^{-2\lambda_1 y} - 2\pi(q/\lambda_1)^2 e^{-4\lambda_1 y}]^2$$
$$- 16\pi(q/B)^2 e^{-4\lambda_1 y}[1 + 16m^4/(K\sin\alpha)^2]$$
$$+ 2e^{-2\lambda_1 y}(2F\lambda_1^2/B^2 + F,_r B,_x/B). \qquad (5.6.44)$$

The components of the fluid velocity and of the wave vector of the null radiation are given with respect to the following tetrad:

$$\Theta^1 = du - 4m^2 d\beta, \quad \Theta^2 = BK\, d\alpha,$$
$$\Theta^3 = BK\sin\alpha\, d\beta, \quad \Theta^4 = dx - L\Theta^1. \qquad (5.6.45)$$

They are, respectively:

$$u_{(a)} = [1/(2z), 0, 0, z], \quad k_{(a)} = [1/2, 0, 0, 0], \qquad (5.6.46)$$

where z is defined by:

$$-\kappa(\epsilon + p)z^2/2 = B,_{xx}/B - \lambda_1^2/B^4. \qquad (5.6.47)$$

The fluid has rotation ω given by:

$$(\omega BK\sin\alpha)^2/(16m^4) = (Yz,_u - zY,_u)^2 + (zY,_y - Yz,_y)^2$$
$$+ (Y\lambda_1 K\sin\alpha)^2 z(Lz - 1)/(2B^2 m^4), \qquad (5.6.48)$$

where:

$$Y = 1/(2z) - zL. \qquad (5.6.49)$$

Patel *et al.* did not investigate any properties of this solution. Because of the rotation, it is not a subcase of any other in this family.

Patel and Yadav (1987) considered the Einstein–Maxwell equations for a broad class of metrics, but presented only five solutions explicitly. Their case (i) is a coordinate transform of the Patel–Trivedi (1982) solution. Case (ii) has no perfect fluid component in the source, so it cannot have a FLRW limit. Case (iii) is the limit of vanishing electric (but not magnetic) field of case (iv), and case (v) is an electrovacuum solution. Thus only case (iv) needs be considered here. It will be the subcase of the Patel–Koppar–Yadav (1989) solution (eqs. (5.6.33)–(5.6.49)) defined by:

$$4m^2 = a \cos \alpha, \quad K = a/\sqrt{2}, \quad \lambda = -2/a^2, \quad \lambda_1 = -1/a, \tag{5.6.50}$$

$$B = R/(2K), \quad A^2 = (R/R_0)^2 - A_0/2, \tag{5.6.51}$$

but only if the sign of the right-hand side of (5.6.35) is changed to minus. The limiting transition must be accompanied by the following coordinate transformation performed on Patel and Yadav's case (iv):

$$2 - e^{-2y/R_0} = 2 \sin^2 \alpha,$$
$$\alpha = \pi/2 + \alpha'/2, \quad u = u' - \beta, \quad \beta = R_0 \beta'/2, \tag{5.6.52}$$

the primed variables will then coincide with those of (5.6.33)–(5.6.49). Case (iii) follows when $q = 0$ in addition.

Patel, Koppar and Pandya (1990) published another "electromagnetic generalization" of the Patel–Yadav (1987) solution (iii), but the metric in it is the same as in the limit $q = 0$ of the Patel–Koppar–Yadav (1989) solution. This means that the electromagnetic field in the Patel–Koppar–Pandya solution was introduced simply by reinterpreting the energy-momentum tensor. We shall not count this as a separate solution.

With $F = 0$, Patel and Yadav's (1987) solution (iii) reduces to a Bianchi IX model. It becomes a closed FLRW model when, further, $A_0 = 0$. The flat FLRW model will follow after the reparametrization $R = R_0 \tilde{R}$ as the limit $R_0 = 0$.

Patel and Yadava (1987) published another subcase of the Patel–Koppar–Yadav (1989) solution. It results from (5.6.33)–(5.6.49) when (5.6.50) holds and $q = 0$ in addition. It will reproduce the Patel–Yadav (1987) case (iii) when, further, (5.6.51) holds. However, a misprint in the metric must be corrected, the correct formula is (in the original notation):

$$ds^2 = 2 \, dx \, (du - a \cos \alpha \, d\beta) - m^2(d\alpha^2 + \sin^2\alpha \, d\beta^2)$$

$$- [(l/a)^2 + 2F(r)/\sin^2(\alpha/2)](du - a \cos \alpha \, d\beta)^2. \tag{5.6.53}$$

The paper has $\sin^2\alpha$ in place of $\sin^2(\alpha/2)$; the sign in (5.6.35) has to be changed in this case, too.

Patel and Yadava claimed that when $l = m$, their model will reproduce the Vaidya (1977) solution, but this is not the case. The limit $l = m$ of the Patel–Yadava (1987) solution can reproduce only the limit $R_0 = a$ of the Vaidya (1977) solution. It can be shown that the FLRW limit of the Patel–Yadava (1987) solution is achieved when the radiation density is zero, but the calculation is lengthy.

Singh and Bhamra (1990) presented a solution that belongs to this family because its source includes null radiation. The metric is:

$$ds^2 = D^{-BD}(d + ct)^{BD} dt^2$$

$$- (ar)^{2b} A^{-AD}(d + ct)^{AD}[dr^2 + r^2(d\vartheta^2 + \sin^2\vartheta \, d\varphi^2)], \tag{5.6.54}$$

where a, b, c, d, A, B and D are arbitrary constants. The source in the Einstein equations is a mixture of a perfect fluid, a scalar field (whose gradient is not collinear with the velocity of the fluid), and the null radiation. We shall not present the other formulae because they are lengthy and not very illuminating. Strictly speaking, Singh and Bhamra presented two solutions, but they differ only in the scalar field; the metrics are the same. Because of the scalar field, the Singh–Bhamra solution is not a subcase of any other in this family. When the scalar field is constant and so gives no contribution to the energy-momentum, the null radiation will vanish, too. This shows that the Singh–Bhamra solution does not contain any other from this family. The solution is a pedagogic example because it has $\sigma = \omega = 0 = \dot{u}^{\alpha}$, but is not identical to FLRW because of the nonperfect fluid source. It is not conformally flat. The nonstatic limit of zero radiation (in which the scalar field disappears automatically) occurs when $b=0$ or $b=-2$. In both cases, (5.6.54) reduces to a flat FLRW model with a specific equation of state; the second case is the coordinate transform $r=1/r'$ of the first case.

5.7 Conformally flat solutions with null radiation

Conformally flat spacetimes with null radiation were considered by Vaidya and Shah (1960). They cannot really be called solutions because they are defined by differential equations rather than by explicit algebraic expressions. There are two classes of them. The second (marked II in Figure 5.1) is the more general one. In slightly modified notation, it is:

$$ds^2 = (r/\bar{r})^2(1+Pr+Qr^2)^{-1}dt^2 - (\bar{r}/r)^2(1+Pr+Qr^2)^{-1}dr^2$$

$$- \bar{r}^2(d\vartheta^2 + \sin^2\vartheta \, d\varphi^2), \tag{5.7.1}$$

where:

$$\bar{r} \stackrel{\text{def}}{=} r\left(AQr + \frac{dA}{dm}\right)^{-1} e^{\psi/2} Q e^{(1/2)\int P dm}, \tag{5.7.2}$$

$\psi(t)$ is an arbitrary function, and $P(m)$ and $Q(m)$ are arbitrary functions of the function $m(t,r)$ defined by:

$$\frac{\partial m}{\partial r} = (1+Pr+Qr^2)^{-1},$$

$$\frac{\partial m}{\partial t} = \left(AQr + \frac{dA}{dm}\right)^2 e^{-\psi}(1+Pr+Qr^2)^{-1}Q^{-2}e^{-\int P dm}, \tag{5.7.3}$$

and the function $A(m)$ is defined by:

$$A,_{mm} - (P+Q,_m/Q)A,_m + QA = 0. \tag{5.7.4}$$

The source is a mixture of perfect fluid and null radiation. The coordinates of (5.7.1) are comoving with the fluid, and the source quantities are:

$$\kappa p = p_1(-\psi,_{tt} - 5\psi,_t^2/4) - p_2 - p_3, \tag{5.7.5}$$

$$\kappa \epsilon = 3p_1\psi,_t^2/4 + p_2 + p_3, \tag{5.7.6}$$

$$\kappa T(k^1)^2 = (r/\bar{r})^4 \{(A,_m Q,_m - AQP,_m)r/(AQr + A,_m) + (Q,_m + P,_m/r)\bar{r}^2 \psi,_t/2\}, \quad (5.7.7)$$

where:

$$p_1 = (1 + Pr + Qr^2)(\bar{r}/r)^2, \qquad (5.7.8)$$

$$p_2 = 3\psi,_t [Q(rA,_m - A) - (P/2)(AQr - A,_m)], \qquad (5.7.9)$$

$$p_3 = 3e^{-\psi} Q^{-1} e^{-\int Pdm} (A,_m^2 + QA^2 - AA,_m P). \qquad (5.7.10)$$

Vaidya and Shah included the cosmological constant, but it can be (and was) absorbed into p and ϵ. The conformal flatness of this spacetime is not evident, and was not disclosed in the paper, but it was verified by the program Ortocartan (and was also proved earlier by R. B. Patel (1969) who worked in Vaidya's (1966) parametrization; see below). Note that the second equation of (5.7.3) means that $m,_t = g_{00}$, so necessarily $m,_t \neq 0$.

A necessary condition for the FLRW limit is $\dot{u}^\alpha = 0$, that is, $m,_t = g_{00} = f(t)$ which implies that $m,_{tr} = 0$. Substituting the first equation of (5.7.3) into this we obtain:

$$-(1 + Pr + Qr^2)^{-2}(rP,_m + r^2 Q,_m)m,_t = 0. \qquad (5.7.11)$$

But $m,_t \neq 0$; hence (5.7.11) implies:

$$rP,_m + r^2 Q,_m = 0. \qquad (5.7.12)$$

A consequence of $m,_t \neq 0$ is $\partial(m,r)/\partial(t,r) \neq 0$, so m and r are independent variables. Therefore (5.7.12) implies:

$$P,_m = Q,_m = 0. \qquad (5.7.13)$$

With (5.7.13), the radiation density vanishes, that is, the source becomes a perfect fluid and, moreover, $\sigma = 0$ as can be seen from (5.7.1). Hence, the metric becomes FLRW in this limit. This can be verified by a direct transformation; one has to observe that with P and Q constants, $m = \int(1 + Pr + Qr^2)dr + \int f(t)dt$. All the FLRW models are reproduced, and the sign of $(Q - P^2/4)$ is the sign of k.

The same class of spacetimes, in a slightly different parametrization, was derived and discussed again by Vaidya (1966).

The first class of the Vaidya and Shah (1960) spacetimes is the degenerate case $Q = 0$ of the second class. The case is degenerate because the limit $Q = 0$ is singular; it can be taken in the Einstein equations, but not in the formulae (5.7.1)–(5.7.10). The first class is given by:

$$ds^2 = (r/\bar{r})^2(1 + Pr)^{-1}dt^2 - (\bar{r}/r)^2(1 + Pr)^{-1}dr^2 - \bar{r}^2(d\vartheta^2 + \sin^2\vartheta\, d\varphi^2), \quad (5.7.14)$$

where:

$$\bar{r} \stackrel{\mathrm{def}}{=} r \mathrm{e}^{(\psi - \int P \mathrm{d}m)/2}, \qquad (5.7.15)$$

$\psi(t)$ and $P(m)$ being arbitrary functions, and $m(t,r)$ is defined by:

$$\frac{\partial m}{\partial r} = (1 + Pr)^{-1}, \quad \frac{\partial m}{\partial t} = -\mathrm{e}^{-\psi}(1 + Pr)^{-1}\mathrm{e}^{+\int P \mathrm{d}m}. \qquad (5.7.16)$$

The coordinates of (5.7.14) are again comoving with the fluid, and the formulae for the source quantities are obtained from (5.7.5)–(5.7.10) by the formal substitution $p_3 = 0$, $Q = 0$.

The radiation density will vanish when either $\psi = \mathrm{const}$ or $P = \mathrm{const}$. The first limit is static, the second one leads to the metric:

$$\mathrm{d}s^2 = \mathrm{e}^{-\psi}\phi^{-1}\mathrm{d}t^2 - \mathrm{e}^{\psi}\phi[(1 + Pr)^{-2}\mathrm{d}r^2 + r^2(1 + Pr)^{-1}(\mathrm{d}\vartheta^2 + \sin^2\vartheta \, \mathrm{d}\varphi^2)], \quad (5.7.17)$$

where $\phi(t) \stackrel{\mathrm{def}}{=} P\int \mathrm{e}^{-\psi}\mathrm{d}t$. This can be verified by explicit transformations to be the open (when $P \neq 0$) or flat (when $P = 0$) FLRW metric.

Vaidya and Shah presented some physical discussion. They showed that the metrics (5.7.1) and (5.7.14) can be matched to the FLRW metrics at two boundaries, $r = R_1(t)$ and $r = R_2(t) > R_1(t)$. The two boundaries move away from the centre at the speed of light, and the outer boundary is receding from the inner one. After being matched in this way, each metric thus represents a zone of radiation travelling through a FLRW background, the FLRW curvature behind the zone being smaller than that in front of it. The authors interpreted this configuration as a remnant of a radiating body whose mass was completely converted into radiation. The solutions show that the radiation tends to reduce the curvature of that part of the Universe through which it has passed.

Bayin (1979) found a collection of solutions whose sources are mixtures of perfect fluid and null radiation. His solutions II and III are special FLRW models with no radiation, and solutions IV, V and VI become static in the limit of zero radiation. Therefore, only solution I will be taken into account here. It is, in comoving coordinates:

$$\mathrm{d}s^2 = [C_1 \mathrm{e}^{zr}(zr - 1) + C_2]^{-2}\{A_0 g^2(t)\mathrm{d}t^2 - B_0 k^2(t)[\mathrm{d}r^2 + r^2(\mathrm{d}\vartheta^2 + \sin^2\vartheta \, \mathrm{d}\varphi^2)]\}, \quad (5.7.18)$$

where z, C_1, C_2, A_0 and B_0 are constants, and $g(t)$ and $k(t)$ are arbitrary functions. The solution is manifestly conformally flat. Since it has $\sigma = 0$, it cannot be a subcase of the Vaidya–Shah (1960) solutions: the limit $\sigma = 0$ of the latter are simply the FLRW models (we do not show this, but it is an easy exercise for the reader).

For completeness, the paper by Siemieniec–Oziębło (1983) must be mentioned here. The author assures the readers that the metric

$$\mathrm{d}s^2 = (at^2 + br^2)[\mathrm{d}t^2 - \mathrm{d}r^2 - r^2(\mathrm{d}\vartheta^2 + \sin^2\vartheta \, \mathrm{d}\varphi^2)], \qquad (5.7.19)$$

can be interpreted as a solution of the Einstein equations with a source being a mixture of perfect fluid and null radiation (the coordinates are not comoving). In fact, *every* spherically symmetric metric can be interpreted in this way, unless its special properties force the source to be simpler (e.g. pure perfect fluid, pure null radiation or vacuum). Hence, with no physical or geometrical interpretation of the various functions or constants, such a statement is not very revealing. The null radiation will vanish when $b=0$ or $b=-a$; the metric (5.7.19) then becomes the flat or open FLRW model respectively. This suggests that (5.7.19) may be a subcase of Vaidya and Shah's (1960) first solution, but the calculations required to verify this conjecture are prohibitively complicated, while the answer is not in urgent demand.

This closes the overview of solutions with null radiation.

6

Solutions with a "stiff fluid"/scalar field source

6.1 How to generate a "stiff fluid" solution from a vacuum solution

The subject of this chapter will be solutions of the Einstein equations that are A_2-symmetric and whose source is a "stiff fluid", that is, a perfect fluid with the equation of state:

$$\epsilon = p. \tag{6.1.1}$$

The A_2 symmetry implies that there exist two commuting spacelike Killing vector fields. Since they commute, coordinates may be adapted to them so that the corresponding generators are:

$$k^\alpha_{(1)} \frac{\partial}{\partial x^\alpha} = \frac{\partial}{\partial y}, \quad k^\alpha_{(2)} \frac{\partial}{\partial x^\alpha} = \frac{\partial}{\partial z}. \tag{6.1.2}$$

It is usually assumed in addition that the orbits of the group generated by (6.1.2) are non-null and that the spacetime is orthogonally transitive, that is, that the surfaces tangent to $k^\alpha_{(1)}$ and $k^\alpha_{(2)}$ admit orthogonal surfaces.[45] Then, the other two coordinates (t, x) can be chosen so that the metric form becomes:

$$ds^2 = g_{00}(dt^2 - dx^2) - g_{AB}dx^A dx^B, \tag{6.1.3}$$

where A, $B = 2, 3$, $(x^2, x^3) = (y, z)$ and all the components $g_{\alpha\beta}$ depend only on t and x.

Interest in vacuum solutions with the metric (6.1.3) stems from the fact that there exists an elaborate theory which tells how to generate new vacuum solutions of the form (6.1.3) from known solutions by algebraic operations. That theory is an important part of contemporary relativity, but it is not directly related to relativistic cosmology (as defined in this book), so we refer the readers to other sources for its description. The most recent broad overview of the technique and its applications was given by Verdaguer (1993).

The equation of state (6.1.1) was apparently first proposed by Zeldovich (1961). It should have applied in the early Universe, the justification being the observation that with (6.1.1) the velocity of sound equals the velocity of light, so no material in this Universe could be more stiff.[46] Such stiffness should be conceivable at the very

---

[45] There exist examples of solutions that do not obey this condition, but none of them admits a FLRW limit.

[46] As observed by A. Trautman (private communication), the "stiff fluid" is the modern version of the aether.

high densities occurring just after the Big Bang. However, the real reason for the popularity of (6.1.1) in the literature is the fact that A_2-symmetric solutions with such a source can be relatively simply generated from vacuum solutions with the same symmetry. As explained in our Section 1.2 (eqs. (1.2.19)–(1.2.22)), the energy-momentum tensor of a nonrotating "stiff fluid" is algebraically equivalent to a massless scalar field. The massless scalar field solutions can in turn be interpreted as vacuum solutions in the Brans–Dicke theory (see e.g. Wainwright, Ince and Marshman 1979 or Tabensky and Zamorano 1975).

The possibility of generating A_2-symmetric "stiff fluid" solutions from vacuum solutions was apparently first indicated by Wainwright, Ince and Marshman (1979), who described the algorithm of this procedure.[47] Suppose the metric:

$$ds^2 = e^{2k}(dt^2 - dx^2) - R[f(dy + w\, dz)^2 + (1/f)dz^2] \qquad (6.1.4)$$

is a vacuum solution, where k, R, f and w are functions of t and x. The vacuum Einstein equations then imply that:

$$R_{,tt} - R_{,xx} = 0. \qquad (6.1.5)$$

In addition, it will be assumed that:

$$R_{,t}^2 - R_{,x}^2 \neq 0, \qquad (6.1.6)$$

that is, that the gradient of R is not a null vector. The algorithm is then as follows:

1. Given $R(t,x)$ obeying (6.1.5), find any $\phi(t,x)$ obeying:

$$\phi_{,tt} + (R_{,t}/R)\phi_{,t} - \phi_{,xx} - (R_{,x}/R)\phi_{,x} = 0, \qquad (6.1.7)$$

   and the condition:

$$\phi_{,t}^2 - \phi_{,x}^2 > 0. \qquad (6.1.8)$$

2. Find any $\Omega(t,x)$ obeying:

$$\Omega_{,t}R_{,t} + \Omega_{,x}R_{,x} = R(\phi_{,t}^2 + \phi_{,x}^2), \quad \Omega_{,t}R_{,x} + \Omega_{,x}R_{,t} = 2R\phi_{,t}\phi_{,x}. \qquad (6.1.9)$$

   Equations (6.1.5) and (6.1.7) provide sufficient conditions for the integrability of (6.1.9).

3. The metric (6.1.4) with e^{2k} replaced by $e^{2k+\Omega}$, where Ω is a solution of (6.1.9), will be a solution of the Einstein equations with a stiff perfect fluid source, in which:

$$2\kappa\epsilon = 2\kappa p = e^{-2k-\Omega}(\phi_{,t}^2 - \phi_{,x}^2), \qquad (6.1.10)$$

---

[47] The notations used by different authors are diverse and partly in conflict. In this book, an attempt has been made to make the notations consistent. As a result, several of the original notations have been changed.

and the velocity field of the source is:

$$u_\alpha dx^\alpha = (\phi_{,t}^2 - \phi_{,x}^2)^{-1/2} e^{k+\Omega/2} d\phi. \tag{6.1.11}$$

Note: only the function e^{2k} changes in the procedure; the functions R, f and w are the same for vacuum and for the corresponding stiff fluid. The function ϕ is the massless scalar field in the other interpretation.

Stephani (1988) interpreted this algorithm in terms of the Lie–Bäcklund symmetries of the Einstein equations.

A simple argument based on the existence of the symmetry generated by (6.1.2) allows one to deduce that solutions from this family cannot be more general than class B^3, and cannot be as special as class D in Wainwright's (1979) extrinsic classification, that is, the acceleration and the spatial gradient of the expansion scalar must be collinear and must be shear eigenvectors, while shear cannot vanish (or else the solution degenerates to a FLRW model). Wainwright's intrinsic classes would have to be investigated for each solution separately, and the authors of the papers discussed here did not provide information on this point.

6.2 The allowed coordinate transformations

In comparing different solutions it is important to know how unique the coordinates of (6.1.3) are. The Killing vectors (6.1.2) are determined up to the linear transformations:

$$y = ay' + bz' + a_0, \quad z = cy' + dz' + c_0, \tag{6.2.1}$$

where a, b, c, d, a_0 and c_0 are constants obeying:

$$ad - bc \neq 0. \tag{6.2.2}$$

The individual metric components g_{22}, g_{23} and g_{33} are shuffled among themselves after the transformation (6.2.1), but the determinant of the metric in the (y,z)-surface will only change by the constant factor:

$$g_2' \overset{\text{def}}{=} \det[g_{A'B'}] = (ad - bc)^2 g_2. \tag{6.2.3}$$

Note that the function R obeying (6.1.5) equals $g_2^{1/2}$.

Since the metric components depend only on t and x, the two-dimensional metric of the (t,x)-surface can always be transformed to the form:

$$ds_2^2 = g_{00}(dt^2 - dx^2), \tag{6.2.4}$$

as in (6.1.3), but the coordinates of (6.1.3) are in general not comoving (see eq. (6.1.11)). The (t,x) coordinates are thus determined up to conformal symmetries of the two-dimensional metric (6.2.4). A sufficient condition for a transformation to be a conformal symmetry of (6.2.4) is easy to find. We perform the transformation:

$$t = \xi + \eta, \quad x = \xi - \eta, \tag{6.2.5}$$

on (6.2.4); the result is:

$$ds_2^2 = 4g_{00}(\xi,\eta)d\xi\,d\eta. \tag{6.2.6}$$

Now it is easy to see that any transformation of the form:

$$\xi = f(\xi'), \quad \eta = g(\eta'), \tag{6.2.7}$$

where f and g are arbitrary functions, will be a conformal symmetry. Therefore, any transformation:

$$t = f(t'+x') + g(t'-x'), \quad x = f(t'+x') - g(t'-x') \tag{6.2.8}$$

will preserve the forms (6.2.4) and (6.1.3). That (6.2.8) exhausts all the conformal symmetries of (6.2.4) can be verified by solving the conformal Killing equations for (6.2.4) and finding the transformations corresponding to the generators. They will be of the form (6.2.8). The transformations preserving the form (6.1.3) are composites of (6.2.1) and (6.2.8).

Note that t and x given by (6.2.8) are the general solutions of (6.1.5) in terms of (t',x'), and eq. (6.1.5) is invariant with respect to the transformations (6.2.8). Hence, at every point of the spacetime, coordinates can be chosen so that either $R=t$ or $R=x$, depending on whether the gradient of R is timelike or spacelike respectively. However, the assumption that $R=t$ or $R=x$ everywhere is a limitation. In fact, it may happen that a solution with $R=t$ and another one with $R=x$ will cover two disjoint regions of the same manifold.

Once coordinates have been chosen so that $R=t$ or $R=x$, while the metric has the form (6.1.4), the coordinate transformations that preserve these conditions are (6.2.1) composed with:

$$t = At' + A_0, \quad x = Ax' + B_0, \tag{6.2.9}$$

where $A \neq 0$, A_0 and B_0 are constants.

It may happen that the coordinates of (6.1.4) are comoving, that is, that the scalar field ϕ in these coordinates depends only on time. It is then easy to conclude that in any other coordinate system that is obtained from the comoving one by (6.2.8) (i.e. in any coordinates preserving the form (6.1.4)), the scalar field will obey the equation:

$$[\ln(\phi_{,\xi'}/\phi_{,\eta'})]_{,\xi'\eta'} = 0, \tag{6.2.10}$$

where $t' = \xi' + \eta'$, $x' = \xi' - \eta'$. Equation (6.2.10) is a necessary condition for the existence of coordinates that simultaneously are comoving and obey (6.1.4). It is useful in comparing different solutions.

6.3 The Einstein equations with a stiff fluid source in an A_2-symmetric spacetime

A large part of the literature on the subject has been aimed at disentangling the Einstein equations in such a way that a solution can in principle be found by successive integrations of simple (sets of) differential equations. The papers in which no explicit solutions were derived or discussed are not in fact within the scope of this book. However, they form a clear majority and simply cannot be ignored. The relationships among them are shown in Figure 6.1.

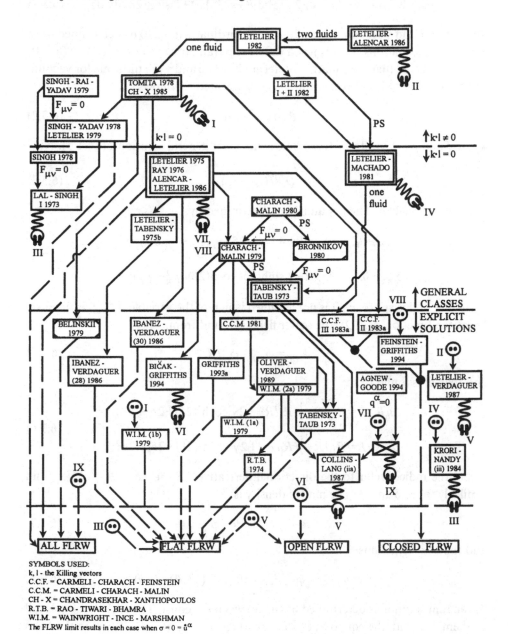

SYMBOLS USED:
k, l - the Killing vectors
C.C.F. = CARMELI - CHARACH - FEINSTEIN
C.C.M. = CARMELI - CHARACH - MALIN
CH - X = CHANDRASEKHAR - XANTHOPOULOS
R.T.B. = RAO - TIWARI - BHAMRA
W.I.M. = WAINWRIGHT - INCE - MARSHMAN
The FLRW limit results in each case when $\sigma = 0 = \dot{\hat{u}}^\alpha$

Figure 6.1. Solutions with a "stiff fluid" source.

The most general case of the Einstein equations for the metric (6.1.3) with the source (6.1.1), and with no further limitations on the Killing vectors, was considered by Tomita (1978). He considered the metric:

$$ds^2 = 2e^{2a}du\,dv - R[\cosh(2\delta)(e^{2\gamma}dy^2 + e^{-2\gamma}dz^2) + 2\sinh(2\delta)\,dy\,dz], \quad (6.3.1)$$

where a, R, δ and γ are functions of the null coordinates (u,v), to be determined from the Einstein equations. This is just a different parametrization of (6.1.3) or (6.1.4). Tomita observed that the equations determining R, δ and γ are the same for vacuum and for the stiff fluid source. Solutions are generated as follows:

$$R = f(v) + g(u), \quad (6.3.2)$$

where f and g are arbitrary functions, is the most general solution of (6.1.5). Now choose R and:

$$Q \stackrel{\text{def}}{=} f(v) - g(u) \quad (6.3.3)$$

as the new variables. Then, γ and δ are determined by:

$$\gamma,_{RR} - \gamma,_{QQ} + \gamma,_R/R + 4\tanh(2\delta)(\gamma,_R\delta,_R - \gamma,_Q\delta,_Q) = 0,$$
$$\delta,_{RR} - \delta,_{QQ} + \delta,_R/R - 2\sinh(2\delta)\cosh(2\delta)(\gamma,_R^2 - \gamma,_Q^2) = 0. \quad (6.3.4)$$

Note that in this approach it is not assumed which of the variables (R,Q) is timelike. Next, the function $\phi(R,Q)$ (the scalar field) is found from:

$$\phi,_{QQ} = \phi,_{RR} + \phi,_R/R, \quad (6.3.5)$$

and then e^a is found from:

$$2a_0 R_0/R = R_{00}/R - \tfrac{1}{2}(R_0/R)^2 + 2\gamma_0^2\cosh^2(2\delta) + 2\delta_0^2 + 2\phi_0^2,$$
$$2a_1 R_1/R = R_{11}/R - \tfrac{1}{2}(R_1/R)^2 + 2\gamma_1^2\cosh^2(2\delta) + 2\delta_1^2 + 2\phi_1^2, \quad (6.3.6)$$

where the indices refer to the directional derivatives, $R_0 \stackrel{\text{def}}{=} e^{-a}R,_u$, $R_1 \stackrel{\text{def}}{=} e^{-a}R,_v$, and similarly for γ, δ and ϕ. The matter-density is:

$$\kappa\epsilon = \kappa p = 2e^{-4a}\phi,_u\phi,_v, \quad (6.3.7)$$

and the velocity field is:

$$u_i = e^a\phi,_i/(2\phi,_u\phi,_v)^{1/2}, \quad i = 0, 1; \quad (x^0, x^1) = (u,v). \quad (6.3.8)$$

A vacuum solution is determined by (6.3.6) when $\phi = $const. The FLRW models (all of them, but with the equation of state $\epsilon = p$) result from (6.3.1)–(6.3.8) in the form (1.3.13)–(1.3.14) when:

$$v = (t-x)/\sqrt{2}, \quad u = (t+x)/\sqrt{2}, \quad \delta = 0, \quad (6.3.9)$$

and:

$$[e^{2a}, e^{2\gamma}, R] = \begin{cases} [\tfrac{1}{2}\sin(2t), \cot x, \tfrac{1}{2}\sin(2t) \sin x \cos x], \\ [\tfrac{1}{2}\sinh(2t), \coth x, \tfrac{1}{2}\sinh(2t) \sinh x \cosh x], \\ [t, 1/x, tx], \end{cases} \qquad (6.3.10)$$

for the closed, the open and the flat FLRW model respectively.[48]

Tomita (1978) used these solutions to consider the influence on the spacetime of an initial singularity that is partly timelike and partly spacelike. It turns out that boundary points between the two kinds of singularity initiate discontinuities in matter-density and curvature that propagate through the spacetime with the speed of light. The null surfaces of discontinuity are not singular. Tomita investigated the same problem in perfect fluids with $p < \epsilon$, again without deriving solutions (the discontinuities are then timelike hypersurfaces). He also considered properties of the spacetime in the neighbourhood of the singularity, but only approximately.

The set of equations equivalent to (6.3.1)–(6.3.8) was rederived by Chandrasekhar and Xanthopoulos (1985), in different variables. Along the way, these authors rediscovered a few properties of such spacetimes (e.g. that only one metric function is different than in vacuum). This work was a preliminary to consideration of a collision of plane impulsive gravitational waves. Chandrasekhar and Xanthopoulos assumed that the waves propagate into a flat spacetime and are followed by trains of null radiation. Then they collide head-on and create the "stiff fluid" in the wake of the collision. However, the explicit solution considered by these authors has no FLRW limit.[49]

The relationship between the approaches of Tomita (1978) and Chandrasekhar–Xanthopoulos (1985) was described in more detail by Griffiths and Ashby (1993; in fact the authors rediscovered Tomita's approach). This last paper also contains a detailed description of the algorithm for generating stiff fluid solutions from vacuum solutions, and a prescription for superposing different solutions.

6.4 Subcases of the Tomita (1978) class

The Tomita (1978) equations (6.3.1)–(6.3.8) involve no assumptions in addition to the A_2 symmetry and the stiff fluid source. Simpler subcases have been investigated

---

[48] The investigation of the FLRW limits of the metrics from this chapter is a rather laborious process because of the coordinates being in most cases noncomoving. We shall ignore details of the calculation, but the readers should know that the conclusions are not obvious or immediate.

[49] The same team of authors produced several other papers in which similar configurations were considered, but none of the explicit solutions they discussed allow a FLRW limit. Note the criticism of the Chandrasekhar–Xanthopoulos approach by Feinstein, MacCallum and Senovilla (1989). The Einstein equations should in fact be considered together with the field equations for the null fluid, and with the boundary conditions for the latter. Then it turns out that the outcome of the collision is much less arbitrary; in particular the two interacting null fluids will not create a stiff perfect fluid.

by several other authors; some of them are defined by invariant properties. In this section we shall mention only those papers in which the field equations have been investigated; explicit solutions will be discussed separately.

Singh and Yadav (1978a,b; the two papers seem to be literally identical) worked out the same case as did Tomita (1978), in another parametrization, but assumed $R = x \overset{\text{def}}{=} u - v$. As explained in Section 6.2, this implies that the gradient of R is a spacelike vector. The solutions obtained from that approach would thus cover only part of the Tomita manifold. As can be seen from (6.3.9)–(6.3.10), even in the FLRW limit the gradient of R is timelike in one part of the FLRW manifold, spacelike in another part and null on the boundary between them. However, Singh and Yadav also presented the metric in the comoving coordinates in which it can be analytically extended to the other part of the manifold.

Singh and Yadav observed that (6.3.5) is the two-dimensional wave equation in cylindrical coordinates, and its separable solutions are expressed through the Bessel functions of the first kind with the index zero:

$$\phi = M\,J_0(kR)\,\cos(kQ), \tag{6.4.1}$$

where M and k are constants and J_0 is the Bessel function. The general solution of (6.3.5) may thus be represented as:

$$\phi = M\!\int_0^\infty\! e^{-ak} J_0(kR)\,\cos(kQ)\,\mathrm{d}k, \tag{6.4.2}$$

where a is a constant. Singh and Yadav gave a few simple examples of explicit solutions, but these were not discussed further.

Letelier (1979) discussed exactly the same problem as did Singh and Yadav (1978), and gave other simple explicit examples without discussing them.

Lal and Singh (1973) worked out three cases, but each one with additional simplifying assumptions. The additional assumptions exclude a FLRW limit in cases II and III. Case I is the limit of the Singh–Yadav (1978) class corresponding to the Killing vectors being orthogonal (i.e. $R = x = u - v$, $\delta = 0$ in the Tomita parametrization (6.3.1)), but with the additional assumption that (again in the Tomita variables) $e^{2\gamma}/R$ is a function of ϕ only. This additional assumption has the result that only the flat FLRW limit is contained in their class. The paper is remarkable because of the year of its publication: it was contemporary to the paper by Tabensky and Taub (1973, see later in this section) that initiated the whole activity.

The subcase $\delta = 0$ of the Tomita (1978) class, that is, the case when the two Killing vectors are orthogonal, was considered by Letelier (1975), in a still different parametrization. He showed that the subcase corresponding to Tomita's $R = tx$ has a velocity-dominated singularity in the sense of Eardley, Liang and Sachs (1972, see our Section 3.6).

Ray (1976) corrected a misprint in the Letelier (1975) paper and showed how the solution of (6.3.5) is represented through a series in the Bessel functions.

Alencar and Letelier (1986) considered the same problem as did Letelier (1975) and Ray (1976), but in such a parametrization that clearly shows that the general case may be interpreted as a perturbation superimposed on a FLRW background. As examples of explicit solutions they gave the flat-background Belinskii (1979) solution (it will be mentioned in our Section 6.7), and another solution that is a non-solitonic perturbation of theFLRW background. The second solution was not discussed at all.

Letelier and Tabensky (1975b) worked out the subcase of the Letelier (1975) class defined by $\delta=0$ and $e^{2\gamma}=x=u-v$ in Tomita's notation. This is a genuine limitation of generality (with no invariant definition). As a result, only the flat FLRW limit is contained in this class. In another paper, the same authors (Letelier and Tabensky 1975a) investigated the properties of singularities in their 1975b class, and in the subcase of the Letelier (1975) class that corresponds to $\delta=0$, $R=x=u-v$ in (6.3.1). It was found that the singularities are velocity-dominated and that initial data given at the singularity are not sufficient to determine the evolution of the spacetime for all times.

Charach and Malin (1979) worked out the subcase of the Letelier (1975) class corresponding to $\delta=0$, $R=t=u+v$ in (6.3.1). Compared to Letelier's paper, the only additional limitation is $R=t$, and this, as explained before, only means that the consideration applies to that coordinate patch of the manifold on which the gradient of R is timelike. This class still contains all the FLRW models, although we cannot see this from Figure 6.1.

Charach and Malin interpreted this class of spacetimes as representing gravitational and scalar waves travelling on a spatially homogeneous background. This makes the spacetimes analogous to the Gowdy (1974) vacuum solutions which are contained here as a subcase of the limit $\phi=0$. The four examples of more special solutions considered in the paper are: the Kasner (vacuum) solution, the Belinskii–Khalatnikov solution of Bianchi type I, the Tabensky–Taub (1973) class (see below) and the vacuum Gowdy solution.

Matzner, Rosenbaum and Ryan (1982) investigated the solvability of the Charach–Malin (1979) equations in the formalism of the initial value problem, and showed how the solution can be constructed as a Fourier integral of separable solutions.

Tabensky and Taub (1973) worked out the plane symmetric limit of the Letelier (1975) class, which results from (6.3.1) when $\delta=\gamma=0$, $R=t=u+v$. Hence, the Tabensky and Taub class follows as the plane symmetric limit of the Charach–Malin (1979) class. However, Tabensky and Taub also presented the equations in the comoving coordinates. Only the flat and the open FLRW models are contained in this class. This is the oldest entry in Figure 6.1 that initiated this field of activity.

Tabensky and Taub discussed the equations they derived from the point of view of the Cauchy problem, and presented a few explicit solutions. The first of them was static; the second was the flat FLRW model; the third was an inhomogeneous generalization of the flat FLRW model; and the fourth was a shock wave propagating into a FLRW background. The first two examples are not within the scope of this book, and in the fourth one only the solution for the scalar field was provided. The third example will be mentioned in Section 6.7.

Tabensky and Zamorano (1975) considered the same class as did Tabensky and
Taub (1973), but without the assumption $R=t$. However, the case worked out in
detail has no FLRW limit.

Carot and Ibanez (1985) showed that the Tabensky–Taub (1973) class of solutions may be interpreted as having a viscous and heat-conducting fluid source whose
viscosity and heat-conduction terms cancel out. In their paper, the same reinterpretation is done for other solutions, but they are static.

Li (1992) claimed to have proved that a plane symmetric metric with a "scalar
wave" source must be either static or spatially homogeneous. On closer inspection,
though, it turns out that Li defined the "scalar wave" to be a scalar field that obeys
eq. (6.1.5) (the flat space wave equation) in addition to (6.1.7). Hence, the result is
rather trivial and of unknown physical significance.

For completeness, let us note that Lapedes (1977) considered the ADM quantization in the Tabensky–Taub (1973) spacetimes and in the subcase $\{R=x=u-v,\ \delta=0\}$
of the Tomita (1978) spacetimes. His results are beyond the scope of this book.

6.5 Electromagnetic generalizations of the stiff fluid spacetimes

Singh, Rai and Yadav (1979) investigated the Einstein–Maxwell equations for the
Singh and Yadav (1978) metric under the additional assumption that the electromagnetic field inherits the symmetry of the metric. Singh *et al.* found that two components of the electromagnetic four-potential may be nonzero; these would be A_y
and A_z in the Tomita (1978) parametrization of (6.3.1). Just as in the perfect fluid
case, all the equations except those determining e^a in (6.3.1) are independent of the
scalar field. However, the equations to determine the metric components will be
modified by the electromagnetic field. Singh *et al.* were interested in generating the
solutions in question from electrovacuum solutions, and so they only presented the
equations determining e^a in (6.3.1). Hence, assembly work spread over at least two
papers is required to give a full prescription in analogy with (6.3.1)–(6.3.8). We shall
leave this task to the interested reader. The main result of the paper is as follows.

If a metric of the form (6.3.1) with $R=x=u-v$ is a solution of the electrovacuum
Einstein–Maxwell equations, then a metric obeying the Einstein–Maxwell equations with a stiff fluid source is obtained when the function a in (6.3.1) is replaced
by $(a+a^\phi)$, where a^ϕ is a solution of:

$$(a^\phi)_{,x}=\tfrac{1}{2}\kappa x(\phi_{,t}^2+\phi_{,x}^2), \quad (a^\phi)_{,t}=\kappa x\ \phi_{,t}\phi_{,x}, \qquad (6.5.1)$$

the function ϕ being the scalar field determined by (6.3.5). These equations follow
from (6.1.9) when $R=x$. The explicit example of a solution given in the paper is
static.

Singh (1978) worked out the subcase of the Singh–Rai–Yadav (1979) class corresponding to the Killing vectors being orthogonal, and the electromagnetic field
obeying additional simplifying assumptions of unexplained meaning. Singh showed
that his equations reduce to those of Lal and Singh (1973) when $F_{\mu\nu}=0$.

Charach and Malin (1980) presented the set of Einstein–Maxwell equations generalizing the set that they derived before (Charach and Malin 1979). Since this seems to be the most general invariantly defined case of this family that is currently available in the literature, we shall reproduce their equations here, in a slightly modified notation. The metric is:

$$ds^2 = e^{2(\gamma - \psi)}(dt^2 - dx^2) - t^2 e^{2\psi} dy^2 - e^{-2\psi} dz^2. \tag{6.5.2}$$

Note that this follows from (6.3.1) as the subcase $\delta = 0$, $R = t = u + v$; the other differences are just renamings of the functions. The nonzero components of the electromagnetic four-potential are:

$$A_y = \chi, \quad A_z = \omega. \tag{6.5.3}$$

The scalar field is determined by:

$$\phi_{,tt} + \phi_{,t}/t - \phi_{,xx} = 0. \tag{6.5.4}$$

The other Einstein–Maxwell equations are coupled together, and they are:

$$\omega_{,tt} + \omega_{,t}/t - \omega_{,xx} = 2(\omega_{,t}\psi_{,t} - \omega_{,x}\psi_{,x}), \tag{6.5.5}$$

$$\chi_{,tt} - \chi_{,t}/t - \chi_{,xx} = -2(\chi_{,t}\psi_{,t} - \chi_{,x}\psi_{,x}), \tag{6.5.6}$$

$$\omega_{,t}\chi_{,t} = \omega_{,x}\chi_{,x}, \tag{6.5.7}$$

$$\psi_{,tt} + \psi_{,t}/t - \psi_{,xx} = (e^{2\psi}/t^2)(\chi_{,t}^2 - \chi_{,x}^2) - e^{-2\psi}(\omega_{,t}^2 - \omega_{,x}^2), \tag{6.5.8}$$

$$\gamma_{,x} = 8\pi t\phi_{,t}\phi_{,x} + 2t\psi_{,t}\psi_{,x} + 2te^{-2\psi}\omega_{,t}\omega_{,x} + (2/t)e^{2\psi}\chi_{,t}\chi_{,x}, \tag{6.5.9}$$

$$\gamma_{,t} = 4\pi t(\phi_{,t}^2 + \phi_{,x}^2) + t(\psi_{,t}^2 + \psi_{,x}^2) + te^{-2\psi}(\omega_{,t}^2 + \omega_{,x}^2) + (e^{2\psi}/t)(\chi_{,t}^2 + \chi_{,x}^2). \tag{6.5.10}$$

The explicit example of a solution discussed in the paper has the property that $\psi = $ const in the limit $F_{\mu\nu} = 0$. Hence, it does not contain any FLRW model because $\sigma = 0$ then implies that $\theta = 0$ (see the paragraph containing eq. (1.3.6)).

Ray (1982) reformulated the Charach and Malin (1980) equations in a somewhat simpler way. He also indicated how the A_2-symmetric solutions discussed here may be obtained from the stationary axisymmetric ones by the formal complex transformation $t = iy'$.

Bronnikov (1980) worked out the Einstein–Maxwell equations in the plane symmetric subcase of the Charach–Malin (1980) class (but without the coordinate condition $g_2^{1/2} = t$ assumed by the latter). His equations reproduce those of Tabensky and Taub (1973) in the limit $F_{\mu\nu} = 0$. The paper contains more results. Bronnikov rederived the Letelier (1975) and Ray (1976) formulae and used them to study perturbations (within the exact theory) of a static cylinder and of a Bianchi I spacetime. The conclusions were that perturbations of a static cylinder that preserve the A_2 symmetry do not cause radiation of gravitational waves, and that the Kasner model is unstable with respect to the clustering of matter.

6.6 Two- and many-fluid generalizations of the stiff fluid spacetimes

Letelier (1980) discussed the hydrodynamics of a mixture of two perfect fluids, and thereby initiated a new line of activity. For this book, the most useful part of his paper are the formulae showing how such a mixture can be reinterpreted as a single fluid with anisotropic pressure. Since a FLRW metric defines the energy-momentum tensor of a single (isotropic) perfect fluid, solutions obtained in the Letelier scheme will have a FLRW limit only if they admit a subcase that is a single perfect fluid. This happens when any of the conditions 1, 2, 3 or 5 listed below holds. The statements can easily be read out from Letelier's formulae.

1. One of the two perfect fluids degenerates to a "cosmological fluid" with $\epsilon+p=0$. The second pressure can then be incorporated into the first one. In particular, this covers the case when one of the fluids degenerates into vacuum, $\epsilon=p=0$.
2. The two velocities are collinear or anti-collinear. The anti-collinear case should be discarded on physical grounds because one of the velocities would then be past-pointing.
3. A mixture of a perfect fluid and a null fluid will become a single perfect fluid only when the null fluid density is zero.

The following other situations can occur:

4. A mixture of two null fluids can never degenerate into a single perfect fluid.
5. A mixture of two scalar fields will become a single perfect fluid when one of the fields is either constant or is a function of the other. In the first case, the mixture becomes a single scalar field.

The explicit example considered in Letelier's paper is a mixture of two null fluids.

Ferrando, Morales and Portilla (1990a) formulated invariant criteria for interpreting a given energy-momentum tensor as a mixture of two perfect fluids. In another paper, the same authors (1990b) proved that a given energy-momentum tensor corresponding to two perfect fluids can be decomposed into single perfect fluids up to a one-parameter family of solutions of an algebraic equation. It follows that the energy-momentum tensors of the electromagenetic field and of null radiation cannot be decomposed into two perfect fluids. Then Ferrando *et al.* discussed special cases of this decomposition corresponding to different Segre types.

Since the activity considered here has not so far produced results of importance to cosmology, we shall only briefly describe the papers in this category, without quoting details.

Letelier (1982) worked out in detail the Einstein equations for the case when the source is a mixture of two scalar fields. The examples given in the paper are in fact whole classes rather than explicit solutions; for these, only the functions $R=t$, γ and δ from (6.3.1) are specified; the scalar fields and the function a are general. The functions γ and δ are such that the Killing vectors can become orthogonal only if the metric simultaneously becomes plane symmetric. Hence, in the one-fluid limit the two examples can reproduce only the Tabensky–Taub (1973) class.

The plane symmetric subcase of the Letelier (1982) class was worked out by Letelier and Machado (1981). It is contained within Letelier's (1982) two examples, and in the limit of one scalar field it reduces to the Tabensky–Taub (1973) class.

Letelier and Alencar (1986) considered a far-reaching generalization of the whole class, to the case when the source is a mixture of an arbitrary number of perfect fluids.

This closes the discussion of general classes. The explicit solutions will be discussed in the next section. However, for better clarity of presentation, the explicit solutions with many-fluid sources will be mentioned here.

Letelier and Verdaguer (1987) used Letelier and Alencar's (1986) approach to study a metric resulting from the Charach–Malin (1979) class by Belinskii and Zakharov's generating technique, whose source is a mixture of three scalar fields. That solution does not contain any other from Figure 6.1 as a subcase, and in the limit of one scalar field it produces an inhomogeneous generalization of the flat and open FLRW models.

Krori and Nandy (1984) presented five explicit solutions, each with two scalar fields as a source. The first example is a Bianchi I spacetime. The second becomes static in the limit of zero shear. The third and fourth have no FLRW limit. Hence, only the last example (case (iii) of section III) qualifies for this book. It is plane symmetric, so it is a subcase of the Letelier–Machado (1981) class, and its one-fluid limit is within the Tabensky–Taub (1973) class. It can reproduce only the flat FLRW limit.

6.7 Explicit solutions with a stiff perfect fluid source

The foregoing sections of Chapter 6 have presented papers in which the field equations were analysed, usually without deriving explicit solutions. In this section we shall discuss the explicit solutions.

The most elaborate solution, representing a soliton wave travelling in a FLRW background, was found by Belinskii (1979). It is:

$$ds^2 = a_0^2 rs^{-2}k^{-1}\sin(2kt)QL^{-1}(dt^2 - dx^2)$$

$$-(2kQ)^{-1}\sin(2kt)\{[2(a_0/k)^2 L \sin^2(kx) + \Sigma\mu \cos^2\gamma + (\Sigma\mu/s^2)R \sin^2\gamma]dz^2$$

$$+(s^2/r)[2a_0^2 L \cos^2(kx) - \Sigma k^2\mu \sin^2\gamma - \Sigma(k/s)^2\mu R \cos^2\gamma]dy^2$$

$$+\Sigma k^2(\mu/s)(R-s^2)\cos(2\gamma)dz\, dy\}, \tag{6.7.1}$$

where:

$$Q = s^2\sin^2(kt) + R \cos^2(kt), \tag{6.7.2}$$

$$L = r \sin^2(kt) + R \cos^2(kt), \tag{6.7.3}$$

$$\mu = \tfrac{1}{2}(a_0/k)^2 \big\{ \cos)(2\gamma) + \cos(2kt)\cos(2kx)$$
$$-[(\cos(2\gamma) + \cos(2kt)\cos(2kx))^2 - \sin^2(2kt)\sin^2(2kx)]^{1/2} \big\}, \qquad (6.7.4)$$

$$R = k^6 a_0{}^{-4}\cos^{-2}(2\gamma)[a_0{}^2k^{-4}\tan^2(kt)(\cos(2\gamma) - \cos(2kx)) - \mu k^{-2}\cos^{-2}(kt)]^2, (6.7.5)$$

a_0, s, k, Σ, r and γ are constants connected by:

$$r = -k^{-2}\tan^2(2\gamma), \quad \Sigma = s^2 - r, \qquad (6.7.6)$$

the energy-density is:

$$\kappa\epsilon = \kappa p = 3a_0{}^{-2}s^2r^{-1}k^3LQ^{-1}\sin^{-3}(2kt), \qquad (6.7.7)$$

and the scalar field is:

$$\phi = (3/2)^{1/2}\ln[k^{-1}\tan(kt)]. \qquad (6.7.8)$$

The last formula implies that the coordinates of (6.7.1) are comoving. The closed FLRW model results from here when $\Sigma = 0$; the open FLRW model results after the reparametrization $k = ik'$ with $\Sigma = 0$. In order to obtain the flat FLRW limit, the following reparametrization is necessary:

$$a_0{}^2 = 1/2, \quad s = 2s', \quad \Sigma = 4q, \quad \cos(2\gamma) = -1 - 2k^2l^2, \qquad (6.7.9)$$

after which the limit $q = k = 0$ has to be taken. Since the formulae (6.7.1)–(6.7.8) are rather difficult to manipulate, let us note the result of (6.7.9) combined with the limit $k = 0$, that is, a soliton wave on a flat FLRW background:

$$ds^2 = (l/s)^2 t[(st)^2 + (t^2 + \mu)^2][(lt)^2 + (t^2 + \mu)^2]^{-1}(dt^2 - dx^2)$$
$$- t[(st)^2 + (t^2 + \mu)^2]^{-1}\big\{ [(stx)^2 + x^2(t^2 + \mu)^2 + qx^2(t^2 + \mu) - q^2\mu]dz^2$$
$$+ [(st)^2 + (t^2 + \mu)^2 - q(t^2 + \mu)]dy^2 + 2qs\mu \, dz \, dy \big\}, \qquad (6.7.10)$$

where:

$$q = s^2 - l^2, \qquad (6.7.11)$$

$$\mu = -\tfrac{1}{2}(l^2 + t^2 + x^2) + \tfrac{1}{2}[(l^2 + t^2 + x^2) - 4(tx)^2]^{1/2}, \qquad (6.7.12)$$

and the matter-density is:

$$\kappa\epsilon = \tfrac{3}{4}[s/(lt)]^2[(lt)^2 + (t^2 + \mu)^2]/[(st)^2 + (t^2 + \mu)^2]. \qquad (6.7.13)$$

In Belinskii's paper, the evolution of the model is described qualitatively and is illustrated by graphs. In short, at $t = 0$ the perturbation of the FLRW density is concentrated around the x-axis. As it wanders away, it evolves into a wave. In the flat and open backgrounds, the wave recedes to infinity with a gradually decreasing amplitude, leaving in its wake the FLRW background with a modified curvature. In the closed background, the evolution of the shape of the wave is time-symmetric with respect to the moment of maximal expansion; the wave has just enough time to pass from $x = \pi/2$ at $t = 0$ to $x = 0$ at $t = \pi/2$.

In comparing the Belinskii (1979) solution to the other explicit solutions, the following properties are important:

The Belinskii solution is the only explicit solution in the $\epsilon=p$ family that can reproduce all the FLRW models, so it will not be a subcase of any other.

The determinant of the two-dimensional metric in the (z, y)-space is given by:

$$g_2^{1/2}=\tfrac{1}{2}(a_0/k)^2\sin(2kt)\sin(2kx). \tag{6.7.14}$$

Gleiser, Diaz and Grosso (1988a,b) discussed the Belinskii solution with $k=0$, eqs. (6.7.10)–(6.7.13), further. They showed that it can be analytically continued to negative values of l^2. In the 1988a paper, the continuation is of class C^1 and results in a null shock front that propagates into a mixture of ingoing and outgoing null radiation. Behind the shock, the "stiff fluid" is created that evolves towards a FLRW spacetime. This is a similar configuration to the one considered by Chandrasekhar and Xanthopoulos (1985). In the 1988b paper, the same extension was discussed again, and, in addition, Gleiser *et al.* showed that a C^∞ extension is also possible. In the latter case, the matching surface separates an ordinary stiff fluid from a tachyonic stiff fluid, moving with a superluminal velocity.

Ibanez and Verdaguer (1986) considered a novel application of the generating technique that leads to solutions with anisotropic pressure. Those solutions will be discussed in Section 7.8. In passing, Ibanez and Verdaguer presented two stiff fluid solutions (their eqs. (28) and (30)) that result by a conformal transformation of the former solutions. The first of them is:

$$ds^2=g_{00}(dt^2-dx^2)-t(x^2dy^2+dz^2), \tag{6.7.15}$$

where:

$$g_{00}=Ct^{-1}[\sigma_1\sigma_2(x\sigma_1+t)(x\sigma_2+t)]^2[x(\sigma_1\sigma_2-1)]^{-2}$$

$$\times[(\sigma_1^2-1)(\sigma_2^2-1)(\sigma_1-\sigma_2)^{-1/2}]^{-4/3}, \tag{6.7.16}$$

$$\sigma_1=\sigma_1^{(-)}, \quad \sigma_2=\sigma_2^{(+)}, \tag{6.7.17}$$

$$\sigma_i^{(\pm)}=(2xt)^{-1}\{\pm[(l_i^2+t^2+x^2)^2-4(tx)^2]^{1/2}-(l_i^2+t^2+x^2)\}, \tag{6.7.18}$$

l_1, l_2 and $C\overset{\text{def}}{=}(l_1^2-l_2^2)^2/l_1^4$ are constants, and the scalar field is:

$$\phi=(3/2)\ln[t(\sigma_1\sigma_2)^{2/3}]. \tag{6.7.19}$$

This solution is in the Letelier and Tabensky (1975b) class, and reduces to the flat FLRW model when shear becomes zero.

The second of Ibanez and Verdaguer's stiff fluid solutions (1986, eq. (30) in the paper) is:

$$ds^2=g_{00}(dt^2-dx^2)-t[(\sigma_1\sigma_2)^{-1}x^2dy^2+\sigma_1\sigma_2dz^2], \tag{6.7.20}$$

where:

$$g_{00} = C\sigma_1\sigma_2[tx^2(\sigma_1\sigma_2-1)^2]^{-1}[(\sigma_1^2-1)(\sigma_2^2-1)(\sigma_1-\sigma_2)^{-1/2}]^{-4/3}, \qquad (6.7.21)$$

the definitions of σ_1 and σ_2 being the same as before. A suspected misprint in the exponent of $(\sigma_1-\sigma_2)$ has been corrected in (6.7.21). The scalar field is:

$$\phi = (3/2)^{1/2}\ln[t(\sigma_1\sigma_2)^{-1/3}], \qquad (6.7.22)$$

and the definition of C is presumably $C=(l_2^2-l_1^2)^2$. This solution is in the Letelier (1975) class, and again becomes the flat FLRW model when shear is zero.

The solution (6.7.15)–(6.7.19) describes solitonic density waves propagating through a FLRW background that are not accompanied by gravitational waves (the metric in the subspace $\{t=\text{const}, x=\text{const}\}$ is the same as in the background). This is in contrast to the Belinskii (1979) solution in which the density wave was coupled to a gravitational wave. In the solution (6.7.20)–(6.7.22) both kinds of waves are present.

Griffiths (1993a) considered a gravitational wave propagating into the flat FLRW background with $\epsilon=p$, leaving behind it a Petrov type I fluid with $\epsilon=p$. The metric (in notation adapted to ours) is:

$$ds^2 = t\,e^\Omega\,du\,dv - t[f\,dy^2+(1/f)dz^2], \qquad (6.7.23)$$

where $t=u+v$, and f is given by:

$$\ln f = \sum_{n=1}^{\infty} c_n R_n(u/v)v^{n-1/2}, \qquad (6.7.24)$$

where c_n is an infinite set of arbitrary constants, and the functions R_n are related to the hypergeometric function by:

$$R_n = x^n F(1/2,\ 1/2+n,\ 1+n,\ -x). \qquad (6.7.25)$$

The function Ω is then determined by:

$$\Omega_{,v} = (u+v)(\ln f)_{,v}^2/2, \qquad \Omega_{,u} = (u+v)(\ln f)_{,u}^2/2. \qquad (6.7.26)$$

In this solution, the Killing vectors are orthogonal, and in the limit of plane symmetry it reduces to the flat FLRW model. It is within the Charach–Malin (1979) class. The result (6.7.23)–(6.7.26) is not in fact an explicit solution, but goes a long way towards one, and therefore the paper is displayed in Figure 6.1 among the explicit solutions.

In a subsequent paper (Griffiths 1993b), the author considered a generalization of the model described above to the situation when two waves, each obeying (6.7.23)–(6.7.26), propagate into the flat FLRW background in opposite directions, then collide and interact. Bičak and Griffiths (1994) modified the results of both Griffiths (1993a,b) papers to describe the case when the background region into which the waves propagate is the open FLRW model with $\epsilon=p$. This case is still within the Charach–Malin (1979) class. Feinstein and Griffiths (1994) found the corresponding result for the closed FLRW background with $\epsilon=p$. Their result is in the Letelier (1975) class.

In discussing the solutions by Wainwright, Ince and Marshman (WIM, 1979) and by Carmeli, Charach and Malin (1981) we will depart from the "general to special" order adopted in this book because the more general solutions of these are constructed with reference to the more special ones.

WIM presented several examples of explicit solutions to illustrate their generating algorithm. Three of the examples are described below, the other two have no FLRW limit. The formulae (6.7.27)–(6.7.33) given below refer to the metric (6.1.4)–(6.1.11). Example 1(a) is:

$$R=t, \quad \phi=-\alpha \ln t+\beta x, \quad \Omega=\alpha^2\ln t+\tfrac{1}{2}\beta^2t^2-2\alpha\beta x, \qquad (6.7.27)$$

$$f=t^m e^{2nx}, \quad w=0, \quad 2k=\tfrac{1}{2}(m^2-1)\ln t+n^2t^2+2mnx, \qquad (6.7.28)$$

where m, n, α and β are arbitrary constants. This solution reproduces the flat FLRW model when shear and acceleration vanish, that is, when $\beta=m=n=0$, $\alpha^2=3/2$. It is within the Charach and Malin (1979) class.

Example 1(b) has R, ϕ and Ω given by (6.7.27), and:

$$f=t^{n-1}/(1+b^2t^{2n}), \quad w=2bnx,$$

$$2k=(n^2/2-n)\ln t+\ln(1+b^2t^{2n}), \qquad (6.7.29)$$

where b is another constant. This solution reproduces the flat FLRW limit when $\beta=b=0$, $n=1$, $\alpha^2=3/2$. It is within the Tomita (1978) class.

WIM's example 2(a) is:

$$R=e^{-2qx}\sinh(2qt), \quad e^\phi=[\tanh(qt)]^{-\alpha}[\sinh(2qt)]^{-\beta}e^{2\beta qx}, \qquad (6.7.30)$$

$$e^\Omega=[\tanh(qt)]^{2\alpha\beta}[\sinh(2qt)]^{\alpha^2+\beta^2}e^{2q(\alpha^2-\beta^2)x}, \qquad (6.7.31)$$

$$f=R^n[\tanh(qt)]^m, \quad w=0, \qquad (6.7.32)$$

$$e^{2k}=[\tanh(qt)]^{mn}[\sinh(2qt)]^{(m^2+n^2-1)/2}e^{(m^2-n^2-3)qx}, \qquad (6.7.33)$$

where q is a new constant. This example is within the Charach and Malin (1979) class. It reproduces the open FLRW model when $\beta=m=n=0$, $\alpha^2=3/2$, in a form similar to (1.3.15). The flat FLRW model results when $\alpha=\beta=(3/8)^{1/2}$, $m=-n=\pm1/2$ (the $+$ and $-$ are equivalent under the interchange of y and z). The resulting metric is:

$$ds^2=\sqrt{2}\,\sinh(qt)e^{-3qx}[dt^2-dx^2-e^{2qx}dy^2-2\cosh^2(qt)dz^2]. \qquad (6.7.34)$$

This will be transformed to a form similar to (1.3.15) by:

$$t=(2q)^{-1}\ln[(x'+t')/(x'-t')],$$
$$\qquad (6.7.35)$$
$$x=-q^{-1}\ln x'-(2q)^{-1}\ln(1-t'^2/x'^2),$$

and then it can be recognized as the flat FLRW model.

Example 1(a), that is, the solution given by (6.7.27)–(6.7.28), results from example 2(a) (eqs. (6.7.30)–(6.7.33)) as the limit $q \to 0$, but only after a rather elaborate reparametrization that makes the limit nonsingular. In the first step we transform:

$$x = x' + C,$$

$$y = 2^{-(n+1)/2} q^{-(m+n+1)/2} e^{2q(n+1)C} y', \qquad (6.7.36)$$

$$z = 2^{(n-1)/2} q^{(m+n-1)/2} e^{2q(1-n)C} z',$$

where the constant C is chosen so that (6.7.36) changes every occurrence of $\sinh(qt)$ into $[q^{-1}\sinh(qt)]$; it is:

$$C = q^{-1}(2\alpha^2 - 2\beta^2 + m^2 + n^2 - 3)^{-1}\left\{[-(\alpha+\beta)^2 - (m+n)^2/2 + \tfrac{1}{2}]\ln q \right.$$
$$\left. -[\alpha^2 + \beta^2 + (m^2+n^2)/2 - \tfrac{1}{2}]\ln 2\right\}. \quad (6.7.37)$$

Then we define the new constants A, B, M and N by:

$$\alpha = A/2 - B/(2q), \quad \beta = A/2 + B/(2q),$$

$$m = M/2 + N/q, \quad n = M/2 - N/q, \qquad (6.7.38)$$

and observe that:

$$\lim_{q \to 0}\{q^{-1}\sinh(qt), [\cosh(qt)]^{\gamma/q^2}\} = \{t, e^{\gamma t^2/2}\}, \qquad (6.7.39)$$

$$\lim_{q \to 0}\{[\cosh(qt)]^{\delta/q}, [\cosh(qt)]^{\varepsilon}\} = \{1, 1\}, \qquad (6.7.40)$$

where γ, δ and ε are arbitrary constants. Knowing all this, one can verify that the metric of example 2(a), transformed by (6.7.36) with (6.7.37)–(6.7.38), reduces to the metric of example 1(a) in the limit $q \to 0$. After taking the limit, the constants $\{A, B, M, N\}$ should be identified with $\{\alpha, \beta, m, n\}$ from (6.7.27)–(6.7.28) respectively.

McIntosh (1978) showed that the WIM solutions admit a homothetic motion, that is, a conformal symmetry with $\pounds_k g_{\alpha\beta} = 2\psi g_{\alpha\beta}$, where $\psi = $ const. For example 1(a) (eqs. (6.7.27)–(6.7.28) above) the generator is:

$$k^\alpha \frac{\partial}{\partial x^\alpha} = \frac{\partial}{\partial x} + (mn - \alpha\beta - n)y\frac{\partial}{\partial y} + (mn - \alpha\beta + n)z\frac{\partial}{\partial z}, \qquad (6.7.41)$$

and $\psi = mn - \alpha\beta$. The whole group of conformal symmetries is in this case of Bianchi type I when $n = \beta = 0$, of type V when $n = 0 \neq \beta$, of type III when $n = \pm(mn - \alpha\beta) \neq 0$ and of type VI otherwise. The examples 1(b), 2(a) and 2(b) also admit "obvious" (so says the author) nontrivial homothetic motions.

Diaz, Gleiser, Gonzalez and Pullin (1989) showed that the solution (6.7.30)–(6.7.33), that is, example 2(a) of WIM, is geodesically incomplete when $m^2 < 2$ because then the spurious infinity at $\{y \to \infty, z \to \infty\}$ is in fact only a finite geodesic distance away from points in the spacetime. When $m^2=0$, $m^2=1$ and $3/2 \leq m^2 < 2$, the spacetime can be extended beyond that coordinate boundary, and Diaz *et al.* found the extension.

The paper by Carmeli, Charach and Malin (1981) is a large review which covers several topics omitted in this book (a bibliography on the Bianchi models, a discussion of vacuum, electrovacuum and scalar field Bianchi models, a discussion of vacuum solutions with wave features, and a brief summary of linear perturbations of the FLRW models) and one topic dealt with here (an overview of those solutions of the Einstein–Maxwell equations in which the source is the scalar field and the electromagnetic field). Carmeli *et al.* presented a new class of solutions generalizing Wainwright, Ince and Marshman's (1979) example 2(a), eqs. (6.7.30)–(6.7.33) above. They first observed that the transformation:

$$T=\sinh(2qt)e^{-2qx}, \quad \zeta=\cosh(2qt)e^{-2qx}, \tag{6.7.42}$$

transforms the WIM metric to the form:

$$ds^2=e^{2k+\Omega}(dT^2-d\zeta^2)-T[f\,dy^2+(1/f)dz^2], \tag{6.7.43}$$

where:

$$f=T^n[\zeta/T+(\zeta^2/T^2-1)^{1/2}]^{-m}, \tag{6.7.44}$$

and the scalar field to the form:

$$(4\pi)^{1/2}\phi=-\beta \ln T+\alpha \ln[\zeta/T+(\zeta^2/T^2-1)^{1/2}]. \tag{6.7.45}$$

These formulae can now be generalized as follows:

$$\ln f=n \ln T-m\int_0^{\xi-T}[(\zeta-\lambda)^2-T^2]^{-1/2}g_1(\lambda)d\lambda, \tag{6.7.46}$$

$$(4\pi)^{1/2}\phi=-\beta \ln T+\alpha\int_0^{\xi-T}[(\zeta-\lambda)^2-T^2]^{-1/2}g_2(\lambda)d\lambda, \tag{6.7.47}$$

where $g_1(\lambda)$ and $g_2(\lambda)$ are arbitrary functions. The generalized formulae still define a stiff perfect fluid solution, and the WIM example 2(a) is recovered when $g_1=g_2=1$. Carmeli *et al.* discussed the solutions with travelling shock waves, in which:

$$g_i(\lambda)=\begin{cases}1 \text{ for } 0<\lambda<M_i=\text{const} \\ 0 \text{ for } \lambda>M_i, \quad i=1, 2.\end{cases} \tag{6.7.48}$$

The shock wave defined by g_1 is gravitational; the other one is a density wave. A pulse of finite duration is obtained when $g(\lambda)=1$ for $0<\lambda\leq M$, $g(\lambda)=0$ for $\lambda>M+1$ and $g(\lambda)=1-G(\lambda-M)/G(1)$, where:

$$G(\lambda)=\int_0^\lambda \exp[-\mu^{-2}-(\mu-1)^2]d\mu \tag{6.7.49}$$

for $M<\lambda<M+1$. The gravitational and density pulses propagate into the WIM 2(a) space, and leave behind them a Bianchi space of type III (when $n=\pm 1$), V (when

$n=0$) or VI$_h$ (otherwise). Between the pulses, the space is still something else (see paper).

The Carmeli–Charach–Malin (1981) solution described above is within the Charach–Malin (1979) class, and by construction it generalizes the WIM example 2(a).

Oliver and Verdaguer (1989) found a solution which can be obtained from (6.7.43)–(6.7.45) (i.e. from the Carmeli–Charach–Malin 1981 representation of the WIM example 2(a)) by the analytic continuation to the values of $|\zeta/T|\leq 1$, using the identity:

$$\text{arc } \cos(\zeta/T)=-i \text{ ar } \cosh(T/\zeta)\equiv -i \ln[T/\zeta+(T^2/\zeta^2-1)^{1/2}]. \quad (6.7.50)$$

In order to cancel the imaginary unit, the constants α and m should be redefined by:

$$\alpha=-ib, \quad m=ih. \quad (6.7.51)$$

With (6.7.50) and (6.7.51), the Oliver–Verdaguer (1989) solution results from (6.7.43)–(6.7.45). We thus propose to interpret the Oliver–Verdaguer solution as being determined on another region of the same manifold that underlies the WIM example 2(a). This is why the two solutions are placed in the same rectangle in Figure 6.1.

Rao, Tiwari and Bhamra (1974) found two solutions of the vacuum Brans–Dicke equations that define, by correspondence, solutions of the Einstein equations with a scalar field source. The second of them has no FLRW limit. The first one is:

$$ds^2=f(t,x) (dt^2-dx^2)-x^{2(1-k)}(at+b)^{1-2k/a}dy^2-x^{2k}(at+b)^{1+2k/a}dz^2, \quad (6.7.52)$$

where a, b and k are constants, and:

$$f(t,x)=x^{2(2k^2-k^2a+ka-k)}(at+b)^{2(2k^2-k^2a+ka-k/a)+1}[x^{-2}-a^2/(at+b)^2]^{-k^2a+k^2+ka}. \quad (6.7.53)$$

The scalar field is:

$$\phi=(1/\sqrt{2}) (\omega+3/2)^{1/2} \ln (at+b), \quad (6.7.54)$$

that is, the coordinates of (6.7.52) are comoving. The constant ω is the Brans–Dicke parameter. The constants a and k are connected by:

$$(k/a) (a-1) [a-k(a-2)]=0, \quad (6.7.55)$$

which implies $k=0$ or $a=1$ or $k=a/(a-2)$. With $k=0$, the solution is just the flat FLRW model. With $k=a/(a-2)$, the limit $\sigma=0$ requires $a=0$, which is a vacuum solution. Hence, only the case $a=1$ is nontrivial and has a FLRW limit. It can be shown to be the subcase of the WIM example 2(a) defined by:

$$m=-1/2, \quad n=1/2-2k, \quad \alpha=\beta=\pm(3/8-k/2)^{1/2}. \quad (6.7.56)$$

The (flat only) FLRW limit results when $k=0$.

The explicit example of a solution given by Tabensky and Taub (1973) is the subcase $d=h=0$ of the Oliver–Verdaguer (1989) solution, that is, the subcase $m=n=0$ of (6.7.43)–(6.7.45) with (6.7.50) and (6.7.51). It reproduces the flat FLRW limit when further $\alpha=0$, $\beta^2=6\pi$.

A more elaborate family of explicit solutions from the Tabensky–Taub (1973) class was presented by Collins and Lang (1987, their class (ii)). The authors included $\Lambda\neq0$, but $\Lambda=0$ may be assumed by redefining ϵ and p. The metric is (in the original notation except for the signature):

$$ds^2 = A^{2(D+1)}B^{2(1-D)}(\delta A^4+\varepsilon)^{-1}dA^2$$

$$-A^{2D}B^{2(2-D)}(\delta B^4+E)^{-1}dB^2 -A^2B^2(dy^2+dz^2), \qquad (6.7.57)$$

where (A, B, y, z) are the coordinates, and the constants D, E, δ and ε obey one of the following three sets of conditions:

(a) $D\neq3/2$, $E=0$, $\delta=1$, $\varepsilon=\pm1$;
(b) $D=3/2$, $E-$arbitrary, $\delta=1$, $\varepsilon=\pm1$;
(c) $D=3/2$, $E=\varepsilon=1$, $\delta=0$.

The coordinates of (6.7.57) are comoving, and the matter-density is:

$$\kappa\epsilon = M_0 B^{2(D-1)}/A^{2(D+2)}, \qquad (6.7.58)$$

where M_0 is one more constant. This solution is within the Tabensky–Taub (1973) class because of plane symmetry. The limit of zero shear requires $D=1$, and so only the set (a) will have a FLRW limit. The open FLRW model results when $D=\delta=1$, $E=0$; the flat FLRW model results as the formal limit $\delta=0$ of the case $D=1$, $E=0$ after the transformation $B=\exp(\delta^{1/2}x)$ is performed.

By introducing the coordinates of (6.1.4) with $R=t$ (which are unique up to (6.2.9)) one can verify that the sets (a) and (c) defined above are subcases of the WIM example 2(a), while the set (b) is not. The proof is left as a (rather laborious) exercise to the reader. Hence, all those Collins–Lang class (ii) solutions that have a FLRW limit are within the WIM family. The class (i) solutions from this paper have no FLRW limit; the class (iii) solutions will be described in Section 7.1.

Carmeli, Charach and Feinstein (1983a) studied generating stiff fluid solutions from vacuum solutions, taking special care about using such coordinates in which the gradient of R in (6.1.4) is allowed to change from timelike to spacelike over the manifold. They gave three explicit examples of solutions. The first of these (the generalized Taub solution) has no limit $\dot{u}^\alpha=0$, and so will not be discussed here. The second one (the generalized Bianchi IX Universe) has the metric:

$$ds^2 = e^{2k+\Omega}(dt^2-dx^2) - [(I_1^2\sin^2x+I_3^2\cos^2x)dy^2+I_3^2dz^2+2\,I_3^2\cos x\,dy\,dz], \quad (6.7.59)$$

where:

$$I_3^2 = 2A/\{[\tan(t/2)]^{2A/k} + [\cot(t/2)]^{2A/k}\}, \qquad (6.7.60)$$

$$I_1^2 = k^2 I_3^{-2}/[\tan(t/2) + \cot(t/2)]^2, \qquad (6.7.61)$$

$$2k + \Omega = 2 \ln I_1 + 9\pi A_2^2 \sin^2 t \cos^2 t \sin^2 x\, (1 + 3\cos^2 x)$$
$$+ (9/2)\pi A_2^2 \sin^2 t\, (1 - 3\cos^2 t) \sin^4 x - 48\sqrt{2}\pi(MA_2/k) \cos t \sin^2 x, \quad (6.7.62)$$

and the scalar field is:

$$\phi = (\sqrt{2}M/k) \ln [\tan(t/2)] + (A_2/16)(3\cos^2 t - 1)(3\cos^2 x - 1). \qquad (6.7.63)$$

The constants A, M and k are connected by:

$$A^2 + 8\pi M^2 = k^2, \qquad (6.7.64)$$

and A_2 is an arbitrary constant (the paper implies that the numerical coefficient of A_2 in (6.7.63) is (1/4), but the value given here is consistent with the other formulae in the paper). By rather complicated calculations it can be shown that the (closed only) FLRW limit results when $A_2 = 0$, $k = 2A$. In general, the solution is in the Tomita (1978) class.

The third solution presented by Carmeli, Charach and Feinstein (1983a) (a generalized FLRW Universe) has the metric:

$$ds^2 = e^{2k+\Omega}(dt^2 - dx^2) - \sin t \sin x\, [f\, dy^2 + (1/f)\, dz^2], \qquad (6.7.65)$$

where:

$$2k + \Omega = \ln [(k/2) \sin t] - 2A_1 \cos t - 12\sqrt{3}\pi\, A_2 \cos t \sin^2 x$$
$$+ \tfrac{1}{2}A_1^2 \sin^2 t \sin^2 x + 9\pi A_2^2 \sin^2 t \cos^2 t \sin^2 x\, (1 + 3\cos^2 x)$$
$$+ \tfrac{9}{2}\pi A_2^2 \sin t\, (1 - 3\cos^2 t) \sin^4 x, \qquad (6.7.66)$$

$$f = -\cot (t/2) + \exp (2A_1 \cos t \cos x), \qquad (6.7.67)$$

$$\phi = \tfrac{1}{4}(3/\pi)^{1/2} \ln [\tan(t/2)] + (A_2/16)(3\cos^2 t - 1)(3\cos^2 x - 1), \qquad (6.7.68)$$

where k, A_1 and A_2 are arbitrary constants. This solution is in the Letelier (1975) class, and its (closed only) FLRW limit results when $A_1 = A_2 = 0$.

Carmeli *et al.* showed that their solutions describe the (power law) growth of density perturbations on a Bianchi IX background, and the growth may be faster than in the case of an isotropic FLRW background. The first two solutions were published separately in another paper (Carmeli, Charach and Feinstein 1983b).

Agnew and Goode (1994) found all those solutions with a heat-conducting "stiff fluid" source for which the following conditions are obeyed:

1. The metric is invariant under the A_2 group, and the two Killing vectors are mutually orthogonal.
2. The coordinates of (6.1.4) are comoving.
3. The metric components are separable in those coordinates.

The solutions thus defined were obtained explicitly. They are not reproduced here because of the large number of formulae necessary to present them. Of the 14 cases found, two (M12 and M14) have no FLRW limits; the others cover all the three FLRW models with $\epsilon = p$. The limit of zero heat-flow of those solutions is in the Letelier (1975) class, and it contains the Collins–Lang (iia) (1987) solutions as subcases.

This completes the overview of solutions with a "stiff fluid" source. Readers interested in more details may wish to consult the reviews by Carmeli and Charach (1984), Kitchingham (1984) and MacCallum (1984), and the paper by Kitchingham (1986). In particular, the first paper by Kitchingham (1984) focuses on revealing which solutions arise from the same seed solution by different generating techniques. In the second paper (Kitchingham 1986), the author shows which solutions can be generated from the Minkowski spacetime (Belinskii 1979 is among them). Also, the global geometry of such spacetimes is investigated, in which the gradient of R is timelike in one part of the manifold and spacelike elsewhere. The latter consideration may be useful in extending some of the solutions from this section. Very detailed overviews of the generating techniques, though strongly concentrated on vacuum solutions, were given by Verdaguer (1985 and 1993). Jantzen (1980) showed how the Belinskii–Zakharov generating technique may be applied to spatially homogeneous and spatially self-similar spacetimes to obtain inhomogeneous cosmological solutions. However, no explicit solution was actually presented.

7

Other solutions

In this chapter, those solutions will be described that could not be assembled into large interconnected families. Since most of the solutions have little in common with one another, they are usually presented here in the original notations.

7.1 Other A_2-symmetric perfect fluid solutions

These are displayed in the upper part of Figure 7.1. They are A_2-symmetric, but do not obey the "stiff" equation of state $\epsilon = p$, and therefore did not fit into Figure 6.1 and Chapter 6. Some of them obey $p = (\gamma - 1)\epsilon$ with $2 \neq \gamma = \text{const}$, some others do not obey any particular equation of state. Though there are quite a number of them, only two can be reproduced from others by limiting transitions; all the others are simply independent.

Since these solutions have the same symmetry as those from Chapter 6, the same argument applies here: their extrinsic class in the Wainwright (1979) classification cannot be more general than B_3. They cannot be of class D or more special because then they would belong to the Stephani–Barnes family (see Chapter 4). Insufficient information for the intrinsic classification was given in the papers.

Wils (1990) found a large collection of perfect fluid solutions with the metric:

$$ds^2 = e^{2k}(dt^2 - dx^2) - R[f\,dy^2 + (1/f)dz^2], \qquad (7.1.1)$$

where the coordinates of (7.1.1) were assumed comoving, and the functions f, R and e^{2k} were assumed to separate as follows:

$$F_1 = T^{m_i}\,D^{n_i}\,S^{p_i}\,C^{q_i}, \quad i=1, 2, 3, \qquad (7.1.2)$$

with $(F_1, F_2, F_3) = (f, R, e^{2k})$, and:

$$
\begin{array}{llll}
T = \beta t, & D = 1 & \text{when } \varepsilon = 0, & \\
T = \sinh(\beta t), & D = \cosh(\beta t) & \text{when } \varepsilon = +1, & \\
T = \sin(\beta t), & D = \cos(\beta t) & \text{when } \varepsilon = -1, & (7.1.3) \\
S = x, & C = 1 & \text{when } \delta = 0, & \\
S = \sinh x, & C = \cosh x & \text{when } \delta = +1, & \\
S = \sin x, & C = \cos x & \text{when } \delta = -1.
\end{array}
$$

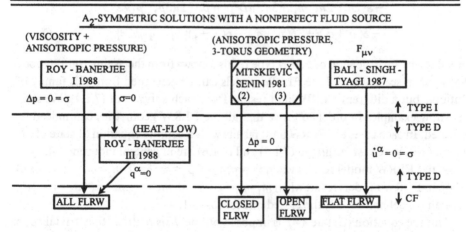

Figure 7.1. A_2-symmetric solutions with a perfect fluid source (upper part) and A_2-symmetric solutions with a nonperfect fluid source (lower part).

The constants $\varepsilon=0$, ± 1; $\delta=0$, ± 1; β, m_i, n_i, p_i and q_i obey the following equations resulting from the Einstein equations:

$$m_1(m_2-1)=p_1(p_2-1)=(m_2-m_3)(m_2-1)=0,$$
$$n_1(n_2-1)=q_1(q_2-1)=(n_2-n_3)(n_2-1)=0,$$
$$p_1^2=p_2p_3+p_2+p_3, \quad q_1^2=q_2q_3+q_2+q_3,$$
$$m_1p_1+m_2p_2=m_2p_3+m_3p_2, \quad n_1p_1+n_2p_2=n_2p_3+n_3p_2, \qquad (7.1.4)$$
$$m_1q_1+m_2q_2=m_2q_3+m_3q_2, \quad n_1q_1+n_2q_2=n_2q_3+n_3q_2,$$
$$\varepsilon\beta^2[m_1(n_2+1)+n_1(m_2+1)]=\delta(p_1+q_1)(p_2+q_2),$$
$$\varepsilon\beta^2(m_2+n_2)(m_3-m_2+n_3-n_2)=\delta[(p_1+q_1)^2-(p_2+q_2)(p_3+q_3)].$$

Hence, the whole collection of solutions splits in fact into several disjoint subsets. The matter-density and pressure are:

$$4\kappa e^{2k}p=\beta^2[(2m_2+2m_3-m_1^2-m_2^2)/T^2-\varepsilon(m_1+n_1)^2-\varepsilon(m_2+n_2)^2$$
$$-\varepsilon(2n_2+2n_3-n_1^2-n_2^2)/D^2]+(p_2-1)(p_2+p_3)/S^2$$
$$+\delta(p_1+q_1)^2+\delta(p_2+q_2)^2-\delta(q_2-1)(q_2+q_3)/C^2, \qquad (7.1.5)$$

$$4\kappa e^{2k}\epsilon=\beta^2(2m_3-2m_2-m_1^2+3m_2^2)/T^2-\varepsilon(m_1+n_1)^2+3\varepsilon(m_2+n_2)^2$$
$$-\varepsilon(2n_3-2n_2-n_1^2+3n_2^2)/D^2]-[(p_2-1)(3p_2-p_3)/S^2$$
$$-\delta(p_1+q_1)^2+3\delta(p_2+q_2)^2-\delta(q_2-1)(3q_2-q_3)/C^2]. \qquad (7.1.6)$$

The solutions are in general of Petrov type I; this follows from the Wainwright–Goode (1980) paper, see below. All the FLRW models can be generated from this family of solutions, but a different FLRW limit results from each subcase of (7.1.4). We shall only show examples in order to prove that all the FLRW geometries can indeed be recovered. In each case, the FLRW limit results with a definite equation of state of the type $p=(\gamma-1)\epsilon+\text{const}$, although (7.1.5) and (7.1.6) are in general independent.

The flat FLRW model results when $m_1=p_1=p_2=p_3=\varepsilon=\delta=0$, $m_2=m_3$. The closed and open FLRW models result when $m_1=p_3=q_3=\varepsilon=0$, $m_2=m_3$, $p_1=p_2=q_2=1=-q_1$; the open model has $\delta=+1$, the closed one has $\delta=-1$.

When the equation of state $\epsilon=p$ is imposed on the Wils metrics, they trivialize to either the Minkowski metric, or a static solution, or a spatially homogeneous solution.[50]

---

[50] There are fewer solutions in the set (7.1.1)–(7.1.4) than it seems at first sight. A. Feinstein, E. Ruiz and J. M. M. Senovilla verified that all the inhomogeneous Wils solutions are contained either in the family of Wainwright and Goode (1980, see next paragraph) or in the family of Ruiz and Senovilla (1992, see later in this section). I am grateful to J. M. M. Senovilla for informing me about this unpublished result.

Wainwright and Goode (1980) obtained a solution very similar in form to those of Wils:

$$ds^2 = S^{-2m}C^{-2m-2}(dt^2 - dx^2) - SC^{\alpha}(T^n dy^2 + T^{-n}dz^2),\qquad(7.1.7)$$

where:

$$S = \sinh(2qt),\quad C = \cosh(2qx/\alpha),\quad T = \tanh(qt),$$

$$\alpha = -(2m+2)/(2m+1),\qquad(7.1.8)$$

q, m and n being arbitrary constants. The matter-density and pressure are:

$$\kappa\epsilon = q^2 S^{2m}C^{2m}[(1-4m-n^2)C^2/S^2 - (4m+3)(m-1)/(m+1)],\qquad(7.1.9)$$

$$\kappa p = q^2 S^{2m}C^{2m}[(1-4m-n^2)C^2/S^2 - (4m+3)],\qquad(7.1.10)$$

and the coordinates of (7.1.7) are comoving. The (flat only) FLRW limit results when $n=0$ and $m=-1/2$ (then $\alpha\to\infty$). Contrary to Wils' (1990) claim, the solution (7.1.7)–(7.1.8) is not contained in his class unless $\alpha=2^{m+1}q$; in that case the solution (7.1.7)–(7.1.8) results when the Wils parameters have the values $m_2=n_2=\varepsilon=\delta=1$, $p_1=p_2=p_3=q_1=0$, $\beta=\alpha/2$, $m_1=-n_1=n$, $m_3=n_3=-2m$, $q_2=\alpha$, $q_3=-2m-2$. The solution is in general of Petrov type I; when $mn=0$ it degenerates to type D. The slices t=const of the spacetime are conformally flat, and their Ricci tensor has an eigenvalue of multiplicity 2. However, the shear tensor has in general distinct eigenvalues. The acceleration is nonzero. When $m\geq-1$, the slices t=const are geodesically incomplete: they are finite in the x-direction and bounded by curvature singularities on both edges. When $m<-1$, the curvature singularities disappear and the spacetime becomes geodesically complete. The solution does not tend to spatial homogeneity either at the Big Bang or in the asymptotic future.

Through a somewhat misleading title, Wainwright and Goode suggested that the solution obeys the equation of state $p=(\gamma-1)\epsilon$. In fact, as explained in the paper, this happens only when $1-4m-n^2=0$, then $\gamma=2m/(m-1)$. Nevertheless, Grøn (1985) let himself be misled and discussed the case $m=0$ without taking $n^2=1$ into account, claiming that this was the "vacuum fluid" of inflationary models with $\epsilon=-p$. Barrow and Grøn(1986) interpreted the Wainwright–Goode (1980) solution as a mixture of a stiff perfect fluid and a barotropic perfect fluid with $p=(\gamma-1)\epsilon$, the velocity fields of both components being the same. In the case $m=0$, the second component degenerates to the cosmological term, and the mixture evolves asymptotically towards the de Sitter spacetime.

Roy and Prasad (1989) found two solutions, of which only case (a) has a FLRW limit. It is:

$$ds^2 = e^{F(t,x)}(dt^2 - dx^2) - e^{G(t,x)}dy^2 - e^{H(t,x)}dz^2,\qquad(7.1.11)$$

where:

$$F(t,x) = k + [m^2 + (1-h)^2]t/m + [ne^{(1+h)x} - S(1-h)/(1+h)]e^{-mt},\qquad(7.1.12)$$

$$G(t,x)=[m^2+(1-h)^2]t/m-Se^{-mt}+2x, \qquad (7.1.13)$$

$$H(t,x)=[m^2-(1-h)^2]t/m+Se^{-mt}+2hx, \qquad (7.1.14)$$

k, m, n, S and $h\neq-1$ are constants. We shall not quote the rather lengthy formulae for the source quantities. The coordinates of (7.1.11) are comoving. The open FLRW limit results when $n=S=0$, $h=1$, the flat FLRW limit can then be obtained after a reparametrization.

In another paper, the same authors (Roy and Prasad 1991) found a collection of solutions of the kind considered here. Since the journal is easily accessible while the solutions are rather complicated and are not discussed physically, we shall not reproduce the formulae. Of the cases considered in the paper, cases II, III and IV have no FLRW limit. Cases Ii, Iii(a) and Iii(b) reproduce the open FLRW model when $\sigma=0$, and can be reparametrized to reproduce the flat FLRW model, too. These three cases are listed in one rectangle in Figure 7.1.

Ruiz and Senovilla (1992) investigated perfect fluid solutions of the Einstein equations for the following metric:

$$ds^2=T^{2m}F^2(dt^2-H^2dx^2)-TG\,(T^nP\,dy^2+T^{-n}P^{-1}dz^2), \qquad (7.1.15a)$$

where the functions $T(t)$, $F(x)$, $H(x)$, $G(x)$ and $P(x)$ are to be found from the Einstein equations, and m, n are constants (the coordinates are assumed comoving). It is a subclass of metrics that are separable in the comoving coordinates. Partial integration of the Einstein equations leads to:

$$F^2=G^{1-2m}P^n, \quad T_{,tt}/T=\varepsilon a^2, \qquad (7.1.15b)$$

where $\varepsilon=0$, ±1 and a is an arbitrary constant.

Ruiz and Senovilla calculated the hydrodynamical and kinematical quantities and the Weyl tensor at this level of generality, and investigated some properties of the metrics (they are in general of Petrov type I). Here we shall be interested only in the explicit solutions. Four classes of them were found, two of which (cases 1 and 4) have FLRW limits.[51]

In case 1, $\varepsilon=0$ so that $T=At+B$, where A and B are constants, and the remaining functions and constants are defined as follows:

$$P^{k+l}=D^{3-4m}+N, \quad G=P^kD^{2(m-1)}, \quad H=sP^{-l}D^{-2m}D_{,x},$$

$$k=[n+(n^2+3-4m)^{1/2}]/(4m-3), \qquad (7.1.16a)$$

$$l=-(n^2+3-4m)^{1/2}/(1-m)-k.$$

---

[51] As given in the paper, the solutions become singular in the limit $\sigma=0$, except in the special case $m=1/2$ discussed separately. The formulae given here result after a reparametrization, and the case $m=1/2$ does not require a separate treatment then. This information was kindly supplied to A.K. by J. M. M. Senovilla.

where N and s are arbitrary constants and $D(x)$ is an arbitrary function. The formulae for matter-density and pressure are given in the paper; they require a small correction: $(2m-1)^2k^2$ should actually read s^2 and C should be replaced by D^{1-2m}. The FLRW limits (all three of them, but with a definite equation of state) result when $n=0$ and $m=1/2$; the sign of $(-N)$ is then the spatial curvature index.

In case 4, all three values of ε are allowed, $m=(n+1)/2$, and the other functions are as follows:

$$P^2 = \varepsilon s^2 a^2 D^{2n} + N D^{2n-1} - K,$$

$$G = P D^{1-n}, \quad H = s P^{-1} D^{n-1} D_{,x}, \tag{7.1.16b}$$

where s, a, n, N and K are arbitrary constants. The FLRW limits (again all three of them) result when $n=0$. In this limit, it can be assumed with no loss of generality that $N \geq 0$, and then the FLRW curvature index is the sign of $(K - \varepsilon s^2 a^2)$.

Case 4 contains Senovilla's singularity-free solution (1990, see the R–S paper for details), which itself does not admit a FLRW limit, so it is not discussed here separately. The R–S paper contains a study of singularities in the general metrics (7.1.15a,b).

We shall not reproduce the solution by Uggla (1992) because of the large number of formulae needed to define it. The author found four solutions, each having an Abelian two-dimensional symmetry group and a three-dimensional homothety group, the velocity field of the perfect fluid sources being tangent to the orbits of the homothety groups. The four solutions are given by eq. (19) in the paper and in the three columns of Uggla's Table 1. Only the solution from the first column of the table has an (open) FLRW limit (when $\sigma=0$), with a definite equation of state.

Davidson (1992) found the following Petrov type I solution:

$$ds^2 = (1+r^2)^{-k(k+1)/(2k+1)} dt^2 - t^{2(3k+2)/(7k+4)}(1+r^2)^{k(3k+1)/(2k+1)} dr^2$$

$$- (1+r^2)^k [t^{2kl(7k+4)} dz^2 + t^{2(3k+2)/(7k+4)} r^2 d\varphi^2], \tag{7.1.17a}$$

where k is an arbitrary constant and the coordinates are comoving. The (closed only) FLRW limit results when $\sigma=0$, that is, when $k=-1$. The reader is referred to the paper for the formulae which define the hydrodynamical quantities.

The solution (7.1.17a) results from Ruiz and Senovilla's (1992) case 4 given by (7.1.16a) and (7.1.16b) through the following specialization of constants and transformation of coordinates:

$$\varepsilon=0, \quad K=N>0, \quad 4s^2=N(2n-1)^2[\tfrac{1}{2}(n+3)]^{-n-1},$$

$$T=t_{R-S}=[\tfrac{1}{2}(n+3)t_D]^{2/(n+3)}, \quad P(x)=N^{1/2}r,$$

$$y=[\tfrac{1}{2}(n+3)]^{-n-1}\phi, \quad z_{R-S}=[\tfrac{1}{2}(n+3)]^{n-1}z_D, \tag{7.1.17b}$$

where the subscripts R–S and D refer to the authors' names. The solution (7.1.17a) is also a subcase of the R–S case 1 corresponding to $2m=n+1$.

The solutions described so far in this section were A_2-symmetric, that is, had only a two-dimensional Abelian symmetry group with spacelike orbits. The solutions described below are plane symmetric, that is, in addition to the A_2 symmetry they are invariant with respect to rotations within the orbits around every point.

Collins and Lang (1987), in the paper already mentioned in Section 6.7, found one such solution. This is their class (iii) given by:

$$ds^2 = A^{2(D+1)}B^{2(1-D)}(EA^4+F_0)^{-1}dA^2 - (AB)^{2D}(EB^{4D}+\varepsilon B^N)dB^2$$

$$-(AB)^2(dy^2+dz^2),\qquad\qquad\qquad (7.1.18)$$

where (A, B, y, z) are the coordinates, and the constants D, E, F_0 and ε obey one of the following sets of equations:

(a) $D\ne-1/2$, $E=1$, $F_0=0$, $\varepsilon=\pm1$;
(b) $D=-1/2$, $E=0$, $F_0=\varepsilon=1$;
(c) $D=-1/2$, $E=1$, F_0 –arbitrary, $\varepsilon=\pm1$,

while $N=4(D^2-D-1)/(D-2)$. The coordinates of (7.1.18) are comoving. For the formulae defining ϵ and p the reader is referred to the paper. When ϵ and p are redefined so that $\Lambda=0$ (the authors allowed for $\Lambda\ne0$), the source will obey $p=(D-2)\epsilon/(D+2)$. The (open) FLRW limit results when $D=E=\varepsilon=1$; the flat FLRW model can be obtained by the formal limit $E+\varepsilon\to0$ of the case $D=1$, after the transformation $B=e^{(E+\varepsilon)x}$.

A solution similar to (7.1.18) may be obtained from the Wainwright–Goode (1980) solution, eqs. (7.1.7)–(7.1.10), as the limit $q=1$, $\sinh(2t)=A^2$, $\cosh(2x/\alpha)=B^{2/\alpha}$; the resulting Collins–Lang constants are $D=-2m$, $E=F_0=1$, $\varepsilon=-1$. It is interesting that the constants then disobey all three sets of the Collins–Lang equations. It happens so because the Collins–Lang solutions were obtained under the assumption of self-similarity, disobeyed by the Wainwright–Goode solution.

Götz (1988) found a few solutions of which only his class D allows for a FLRW limit (achieved when $\sigma=0$). The limit covers both the flat and the open FLRW models. However, Götz's class D is equivalent to the set (a) of the Collins and Lang (1987) class (iii) described above if the Collins–Lang conditions $E=1$ and $\varepsilon=\pm1$ are ignored. Hence, the Götz solution D can be reconstructed from (7.1.18), and is listed accordingly in Figure 7.1. Götz's solution E simply *is* the FLRW model, the open one in general, and the flat one when Götz's $k_2=0$.

Shikin (1981) found four self-similar plane symmetric solutions, but only the first of them allows for a FLRW limit. This is the solution defined by the author's eqs. (2.1)–(2.14), and it is:

$$ds^2 = T^2(z)(dx^0)^2 - X^2(z)(dx^1)^2 - (x^1)^{4/3}Y^2(z)[(dx^2)^2+(dx^3)^2],\qquad (7.1.19)$$

where:

$$z=x^1/x^0,\quad Y=z^{-2/(3\gamma)},\quad T=[Xz^{2/(3\gamma)}]^{\gamma-1},\qquad\qquad (7.1.20)$$

γ is the constant from the equation of state $p=(\gamma-1)\epsilon$, and the coordinates are comoving. The matter-density is:

$$\kappa\epsilon=4\,[9\gamma^2(x^0)^2\mu T^2]^{-1}[\mu+3(\gamma-1)-3\gamma z(X,_z/X)(\mu-\gamma+1)], \qquad (7.1.21)$$

where $\mu\overset{\text{def}}{=}X^{2(2-\gamma)}z^{2(2+\gamma)/(3\gamma)}$, and $X(z)$ is defined by the equation:

$$18\gamma^2\mu(\mu-\gamma+1)\beta\beta,_\mu=16[(2-3\gamma)\mu+(\gamma-1)(3\gamma^2-8\gamma+4)]$$

$$+6\gamma\beta[(3\gamma+2)\mu+(\gamma-1)(10+\gamma)]-9\gamma^2\beta^2(\mu+\gamma-1), \qquad (7.1.22)$$

where $\beta\overset{\text{def}}{=}\mu,_z/\mu$. A suspected misprint has been corrected in (7.1.22); in the paper the second term in the first square bracket has $(-8\gamma-4)$ instead of $(-8\gamma+4)$. The open FLRW model results from the Shikin solution when $\gamma=2/3$ and $X=C/z$ where $C=$const, but it obeys (7.1.22) only if the correction indicated is made. It then obeys the unphysical equation of state $\epsilon+3p=0$. The flat FLRW model will result as the further limit $C\to\infty$, but after a reparametrization. Shikin investigated his solutions by qualitative methods.

Davidson (1988) found the following solution:

$$ds^2=2H\,du^2+2\,du\,dr-ur^{2n}(dx^2+dy^2), \qquad (7.1.23)$$

where:

$$2H=r/u+kr^m u^{(2-m)/(m-1)}, \quad n=m(m-1)/2, \qquad (7.1.24)$$

and k and m are arbitrary constants. The open FLRW model with the equation of state $\epsilon+3p=0$ results when $k=0$. When, further, $n=1/2$, the FLRW limit becomes flat. Hence, the special case $k=0$ discussed separately by Davidson is simply the open FLRW model in odd coordinates.

This closes the overview of A_2-symmetric perfect fluid models. As can be seen, all the solutions in this section were obtained under rather strong (and noninvariant) assumptions. It is now a challenge to find more general classes of solutions from which those described here could be obtained as limiting cases. The interconnections between the solutions of Wils (1990), Wainwright and Goode (1980), Collins and Lang (1987) and Götz (1988) indicated here suggest one direction of search.

The Aguirregabiria, Feinstein and Ibañez (1993) solution belongs in this section, too. As of the moment of writing, it is the only solution with a self-interacting scalar field source that is inhomogeneous and has a FLRW limit. Such a source can be interpreted as a perfect fluid (see Section 1.2, the text after eq. (1.2.22)). The metric of the solution is:

$$ds^2=e^f(dt^2-dz^2)-G\,(e^p dx^2+e^{-p}dy^2), \qquad (7.1.25)$$

where the functions $p\,(t,z), f\,(t,z)$ and $G(t)$ are determined by:

$$p=a\int G^{-1}(t)dt+\gamma z, \qquad (7.1.26)$$

$$f = -k\phi + \ln (G_{,tt}/G) - \ln (2\Lambda), \tag{7.1.27}$$

$$GG_{,tt}^2 - GG_{,t}G_{,ttt} - KG_{,t}^2G_{,tt} + M^2G_{,tt} + A^2G^2G_{,tt} = 0, \tag{7.1.28}$$

the function ϕ is the scalar field:

$$\phi = -\tfrac{1}{2}k \ln G - \frac{\gamma a}{2g} \int G^{-1}(t)dt + gz, \tag{7.1.29}$$

the constants a, γ, k, Λ and g are arbitrary, and K, M and A are constants defined by:

$$K = k^2/4 - 1/2, \quad M^2 = \frac{a^2}{2} + \frac{\gamma^2 a^2}{4g^2}, \quad A^2 = \gamma^2/2 + g^2, \tag{7.1.30}$$

and the self-interaction potential is:

$$V(\phi) = \Lambda e^{k\phi}. \tag{7.1.31}$$

The solution is A_2-symmetric, but is in general not plane symmetric. The flat FLRW limit results when $a = \gamma = g = 0$ and:

$$G_{,t} = C \, G^{3/2 - k^2/4}, \quad C = \text{const}, \quad k^2 \neq 6. \tag{7.1.32}$$

Aguirregabiria *et al.* investigated two examples of explicit solutions analytically, with $G = e^t$ and $G = \sinh(\omega t)$, and the generic case of (7.1.28) by qualitative and numerical methods. They were interested mainly in the inflationary phase of evolution of the model. The results were as follows:

1. With $k^2 < 2$, there are solutions with an inflationary phase.
2. The solutions with arbitrary k do homogenize at late times, but none of them isotropize.
3. For some values of the constants, the inflationary phase can occur several times.

7.2 A_2-symmetric solutions with nonperfect fluid sources

Three unrelated classes of solutions will be listed in this section; they are displayed in the lower part of Figure 7.1.

Roy and Banerjee (1988) obtained a few solutions from *ad hoc* assumptions. The first of them, which we shall label Roy–Banerjee I, has the metric:

$$ds^2 = dt^2 - R^2\{dx^2/(F+K)^4 + (F+K)^{-2}[e^{(1-h)x}dy^2 + e^{(h-1)x}dz^2]\}, \tag{7.2.1}$$

where h is an arbitrary constant and $R(t)$, $K(t)$ and $F(x)$ are arbitrary functions. The solution is of Petrov type I. Roy and Banerjee interpreted the source as a viscous-fluid with anisotropic pressure. The pressure anisotropy vanishes when either $F = \text{const}$ (in which case the model becomes spatially homogeneous) or $K = \text{const}$ and $F_{,xx}/F + F_{,x}^2/F^2 - (h-1)^2/2 = 0$. In the second case, all the FLRW models are reproduced in their full generality, but in nonstandard coordinates.

The second solution, considered in section 3 of the paper, is of Bianchi type VI_0. The third solution is the limit $K=0$ of (7.2.1), and Roy and Banerjee interpreted it as a heat-conducting fluid. It reproduces the FLRW models, all of them in full generality, when $q_\alpha=0$.

Mitskievič and Senin (1981) found three solutions with anisotropic fluid source in which the slices t=const have the local geometry of a three-dimensional torus. The first of them was described in our Section 2.13. The second solution is given in the paper in two coordinate systems; we choose the second:

$$ds^2=dt^2-R^2(t)[dr^2+\sin^2r\,dz^2+(\cos r+A)^2d\varphi^2], \qquad (7.2.2)$$

where A is an arbitrary constant and $R(t)$ obeys:

$$2RR_{,tt}+R_{,t}^2+1=0. \qquad (7.2.3)$$

The matter-density and the two pressures are:

$$\kappa\epsilon=3R_{,t}^2/R^2+(A+3\cos r)R^{-2}/(A+\cos r), \qquad (7.2.4)$$

$$\kappa p_2=AR^{-2}/(A+\cos r)\ \text{(transversal)}, \quad p_1=2p_2\ \text{(radial)}. \qquad (7.2.5)$$

The pressure becomes isotropic when $A=0$, but then the solution reduces to the closed FLRW dust. The slices t=const have the same geometry here as in Senin's (1982) solution, described in our Section 2.8.

The third solution arises from (7.2.2) when sin and cos are replaced by sinh and cosh respectively. The limit $p_1=p_2(=0)$ is then the open FLRW dust.

Bali, Singh and Tyagi (1987), working from a metric ansatz, found a solution of the Einstein–Maxwell equations in which the source is a perfect fluid and a current unrelated to the flow of matter. The metric is A_2-symmetric. We shall not copy the rather lengthy formulae defining the solution. The limit $F_{\mu\nu}=0$ leads to a few subcases, one of which is the flat FLRW model in its full generality, and the others have no FLRW limit. The limit $\dot{u}^\alpha=0$ leads to either the flat FLRW model or to a Bianchi model whose shear may only vanish together with the expansion. Hence, the FLRW limit is invariantly defined by $\dot{u}^\alpha=0$, $\sigma=0\neq\theta$. The second solution discussed in the paper is a subcase of the first one and has no FLRW limit.

7.3 Oleson's Petrov type N solutions

The collection of solutions found by Oleson (1971) is so far the only example of Petrov type N solutions with a perfect fluid source. Oleson assumed that the principal null congruence of the Weyl tensor is geodesic. As a result, the acceleration and shear are connected by:

$$\dot{u}^\alpha\dot{u}_\alpha=-3\sigma^2. \qquad (7.3.1)$$

The solutions have in general no symmetry. These properties allow one to conclude that the solutions are not members of any of the families presented previously, nor do they generalize any of them.

In the Wainwright (1979) classification, these solutions are of extrinsic class B_1, and intrinsic class I, as found by Wainwright (1981) himself.

There are three classes of such solutions. Class I is:

$$ds^2 = -r^{3/2}\left(d\xi - 2r^{-1/2}\frac{\partial H}{\partial \xi}dv\right)^2 - r^{1/2}\left(d\eta + 2r^{1/2}\frac{\partial H}{\partial \eta}dv\right)^2$$

$$+2H\,dr\,dv - B^{-2}H^2 dv^2,\tag{7.3.2}$$

where $(x^1, x^2, x^3, x^4) = (\xi, \eta, r, v)$, $u_\alpha = B\delta^3_\alpha$ (this holds for the other classes, too), and:

$$B^{-2} = 4\,[a^2(v)r^{1/2} + b^2 r^{3/2}],\tag{7.3.3}$$

$$H = L_1 \cos[a(v)\xi] + L_2 \sin[a(v)\xi],\tag{7.3.4}$$

$$L_\alpha = E_\alpha(v)\cos[b\eta + F_\alpha(v)] \quad \text{when } b \neq 0,$$
$$L_\alpha = E_\alpha(v)\eta + F_\alpha(v) \quad \text{when } b = 0,\tag{7.3.5}$$

b is an arbitrary constant and $a(v)$, $L_\alpha(v)$ and $F_\alpha(v)$ are arbitrary functions, $\alpha = 1, 2$. The pressure and matter-density are:

$$\kappa p = (3/4)r^{-3/2}[a^2(v) - 7b^2 r],\tag{7.3.6}$$

$$\kappa\epsilon = \kappa p + 12b^2/r^{1/2}.\tag{7.3.7}$$

The class IIa solution is:

$$ds^2 = -(k^2 - r^2)\left[R^{-1}\left(d\xi - Rk^{-1}\frac{\partial H}{\partial \xi}dv\right)^2 + R\left(d\eta + R^{-1}k^{-1}\frac{\partial H}{\partial \eta}dv\right)^2\right]$$

$$+2H\,dr\,dv - B^{-2}H^2 dv^2,\tag{7.3.8}$$

where k is another arbitrary constant, H is the same as before, and:

$$B^{-2} = 2(k^2 - r^2)^{1/2}[k^{-1}a^2(v) - b^2(r + k)],\tag{7.3.9}$$

$$R = (k - r)^{1/2}(k + r)^{-1/2},\tag{7.3.10}$$

$$L_\alpha = E_\alpha(v)\cos\{[a^2(v) - 2b^2 k^2]^{1/2}\eta + F_\alpha(v)\} \quad \text{when } a^2 > 2b^2 k^2,$$
$$L_\alpha = E_\alpha(v)\cosh\{[2b^2 k^2 - a^2(v)]^{1/2}\eta + F_\alpha(v)\} \quad \text{when } a^2 < 2b^2 k^2,\tag{7.3.11}$$

and $\alpha = 1, 2$. The pressure and matter-density are:

$$\kappa p = (3/2)(k^2 - r^2)^{-3/2}[b^2(4r^3 - 5k^2 r - k^3) + ka^2(v)],\tag{7.3.12}$$

$$\kappa\epsilon = \kappa p + 12b^2 r(k^2 - r^2)^{-1/2}.\tag{7.3.13}$$

The class IIb solution is:

$$ds^2 = -(r^2 - k^2)\left[R^{-1}\left(d\xi + Rk^{-1}\frac{\partial H}{\partial \xi}dv\right)^2 + R\left(d\eta - R^{-1}k^{-1}\frac{\partial H}{\partial \eta}dv\right)^2\right]$$

$$+2H\,dr\,dv - B^{-2}H^2 dv^2,\tag{7.3.14}$$

where:

$$B^{-2} = 2(r^2 - k^2)^{1/2}[b^2(r+k) - \varepsilon k^{-1}a^2(v)], \quad \varepsilon = \pm 1, \tag{7.3.15}$$

$$R = (r-k)^{1/2}(r+k)^{-1/2}. \tag{7.3.16}$$

For $\varepsilon = +1$, H is the same as before and:

$$L_\alpha = E_\alpha(v) \cosh\{[2b^2k^2 - a^2(v)]^{1/2}\eta + F_\alpha(v)\} \tag{7.3.17}$$

while for $\varepsilon = -1$:

$$H = L_1\cosh[a(v)\xi] + L_2\sinh[a(v)\xi], \tag{7.3.18}$$

$$L_\alpha = E_\alpha(v) \cos\{[a^2(v) + 2b^2k^2]^{1/2}\eta + F_\alpha(v)\}. \tag{7.3.19}$$

The pressure and matter-density are:

$$\kappa p = (3/2)(r^2 - k^2)^{-3/2}[b^2(-4r^3 + 5k^2r + k^3) - \varepsilon ka^2(v)], \tag{7.3.20}$$

$$\kappa\varepsilon = \kappa p + 12b^2r(r^2 - k^2)^{-1/2}. \tag{7.3.21}$$

The rotation is zero in each class. The consideration in the paper shows that $a_{,v} = 0$ is the conformally flat limit of all the solutions. In that case, shear and acceleration vanish automatically, and all the solutions reduce to the FLRW models represented in nonstandard coordinates. The formulae for ϵ and p show that in each case the FLRW limit will have a definite form of the equation of state, with no arbitrary functions. Since $u_\alpha dx^\alpha = B\, dr$, the hypersurfaces $r = $const are orthogonal to the fluid flow, and their curvature determines the limiting FLRW geometry. It can be verified by examples that all three FLRW geometries can be reproduced from class IIb. The open and closed FLRW models result when $a(v)$, $E_\alpha(v)$ and $F_\alpha(v)$ are all constant and $L_2/L_1 = $const; the sign of ε is then the sign of the FLRW spatial curvature. The flat FLRW model results when $a(v) = 0$ in addition. The calculations proving these statements were done by the program Ortocartan.

In special cases, the solutions can acquire a one- or two-dimensional symmetry group, and examples of such limits are given in the paper. Oleson showed that the matching conditions to a vacuum solution can never be fulfilled because the equations $\{p = 0, u^\alpha p_{,\alpha} = 0\}_{\text{on the boundary}}$ lead to a contradiction on any hypersurface. The paper also gives more detailed information on the allowed range of the coordinates and on the properties of the Weyl tensor.

Oleson's paper fulfilled the prediction of Szekeres (1966a), who argued that perfect fluid solutions of Petrov type N might exist except when $p = 0$. For completeness, the Oleson (1971) solutions are displayed in Figure 7.2, although there are no interconnections with other solutions to be shown there.

Another paper to be mentioned here is that by Kundt and Trümper (1962). In it, the authors discussed properties of those solutions of Einstein's equations whose Weyl tensor is of type N. The Oleson solutions are examples to which the following theorems may be applied:

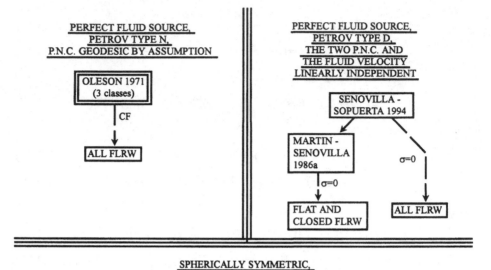

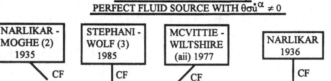

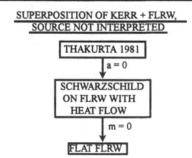

Figure 7.2. Miscellaneous I.

(I) $(C^{\alpha\beta}{}_{\gamma\delta}=0)\Rightarrow(\omega=\sigma=0)\Rightarrow(\epsilon=\epsilon(t)$ in the comoving coordinates)$\Rightarrow$(the FLRW models follow if $p=p(\epsilon)$).

(II) $(\omega=\sigma=0,\ p=p(\epsilon))\Rightarrow(C^{\alpha\beta}{}_{\gamma\delta}=0)$.

(III) There are no type N solutions with $\omega=0=\dot{u}^{\alpha}$.

7.4 Martin and Senovilla's (1986a,b) and Senovilla and Sopuerta's (1994) solutions

Martin and Senovilla's 1986b paper contains an erratum to the 1986a paper. The authors studied generating new solutions from old ones by the Kerr–Schild method, and presented one explicit example.[52] The example is generated from the limit a=const of the class I solution of Oleson (1971, see the preceding section), and so the similarity of the two solutions is not accidental. The solution is:

$$ds^2 = t^{3/2}(dx - 2t^{-1/2}g_{,x}h\,du)^2 + t^{1/2}(dy + 2t^{1/2}gh_{,y}du)^2$$

$$- 2gh\,dt\,du - 2g^2h^2(M + ch/g^3)du^2, \tag{7.4.1}$$

where a, b and c are arbitrary constants, and the functions $g(x)$, $h(y)$ and $M(t)$ are given by:

$$g_{,xx} + a^2g = 0, \quad h_{,yy} + b^2h = 0, \quad M(t) = 2t^{1/2}(a^2 + b^2t). \tag{7.4.2}$$

The pressure, density and velocity field are:

$$\kappa p = (3/4)t^{-3/2}(a^2 - 7b^2t) + 3ch/(8t^2g^3), \tag{7.4.3}$$

$$\kappa\epsilon = \kappa p + 12b^2/t^{1/2}, \tag{7.4.4}$$

$$u_\alpha dx^\alpha = [2(M + ch/g^3)]^{-1/2}dt. \tag{7.4.5}$$

This solution is of Petrov type D and has the one-dimensional symmetry group $u = u' + $const. The velocity field is not spanned on the principal null directions of the Weyl tensor. Martin and Senovilla did not give sufficient information for classifying this solution in Wainwright's (1979 and 1981) scheme. By checking the curvature of the hypersurfaces t=const it may be verified that the limit c=0 of this solution is a closed FLRW model when $ab \neq 0$, and a flat one when $ab = 0$. This calculation was again performed by the program Ortocartan. The solution is displayed in Figure 7.2.

The solution (7.4.1)–(7.4.5) was generalized by Senovilla and Sopuerta (1994). The generalization is reproduced here in the original notation, which is partly in conflict with (7.4.1)–(7.4.5). The metric is:

$$ds^2 = -2G\,dt\,du + 2G^2\tilde{M}\,du^2 + t^{1-c}[dx + (G_{,x}/c)t^c du]^2$$

$$+ t^{1+c}[dy - (G_{,y}/c)t^{-c}du]^2, \tag{7.4.6}$$

where:

$$\tilde{M}(t,x,y) = M(t) + H(x,y), \tag{7.4.7}$$

$$M(t) = 2t(at^c + bt^{-c}), \tag{7.4.8}$$

$$H(x,y) = D[g(x)/h(y)]^{1/c}/G(x,y), \tag{7.4.9}$$

$$G(x,y) = g(x)\,h(y), \tag{7.4.10}$$

a, b, c and D are arbitrary constants and the functions $g(x)$ and $h(y)$ are defined by:

$$g_{,xx} + (2c)^2ag = h_{,yy} + (2c)^2bh = 0. \tag{7.4.11}$$

---

[52] It is rather difficult to assemble the metric using the information given in the paper. This section is partly based on information provided by J. M. M. Senovilla in a letter to A. K., for which this author is very grateful.

This is a perfect fluid solution; the energy-density, pressure and the velocity field are given by:

$$\kappa\epsilon = 3t^{-1}[a(1+c)^2t^c + b(1-c)^2t^{-c}] + (1-c^2)H(x,y)/(2t^2), \qquad (7.4.12)$$

$$\kappa p = -5t^{-1}[a(1+c)(1/5+c)t^c + b(1-c)(1/5-c)t^{-c}] + (1-c^2)H(x,y)/(2t^2), \qquad (7.4.13)$$

$$u_\alpha dx^\alpha = -(2\tilde{M})^{-1/2}dt. \qquad (7.4.14)$$

The rotation is zero, while the shear, acceleration and expansion are nonzero. The formulae are given in Senovilla and Sopuerta's paper. The shear vanishes when $H=0$, that is, $D=0$, and the solution then reduces to the FLRW models. The FLRW limit is flat when $abc=0$, open when $ab<0\neq c$ and closed when $ab>0\neq c$. When $c=-1/2$, the solution (7.4.1)–(7.4.5) is reproduced. The symmetry group is in general one-dimensional ($u=u'+$const). When $h=1$ and $b=0$, the second symmetry $y=y'+$const shows up.

7.5 Spherically symmetric solutions with expansion, shear and acceleration

A few solutions of this kind were derived by various authors from *ad hoc* assumptions. They are shown in Figure 7.2, although there are no interconnections between them. Their extrinsic class in Wainwright's (1979 and 1981) scheme cannot be more general than B_3 because of the symmetry. They cannot be of class C or more special because, otherwise, they would be either in the Szekeres–Szafron or in the Stephani–Barnes families (see Chapters 2 and 4 respectively). Insufficient information was given in the papers to determine Wainwright's intrinsic class.

Narlikar and Moghe (1935), working by guesswork, presented five metrics, none of which were investigated further. The program Ortocartan was used to check whether any of the metrics can have a conformally flat limit. It turned out that the limit is nontrivial only in case 2; in all the other cases the limit either cannot exist or leads to a de Sitter or Minkowski metric. Hence, only case 2 will be presented here. It is:

$$ds^2 = e^\nu dt^2 - e^\mu[dr^2 + r^2(d\vartheta^2 + \sin^2\vartheta \, d\varphi^2)], \qquad (7.5.1)$$

where:

$$\mu = -1/[r(at+b)] + a'(at+b) + b'/(at+b)^2, \quad \nu = -\mu, \qquad (7.5.2)$$

a, b, a' and b' being arbitrary constants. The coordinates are not comoving. The Einstein tensor for this solution has the algebraic form appropriate for a perfect fluid, but no physical quantities were calculated in the paper. The (flat only) FLRW limit is achieved when $a\to\infty$ or $b\to\infty$; the limit must be accompanied by appropriate rescalings of other constants so that the resulting μ is:

$$\mu_{FLRW} = C(At+B) + D/(At+B)^2. \qquad (7.5.3)$$

The FLRW limit (7.5.3) can have a linear barotropic equation of state $p=(\gamma-1)\epsilon$ only if $D=0$, and then $p=-5\epsilon/3$. This is useful when comparing the Narlikar–Moghe solution with other solutions.

Stephani and Wolf (1985) investigated spacetimes which admit flat hypersurfaces (timelike or spacelike), under the additional assumption that the second fundamental tensor of these flat hypersurfaces is proportional to their metric tensor. Only examples of explicit metrics were given. For the class with zero second fundamental form, the example given has no FLRW limit. For the class with $K_{ab}=\lambda g_{ab}$, a, $b=1$, 2, 3, where $\lambda=\lambda(t)$, only the de Sitter and a subcase of the Stephani (1967a) solution were given as examples. Two new perfect fluid solutions were presented in the class where λ was allowed to depend on other coordinates. The first of these is:

$$ds^2=(x^4)^{-2}[\eta_{ab}dx^a dx^b+\varepsilon M^2(dx^4)^2], \tag{7.5.4}$$

where $(\eta_{ab})=\text{diag}(1, 1, -\varepsilon)$; $a, b, m, n=1, 2, 3$; $\varepsilon=\pm 1$ and M is a function of x^4 and of $w\stackrel{\text{def}}{=}\eta_{mn}x^m x^n$ defined by the equation:

$$2(x^4)^2 w M,_{ww}/M+2(x^4)^2 M,_w/M-2\varepsilon M,_w^2/(M^3 M,_{ww})+\varepsilon x^4 M,_4/M^3=0. \tag{7.5.5}$$

The flat hypersurfaces $x^4=\text{const}$ are timelike when $\varepsilon=+1$ and spacelike when $\varepsilon=-1$. The pressure and matter-density of the perfect fluid are:

$$\kappa p=3\varepsilon/M^2+4\varepsilon M,_w^2/(M^3 M,_{ww}), \tag{7.5.6}$$

$$\kappa\epsilon=-3\varepsilon/M^2+4(x^4)^2 w M,_{ww}/M, \tag{7.5.7}$$

and the velocity field of the fluid is:

$$u_a=2\eta_{ab}x^b\{-M,_{ww}/[\kappa(\epsilon+p)M]\}^{1/2},$$

$$(u^4)^2=-\varepsilon(x^4)^2/M^2+4\varepsilon w(x^4)^4 M,_{ww}/[\kappa(\epsilon+p)M^3]. \tag{7.5.8}$$

In the second class, the metric, the variable w and the constant ε are defined as before; the equation defining $M(x^4,w)$ is:

$$\varepsilon m^2(x^4)^2 M,_{ww}/M-4M,_w^2/(M^3 M,_{ww})+2x^4 M,_4/M^3=0. \tag{7.5.9}$$

$m^2=\eta_{ab}m^a m^b$, the m_a are arbitrary constants, the pressure of the fluid is given by (7.5.6), while the density and the velocity field are:

$$\kappa\epsilon=-3\varepsilon/M^2+m^2(x^4)^2 M,_{ww}/M, \tag{7.5.10}$$

$$u_a=m_a\{-M,_{ww}/[\kappa(\epsilon+p)]\}^{1/2},$$

$$(u^4)^2=-\varepsilon(x^4)^2/M^2+\varepsilon m^2(x^4)^4 M,_{ww}/[\kappa(\epsilon+p)M^3]. \tag{7.5.11}$$

The conformally flat limits of both classes were checked using the program Ortocartan. It turns out that the limits are static when $\varepsilon=+1$, that is, when the flat hypersurfaces are timelike. Hence, only $\varepsilon=-1$ will be considered here. Then, the flat hypersurfaces $x^4=\text{const}$ are spherically symmetric, and so is the spacetime. The metric (7.5.4) is conformally flat when $M,_{ww}=0$, and then the Einstein tensor has the algebraic form appropriate for a perfect fluid when $M,_{w}=0$, that is, $M=M(x^4)$. In this limit, (7.5.4) becomes the flat FLRW model in its full generality. The formula defining pressure, eq. (7.5.6), becomes undetermined in this case, and so do the equations defining M, (7.5.5) and (7.5.9). However, they can be used to calculate the FLRW pressure as follows. From (7.5.6) we find that:

$$\lim_{M,_{w}\to 0}[-4M,_{w}^2/(M^3 M,_{ww})]=\kappa p+3/M^2. \qquad (7.5.12)$$

We substitute this into (7.5.5) and obtain:

$$\lim_{M,_{w}\to 0}(\kappa p)=-3/M^2-2x^4 M,_4/M^3. \qquad (7.5.13)$$

This coincides with the FLRW pressure calculated directly from (7.5.4) with $M,_{w}=0$.

The Stephani–Wolf solutions are not contained in the Narlikar–Moghe solution (7.5.1)–(7.5.2) because the latter reproduce the FLRW limit only with a specific equation of state. Conversely, the Narlikar–Moghe solution is not contained in the Stephani–Wolf class because a spherically symmetric hypersurface $t=f(r)$ in (7.5.1)–(7.5.2) can never be flat; this was verified using Ortocartan.

A broad overview of solutions (vacuum and perfect fluid) that can be foliated into flat hypersurfaces, timelike or spacelike, was presented by Wolf (1986; an additional assumption was made in this paper, that the second fundamental tensor of the flat hypersurfaces is covariantly constant over them). Most of the resulting explicit solutions are of Bianchi type I or have zero expansion. A subcase of the Stephani (1967a) solution (see our Section 4.9) shows up as an example in section IIIA of the paper. A new solution with $\theta\sigma\omega\dot{u}^\alpha\neq 0$, eqs. (3.18) in the paper, was found, but its limit $\epsilon,_\mu(\delta^\mu{}_\alpha-u^\mu u_\alpha)=0$, that is, $\dot{u}^\alpha=0$, is a vacuum metric.

McVittie and Wiltshire (1977) found a few solutions which belong in this section, but only one of them (their class A(ii)) allows for a FLRW limit. The class A(i) solution is an open FLRW model itself in noncomoving coordinates, class B does not allow the limit of zero Weyl tensor, the conformally flat limit of class C is the de Sitter metric, and the separable solution has no conformally flat limit.

The class A(ii) solution is:

$$ds^2=z^{4/2}d\eta^2-z^{2/3}[d\xi^2+\xi^2(d\vartheta^2+\sin^2\vartheta\,d\varphi^2)], \qquad (7.5.14)$$

where:

$$z\overset{\text{def}}{=}\varepsilon_1\xi^2/\xi_0^2-\varepsilon\eta/\eta_0, \qquad (7.5.15)$$

ξ_0 and η_0 are arbitrary constants, and $\varepsilon_1=\pm 1=\varepsilon$. The Weyl tensor (calculated using the program Ortocartan) vanishes when $\xi_0\to\infty$; the resulting solution is the flat FLRW model with the equation of state $p=7\epsilon/3$. There is no relationship between (7.5.14)–(7.5.15) and the other solutions from this section because (7.5.14)–(7.5.15) cannot contain a flat hypersurface $\eta=f(\xi)$ (as verified with the help of Ortocartan), and the equations of state in the FLRW limit of the McVittie–Wiltshire and Narlikar–Moghe solutions are mutually exclusive.

Narlikar (1936) showed that the metric:

$$ds^2 = S^2(\chi)dt^2 - R^2(t) [d\chi^2 + \sin^2\chi (d\vartheta^2 + \sin^2\vartheta \, d\varphi^2)] \qquad (7.5.16)$$

will obey the Einstein equations with a perfect fluid source if $R_{,t} = \alpha = $ const and S obeys:

$$4\alpha^2 + 2\alpha^2 SS_{,\chi\chi}/S_{,\chi}^2 + (S^2 S_{,\chi\chi}/S_{,\chi})[\sin^2\chi S_{,\chi\chi} - 2\cos\chi \, S_{,\chi} + 2S] = 0. \quad (7.5.17)$$

The program Ortocartan verified that the Weyl tensor of (7.5.16)–(7.5.17) will vanish when $S = A\cos\chi + B$, where A and B are constants, and then the energy–density ϵ and pressure p will obey $\partial(\epsilon,p)/\partial(t,\chi) = 0$ only if $A = 0$, that is, $S = $ const. The resulting metric is a closed FLRW model with a specific equation of state. Strictly speaking, Narlikar's result does not qualify as a solution (no equation was solved), but is mentioned here as one more evidence that the problem of inhomogeneous cosmological models was appreciated quite early and by several authors.

7.6 Thakurta's (1981) metric

This is included only for completeness. Thakurta set out to construct a superposition of the Kerr and FLRW metrics. He criticized the Vaidya (1977) solution (see our Section 5.6, after eq. (5.6.24)) for not approaching FLRW at large r and not having a perfect fluid source. The first point is a misunderstanding: the solution as it stands is a perturbation of the closed FLRW model only, and simply does not have a spatial infinity. On the second point, Thakurta's metric is hardly an improvement: its energy-momentum tensor has many nonvanishing components which the author did not interpret. The metric is obtained by multiplying the Kerr metric by an arbitrary function of time, $R(t)$, and as such does not in fact qualify as a solution. In the limit $a = 0$ (where a is the usual Kerr parameter), a superposition of the Schwarzschild and the flat FLRW metrics results; the source is then a heat-conducting fluid. When, further, $m = 0$, the flat FLRW model results. The energy-momentum tensor of the Thakurta metric acquires a perfect fluid form only if either $m = 0$ (i.e. in the FLRW limit) or $R(t) = $ const, the latter limit is the Kerr (vacuum) metric. The Thakurta metric is displayed in Figure 7.2.

7.7 Wainwright's (1974) spacetimes and Martin–Pascual and Senovilla's (1988) solutions

Wainwright (1974) considered the class of spacetimes obeying the following conditions:

1. There exists a congruence of expanding, geodesic, shearfree and rotation-free null curves in the spacetime.

2. The field of vectors tangent to this congruence is a repeated principal null direction of the Weyl tensor.
3. The metric of the spacetime obeys the Einstein equations with a perfect fluid source.
4. The repeated principal null vector of the Weyl tensor is parallelly transported along the fluid's flow-lines.
5. The fluid has zero rotation.

The family of metrics obeying these conditions consists of two classes. They are not in fact solutions because they are defined by sets of partial differential equations. However, they do include the FLRW models, and explicit solutions within case 1 were found by Martin–Pascual and Senovilla (1988). Hence, Wainwright's paper has to be included for completeness. The metrics are displayed in the upper part of Figure 7.3. They are in general of Petrov type II, but include subcases of types D and O. They are nonvacuum generalizations of the well-known Robinson–Trautman metrics.

As found by Wainwright (1981), these spacetimes are of extrinsic class A and of intrinsic class I in Wainwright's (1979) scheme, that is, they are the most general on both accounts. In which subcase of class A they are, and whether the Martin–Pascual–Senovilla solutions are more special or not, cannot be decided in any simple way using the information in the papers.

Case 1 is:

$$\mathrm{d}s^2 = -(1/2)P^{-2}\varepsilon r\,\mathrm{d}z\,\mathrm{d}\bar{z} + 2\,\mathrm{d}u(\mathrm{d}r - U\,\mathrm{d}u),\qquad(7.7.1)$$

where:

$$U \overset{\text{def}}{=} r(\ln P)_{,u} + U^0 + erS^0 + 2\varepsilon Gr\ln|r|,\qquad(7.7.2)$$

S^0 and G are arbitrary constants, $\varepsilon \overset{\text{def}}{=} r/|r|$, and the functions $U^0(u,z,\bar{z})$ and $P(u,z,\bar{z})$ are defined by the equations:

$$8P^2(\ln P)_{,z\bar{z}} - \varepsilon\,(\ln P)_{,u} = 2G,\qquad(7.7.3)$$

$$8U^0_{,z\bar{z}} + \varepsilon(U^0/P^2)_{,u} = 0.\qquad(7.7.4)$$

The velocity field of the perfect fluid source is $u_\alpha = (2U)^{-1/2}\delta^2{}_\alpha$, where (x^1, x^2, x^3, x^4) $=(u, r, \mathrm{Re}(z), \mathrm{Im}(z))$, and the pressure and energy-density are given by:

$$\kappa p = r^{-1}[\tfrac{1}{2}(\ln P)_{,u} - 4\varepsilon G - \tfrac{1}{2}\varepsilon S^0] + U^0/(2r^2) - \varepsilon Gr^{-1}\ln|r|,\qquad(7.7.5)$$

$$\kappa\epsilon = \kappa p + (2\varepsilon/r)(4G + S^0) + (4\varepsilon G/r)\ln|r|.\qquad(7.7.6)$$

The metrics of this class have in general nonzero acceleration and shear, and no apparent symmetry. The limit $\sigma = 0$ leads to either the flat FLRW model, the open FLRW model (each with a specific equation of state) or to a static solution. The limit $\dot{u}^\alpha = 0$ reproduces the same two FLRW models, and, in addition, a member of the Kompaneets–Chernov (1964) class (see our Section 2.11) with a specific equation of state. The FLRW models result when $U^0 = 0$ and $\ln P = -2\varepsilon Gu$; the model is open

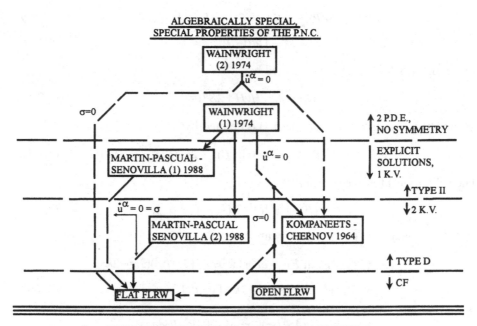

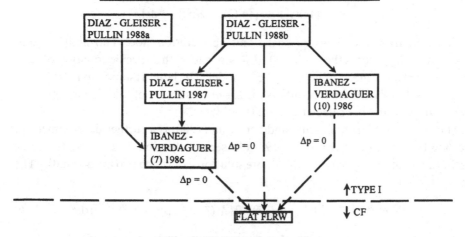

Figure 7.3. Miscellaneous II.

when $G \neq 0$ and flat when $G = 0$. The Kompaneets–Chernov model results when $U = $ const and $P_{,u} = 0 \neq G$. These are only examples; the full collection of FLRW limits is larger. The calculations leading to the limits are rather lengthy.

Wainwright's case 2 is:

$$\mathrm{d}s^2 = -[\chi^2/(2P^2)]\mathrm{d}z\,\mathrm{d}\bar{z} + 2\,\mathrm{d}u\,(\mathrm{d}r - U\,\mathrm{d}u), \qquad (7.7.7)$$

where:

$$\chi^2 = \varepsilon(r^2 - k^2), \qquad (7.7.8)$$

$$U=r (\ln P),_u - \varepsilon K + \chi^2(S^0 + m\int\chi^{-4}dr) + \varepsilon k^2 H, \qquad (7.7.9)$$

$$K=4P^2(\ln P),_{z\bar{z}}. \qquad (7.7.10)$$

The constants S^0, m and H are related by:

$$Hm=0, \quad H(H+2S^0)=0, \qquad (7.7.11)$$

the constant k is arbitrary, and $P(u,z,\bar{z})$ is defined by:

$$P^2[4K,_{z\bar{z}} + \tfrac{1}{2}k^2(P^{-2}),_{uu}] - \varepsilon m (\ln P),_u + 2k^2H (K-k^2H)=0. \qquad (7.7.12)$$

The velocity field of the source is:

$$u_\alpha = [2(U+\chi^2 H)]^{-1/2}(H\chi^2\delta^1_{\ \alpha} + \delta^2_{\ \alpha}), \qquad (7.7.13)$$

and the pressure and energy–density are:

$$\kappa p=(2k^2/\chi^4)[r(\ln P),_u - \varepsilon K + \varepsilon k^2 H] - (2S^0/\chi^2)(3r^2 - 2k^2)$$
$$-(2m/\chi^2)[\varepsilon r/\chi^2 + (3r^2 - 2k^2)\int\chi^{-4}dr], \qquad (7.7.14)$$

$$\kappa\epsilon = \kappa p + (4S^0/\chi^2)(3r^2 - k^2) + 4k^2 H/\chi^2$$
$$+(4m/\chi^2)[\varepsilon r/\chi^2 + (3r^2 - 2k^2)\int\chi^{-4}dr]. \qquad (7.7.15)$$

These metrics have nonzero shear and acceleration, too. The limit of zero acceleration reproduces the flat (only) FLRW model with a specific equation of state or a specific Kompaneets–Chernov (1964) model. The latter results when $H=P,_u=0$, $K=$const$\neq 0$. A FLRW limit results as the $K=m=0$ subcase of this. Again, these are only examples, and the calculations are lengthy.

The paper by Martin–Pascual and Senovilla (1988) is a further development of the one by Martin and Senovilla (1986a,b, see our Section 7.4). Using the same method, the authors generated three more solutions, of which two are nonstatic. The first of them is:[53]

$$ds^2=2e^{\sqrt{2}C(u+v)}(-du\,dv+d\rho\,d\bar{\rho})+2\{M(\bar{\rho})+\bar{M}(\rho)+Ee^{\sqrt{2}C(u+v)}\}du^2, \quad (7.7.16)$$

where C and E are arbitrary constants, ρ is a complex coordinate and $M(\rho)$ is an arbitrary complex function. The velocity field, matter-density and pressure are:

$$u_\alpha dx^\alpha = F(du+dv), \qquad (7.7.17)$$

$$\kappa(\epsilon+p)=2C^2e^{-\sqrt{2}C(u+v)}\{1+E+e^{-\sqrt{2}C(u+v)}[M(\bar{\rho})+\bar{M}(\rho)]\}, \qquad (7.7.18)$$

$$\kappa(\epsilon-p)=4C^2(1+E)e^{-\sqrt{2}C(u+v)}, \qquad (7.7.19)$$

where F should be determined from $u_\alpha u^\alpha = -1$. With $M(\bar{\rho})\neq$const the solution is of

---

[53] This section is again largely based on the letter from J. M. M. Senovilla, for which A.K. is very grateful.

Petrov type II; otherwise it is of Petrov type D. It is the following subcase of Wainwright's (1974) case 1 (our eqs. (7.7.1)–(7.7.6)):

$$G=0, \quad P=\varepsilon=1, \quad S^0=\sqrt{2}C(1+E), \tag{7.7.20}$$

$$r=(\sqrt{2}C)^{-1}e^{\sqrt{2}C(u+v)}, \quad z=2(\sqrt{2}C)^{1/2}\rho, \tag{7.7.21}$$

$$U^0=M(\bar{p})+\bar{M}(\rho). \tag{7.7.22}$$

Still, Martin–Pascual and Senovilla's claim that this is a new solution is justified because their result is explicit, while Wainwright's class is defined by two partial differential equations.

The solution has in general only one Killing vector, $k^\mu \partial/\partial x^\mu = \partial/\partial u - \partial/\partial v$, but the following special cases exist:

(a) If $M(\bar{p})=c\bar{p}+d$, where c and d are constants, then $k^\mu \partial/\partial x^\mu = i\bar{c}\partial/\partial\bar{p} - ic\partial/\partial\rho$ is a second Killing vector.
(b) If $M(\bar{p})=d=$const, then the solution becomes plane symmetric, and acquires the additional Killing vectors $(\partial/\partial\rho + \partial/\partial\bar{p})$, $-i(\partial/\partial\rho - \partial/\partial\bar{p})$ and $i(\bar{p}\partial/\partial\rho - \rho\partial/\partial\bar{p})$.
(c) If $M(\bar{p})=(a/b)\ln(ib\bar{p}+A_2)$ where a and b are real constants, then $k^\mu \partial/\partial x^\mu = (A_2+ib\bar{p})\partial/\partial\bar{p} + (\bar{A}_2-ib\rho)\partial/\partial\rho$ is an additional Killing vector.
(d) If $M(\bar{p})=C(A_1\bar{p}+A_2)^{-2(1+\bar{A}_1/A_1)}-ia(A_1+\bar{A}_1)$ and $E=-1$, then $k^\mu \partial/\partial x^\mu = (A_1+\bar{A}_1)u\partial/\partial u - (A_1+\bar{A}_1)[u\pm(\sqrt{2}C)^{-1}]\partial/\partial v + (A_1\bar{p}+A_2)\partial/\partial\bar{p} + (\bar{A}_1\rho+\bar{A}_2)\partial/\partial\rho$ is an additional Killing vector.

This solution can reproduce only the flat FLRW model with the equation of state $\varepsilon+3p=0$, it results when $M=E=0$.

The second nonstatic solution by Martin–Pascual and Senovilla (1988) is:

$$ds^2=2e^{\sqrt{2}C(u+v)}(-du\,dv+d\rho\,d\bar{p})$$
$$+2[-(\sqrt{2}a/C)e^{-C(u+\rho+\bar{p})/\sqrt{2}}+Ee^{\sqrt{2}C(u+v)}]du^2, \tag{7.7.23}$$

where a, C and E are arbitrary constants. This solution has two Killing vectors:

$$k_1^\mu \partial/\partial x^\mu = -i(\partial/\partial\rho - \partial/\partial\bar{p}),$$

$$k_2^\mu \partial/\partial x^\mu = \partial/\partial u - \partial/\partial v + \tfrac{1}{2}(\partial/\partial\bar{p} + \partial/\partial\rho). \tag{7.7.24}$$

The velocity field is of the form (7.7.17), and the pressure and matter-density are:

$$\kappa p = -C^2(1+E)e^{-\sqrt{2}C(u+v)} - \sqrt{2}aCe^{-2\sqrt{2}C(u+v)-C(u+\rho+\bar{p})/\sqrt{2}}, \tag{7.7.25}$$

$$\kappa\varepsilon = 3C^2(1+E)e^{-\sqrt{2}C(u+v)} - \sqrt{2}aCe^{-2\sqrt{2}C(u+v)-C(u+\rho+\bar{p})/\sqrt{2}}. \tag{7.7.26}$$

This solution is another subcase of Wainwright's (1974) case 1, defined by (7.7.20)–(7.7.21) and:

$$U^0=-(\sqrt{2}a/C)e^{-C(u+\rho+\bar{p})/\sqrt{2}}. \tag{7.7.27}$$

The (flat only) FLRW limit with $\varepsilon+3p=0$ is reproduced when $a=E=0$.

7.8 The anisotropic fluids generated via Kaluza–Klein spaces

Wainwright, Ince and Marshman's (1979) generating algorithm described in Section 6.1 can only produce perfect fluid solutions obeying $\epsilon = p$. Ibanez and Verdaguer (1986) invented a method to bypass this limitation, to some extent. The method is based on the paper by Belinskii and Ruffini (1980) in which the authors showed how the Belinskii–Zakharov generation algorithm (see Verdaguer 1993) may be applied in a five-dimensional space and used to generate solutions of the Einstein–Maxwell equations. The algorithm was later extended by Diaz, Gleiser and Pullin (1987) to include an additional scalar field in the five-dimensional space. Its current version is as follows:

1. As observed by Belinskii and Ruffini (1980), those solutions of the Einstein equations for which the source is of the form:

$$R_{\mu\nu} = \phi^{-1}\phi_{;\mu\nu}, \quad \phi_{;\rho}{}^{;\rho} = 0 \qquad (7.8.1)$$

are vacuum solutions in the five-dimensional Kaluza–Klein theory, with ϕ being the Kaluza–Klein scalar field. There exist perfect fluid solutions for which $R_{\mu\nu}$ has the form (7.8.1); an example is the flat FLRW model with $\epsilon = 3p$, which is the starting point of the procedure.

2. As stated above, the Belinskii–Zakharov generation algorithm also applies in five dimensions. Diaz, Gleiser and Pullin (1987) observed that Wainwright, Ince and Marshman's algorithm (1979, see our Section 6.1) works in five dimensions, too, producing solutions of the five-dimensional Einstein equations with a scalar field source out of vacuum solutions. Applying the two algorithms successively to a five-dimensional vacuum seed metric, a metric is obtained that obeys:

$$R_{AB} = \chi_{,A}\chi_{,B}, \qquad (7.8.2)$$

where χ is the new scalar field and $A, B = 0,...,4$. After reduction to four dimensions, the solution of (7.8.2) will obey the Einstein equations with:

$$R_{\mu\nu} = \phi^{-1}\phi_{;\mu\nu} + \chi_{,\mu}\chi_{,\nu}. \qquad (7.8.3)$$

3. The energy-momentum tensor of any perfect fluid can be represented as a sum of the "stiff" perfect fluid with $\epsilon_s = p_s$ and the radiative perfect fluid with $\epsilon_r = 3p_r$, with $\epsilon_r = 3(\epsilon - p)/2$, $\epsilon_s = 3p/2 - \epsilon/2$. After performing such a decomposition, five-dimensional scalar field solution is found corresponding to the initial metric.

4. The vacuum part of the five-dimensional metric is identified by applying the Wainwright–Ince–Marshman (WIM) algorithm in reverse.

5. The Belinskii–Ruffini generation algorithm is applied to that vacuum part.

6. The WIM algorithm is used to obtain a new scalar field solution from the vacuum solution found in step 5. The resulting scalar field is actually the same one that was discarded in step 4.

7. Reduction from five to four dimensions by the usual Kaluza–Klein procedure yields the final solution.

Ibanez and Verdaguer (1986) applied this method in the subcase $\chi=0$ to the flat FLRW model with $\epsilon=3p$, and obtained the historically earliest solutions of this family. The solutions are solitonic perturbations travelling on the flat FLRW background. They have anisotropic pressure with three different eigenvalues p_1, p_2 and p_3, only one of which obeys $p_1=\epsilon$ (in the direction of the soliton's propagation). One limitation of the WIM algorithm is thus bypassed, but at the price of making the pressure anisotropic, and the perfect fluid limit of all the solutions obtained in this way is only the FLRW model with $\epsilon=3p$. The authors of all the papers described in this section did not provide a hydrodynamical interpretation of the source. The relationships among the different solutions are shown in the lower part of Figure 7.3.

Since all the solutions described in this section have a two-dimensional Abelian symmetry group with spacelike orbits and are orthogonally transitive, they are of extrinsic type no more general than B_3 in the Wainwright (1979) scheme. Not enough information was given in the papers to determine the intrinsic type or delimit the extrinsic type more exactly.

The most elaborate solution of this family was derived by Diaz, Gleiser and Pullin (1988b). It is:

$$ds^2=f(t,r)\,(-dt^2+dr^2)+g_{22}dz^2+g_{33}d\psi^2, \tag{7.8.4}$$

where:

$$f=Ct^{n-2}r^{-2}(\sigma_1-\sigma_2)^{u-2}(\sigma_1\sigma_2)^{-p+2}[(t+\sigma_1r)(t+\sigma_2r)]^Q$$

$$\times[(1-\sigma_1^2)(1-\sigma_2^2)]^{-u/2}(1-\sigma_1\sigma_2)^{-2}, \tag{7.8.5}$$

$$g_{22}=t^n(\sigma_1\sigma_2)^{-q}r^2, \quad g_{33}=t^n(\sigma_1\sigma_2)^{-p}, \tag{7.8.6}$$

where p, q and n are arbitrary constants, and:

$$u=2(q^2+p^2+pq), \quad Q=-(q/2)(3n-2)-(p/2)(3n-4). \tag{7.8.7}$$

The functions σ_1 and σ_2 are defined by:

$$\sigma_k^\pm=-(2tr)^{-1}\{\omega_k^2+t^2+r^2\pm[(\omega_k^2+t^2+r^2)^2-4t^2r^2]^{1/2}\},$$

$$\sigma_1=\sigma_1^-, \quad \sigma_2=\sigma_2^+, \tag{7.8.8}$$

where ω_1 and ω_2 are arbitrary constants. The constant C is:

$$C=(\omega_2^2-\omega_1^2)^2/\omega_1^{2Q}, \tag{7.8.9}$$

and the two scalar fields are:

$$\phi=t^{1-n}(\sigma_1\sigma_2)^{(p+q)/2}, \quad \chi=[3n(2-n)/2]^{1/2}\ln t. \tag{7.8.10}$$

The source of the solution (7.8.4)–(7.8.10) is an anisotropic fluid with the energy-density ϵ and the eigenvalues of pressure P_r, P_z and P_ψ given by:

$$\epsilon=-a^-, \quad P_r=a^+, \quad P_z=\phi_{;zz}/(\phi g_{22}), \quad P_\psi=\phi_{;\psi\psi}/(\phi g_{33}), \tag{7.8.11}$$

where:

$$a^\pm = (2f)^{-1}\left[(\phi_{;rr} - \phi_{;tt})/\phi \pm \{[(\phi_{;rr} + \phi_{;tt})/\phi + \chi_{;t}^2]^2 - 4(\phi_{;tr}/\phi)^2\}^{1/2}\right]. \quad (7.8.12)$$

The formulae given above have been corrected by this author (A.K.) by comparing them with other papers and with other solutions. Several formulae in published articles from this family contain misprints.[54]

The source could become a perfect fluid if $P_r = P_z = P_\psi$. This can happen in two cases: $q = p = 0$ or $\omega_1^2 = \omega_2^2$. In both cases the solution reduces to a flat FLRW model. Hence, this is the only perfect fluid subcase of the solution.

Properties of the travelling solitons were discussed in considerable detail in another paper (Diaz, Gleiser and Pullin 1989).

A solution of a similar kind, describing $4m$ solitons travelling on a flat FLRW background, was found and discussed by Cruzate, Diaz, Gleiser and Pullin (1988). It is rather complicated, and the authors did not comment on its relation to other solutions; thus it was not placed in the diagrams and will not be reproduced here.

Pullin (1990) apparently discussed the Cruzate *et al.* (1988) solution with $m = 1$, and also another one which describes a density wave unaccompanied by a gravitational wave. That new solution is not reproduced here or shown in the diagrams for the same reason.

The Diaz, Gleiser and Pullin (1987) solution is the subcase of (7.8.4)–(7.8.12) defined by $p = q = -2/3$. When further $n = 2$ (i.e. when the second scalar field disappears), the Ibanez–Verdaguer (1986) solution defined by eq. (7) in their paper is obtained. That solution is conformally related to (6.7.15)–(6.7.19). Another solution by Ibanez and Verdaguer (1986, eq. (10) in their paper) results from (7.8.4)–(7.8.12) as the limit $n = 2, p = 4/3, q = -2/3$. This one is conformally related to our (6.7.20)–(6.7.22).

In this family, there is one more solution by Diaz, Gleiser and Pullin (1988a), which is:

$$ds^2 = f(t,r)(-dt^2 + dr^2) + (\sigma_1\sigma_2)^{2(n-1)/3}t^n(dz^2 + r^2d\psi^2), \quad (7.8.13)$$

where:

$$f = Ct^{n-2}r^{-2}(\sigma_1 - \sigma_2)^{2/3}(\sigma_1\sigma_2)^{2(2+n)/3}[(t + \sigma_1 r)(t + \sigma_2 r)]^2$$
$$\times [(1 - \sigma_1^2)(1 - \sigma_2^2)]^{-4/3}(1 - \sigma_1\sigma_2), \quad (7.8.14)$$

σ_1 and σ_2 are defined exactly as before, and C, ω_1 and ω_2 are arbitrary constants. Formulae for matter characteristics are not given in the paper. In the case $n = 2$, this solution reproduces the Ibanez–Verdaguer (1986, eq. (7)) solution mentioned above.

This closes the overview of exact solutions that can be interpreted as cosmological models. However, a certain fundamental problem of gravitation theory, related to the subject of this book, is still unsolved. It will be discussed in the next chapter.

---

[54] Some of the corrections were confirmed by J. Pullin in a letter for which A.K. is very grateful. In general, though, A.K. bears the full responsibility for fallacious corrections, should there be any.

8

Averaging out inhomogeneities of geometry and matter in cosmological models

8.1 Modelling the Universe in relativity

The consequences of the Einstein field equations have been verified experimentally at the scale of planetary systems. The Einstein theory may thus be rather safely assumed to hold at this scale. A mathematical model of the matter distribution in the Universe going down to such a scale would have to be impossibly complicated (and, moreover, there is no observational data on which it could be based). In modelling the Universe, we use continuous functions (such as matter-density, pressure or the kinematical scalars of the velocity field), assuming that they represent "volume averages" of the corresponding fine-scale quantities. This creates additional problems. The scale of averaging (i.e. the diameter of the volume over which the averages are calculated) was never explicitly agreed upon, while the results of averaging in an inhomogeneous medium will obviously depend on the scale. Moreover, a volume average is a well-defined quantity only for scalars. For tensors, it leads to noncovariant quantities, and a more sophisticated definition is required. And the greatest problem is the noncommutation of averaging the metric with calculating the Einstein tensor, which is strongly nonlinear in the metric components. In simple words, the Einstein tensor $\tilde{G}_{\mu\nu}$ calculated from an "averaged" (whatever this means) metric $\bar{g}_{\mu\nu}$ will not be equal to the Einsten tensor $\bar{G}_{\mu\nu}$ which was first calculated from the fine-scale metric $g_{\mu\nu}$ and then averaged. Consequently, if the Einstein equations hold on the scale of planetary systems, then they cannot hold on larger scales that require averaging. Yet, in applying the Einstein equations to models of the Universe we assume precisely the opposite: we take a metric that is supposed to be already averaged, then calculate the corresponding Einstein tensor, and equate it to an energy-momentum tensor – also already averaged. When applied to cosmology, the Einstein equations should in fact be corrected by an additional term compensating for the difference $\gamma_{\mu\nu} \overset{\text{def}}{=} \tilde{G}_{\mu\nu} - \bar{G}_{\mu\nu}$ so that the "cosmological Einstein equations" read:

$$\tilde{G}_{\mu\nu} - \gamma_{\mu\nu} = \kappa \bar{T}_{\mu\nu}. \tag{8.1.1}$$

The term $\gamma_{\mu\nu}$ could then be written on the right-hand side of (8.1.1) and interpreted as a correction to material sources resulting from averaged-out small-scale inhomogeneities of the gravitational field (cosmic gravitational waves). Assuming for a moment that the averaged metric is the FLRW metric and that $\bar{T}_{\mu\nu}$ is still that of a homogeneous and isotropic perfect fluid, the correction $\gamma_{\mu\nu}$ will perturb the energy-density and pressure of the source, and will invalidate the simple FLRW relation between the sign of spatial curvature on the one hand, and the size and lifetime of the Universe on the other (which in any event only holds for dust with zero cosmological constant).

This problem of averaging, and the need for correcting the Einstein equations as in (8.1.1), were brought to general attention by Ellis (1984), although papers on this subject had been published long before. Shirokov and Fisher (1962) seem to have been the first to consider it (see our Section 8.3), and the foregoing consideration is mostly taken from their paper. The most promising advancement towards a solution of the problem was apparently made by Zalaletdinov (1992, see our Section 8.4). The meaning of averaging for cosmology, and the observational issues connected with it, were discussed in much detail by Ellis (1988). However, the problem still remains unsolved – Zalaletdinov presented an axiomatic theory of covariant averaging that has yet to be tested on exact solutions. Since the problem is of fundamental importance for both cosmology and gravitation theory, papers discussing it will be reviewed in the following three sections. They are sorted by the method of averaging proposed.

8.2 Averaging within approximation schemes

In this section we shall consider those papers in which the definition of averaging is either based on or coupled with approximation schemes.

The earliest contribution of this type is the paper by Szekeres (1971). It was not meant to be applied to cosmology, but it inspired several other authors. Szekeres showed that linearized Einstein equations (in which the metric is considered to be a small correction superposed on the Minkowski background) have strong formal similarity to the Maxwell equations. Using this similarity, Szekeres showed that a macroscopic gravitation theory may be built in analogy with the Lorentz theory in electrodynamics. He did not invoke any specific definition of averaging. The theory was then applied to the propagation of gravitational waves through a medium whose molecules are harmonic oscillators. The result was that gravitational waves slow down in such a medium.

The consideration by Sibgatullin (1971) is apparently independent of all the others. The author averaged the Einstein equations after the metric was decomposed into a "background" and a "fluctuation". The result was then calculated approximately under the assumption that the characteristic scale of correlations between matter and geometry is small compared with the scale of variation of the smoothed out geometry. The result is that fluctuations in matter do not influence the equations of zeroth and first order in the small parameter. However, Sibgatullin provided neither a definition of averaging, nor a criterion to separate the metric into the "background" and the "fluctuation".

Bialko (1973) presented another independent approach. The author developed the metric into a FLRW background and a high frequency perturbation. Then he averaged the Einstein equations over spatial volumes, assuming that the characteristic wavelength of the perturbation is small compared with the curvature radius.

The result is that the equations governing the evolution of the averaged perturbation differ from those for linearized perturbations by a logarithmically varying factor.

Noonan (1984) used the weak-field slow-motion approximation to show that when the Einstein equations are averaged by volume, the energy-momentum tensor splits into three parts. The first part can be interpreted as kinematical (due to averaged microscopic motions), the second one as mechanical (due to averaged microscopic stresses), and the third one as gravitational (due to averaged small-scale variations in the gravitational field).

In a subsequent paper (Noonan 1985), the author showed that the time-space components of the macroscopic energy-momentum tensor defined in the earlier paper can be interpreted as the flux of gravitational energy of the microscopic field.

Bildhauer and Futamase (1992) calculated the backreaction of inhomogeneities on the evolution, and in particular on the age of the Universe. Bildhauer and Futamase showed that inhomogeneities slow down the expansion compared to the ordinary Friedmann equation, and so the age of the Universe calculated from the Hubble law is underestimated. The averaging was performed within the perturbative scheme devised by Futamase (1988 and 1989), and the components of tensors are averaged by the spatial volume.

Futamase (1993), working in the same approximation scheme, considered inhomogeneous spacetimes in which preferred slicings exist. Assuming that the preferred slicings go over into the homogeneous spaces of the FLRW models in the limit of zero perturbation, the author calculated the effect on the Friedmann equation of averaging by three-dimensional volumes within the preferred slices. No qualitative conclusions were drawn from the equations.

8.3 Exact noncovariant averaging

In this section we shall consider those papers in which the averaging procedure (by volume in all cases) is defined without recourse to approximations, but where the definition is not covariant. In many cases, though, the definition was applied on top of a perturbative scheme. Readers may thus find the assignment of various papers up to Sections 8.2 and 8.3 disputable.

Shirokov and Fisher (1962) published the earliest contribution to this subject. The consideration presented in our Section 8.1 is mostly taken from their paper. They proposed to define the components of the averaged (macroscopic) metric as the volume-averages of the corresponding components of the small-scale (microscopic) metric. This procedure is not only noncovariant (volume integrals of tensor components do not constitute a tensor), but it has no geometrical interpretation. The metric is a device to calculate, among other things, volumes. Hence, it is not at all clear what is actually represented by a volume-average of a metric component. Assuming, however, that the procedure will find a justification in a future theory, we shall include papers based on such a definition.

Shirokov and Fisher applied this definition to metrics which are small perturbations of the FLRW models. They calculated the Einstein tensor from them, and then
calculated the averages of all components, equating them to the averages of the
appropriate components of the energy-momentum tensor and neglecting terms that
are nonlinear in the small quantities. They obtained a generalization of the FLRW
solutions in this way, with a repulsive term that prevents the singularity for all three
curvatures. At maximal compression, each matter particle fills the interior of a
sphere whose radius equals the particle's gravitational radius.

In a follow-up paper, Shirokov (1967) considered the following deformation of
the FLRW metrics:

$$ds^2 = dt^2 - (1 + \tfrac{1}{4}kr^2)^{-2}G^2(y)f(t,x,y,z)(dx^2 + dy^2 + dz^2), \qquad (8.3.1)$$

where $G(y)$ and $f(t,x,y,z)$ are unknown functions. Shirokov calculated the consequences of averaging the corresponding Einstein tensor without approximations,
but on the assumption that $\tilde{G}_{\mu\nu}$ (see Section 8.1) still obeys the cosmological principle. Shirokov concluded that the term produced by averaging can be interpreted as
a negative contribution to pressure that will prevent the Big Bang. The paper is,
unfortunately, a short conference report on a work in progress and does not present
any further details.

Mitskevich (1969) showed that averaging the Einstein equations by volume
results in an additional term on the right-hand side. The reasoning closely paral-
lelled that of Lorentz in electrodynamics. However, Mitskevich did not investigate
the physical consequences of this additional term.

A similar consideration, but without recourse to any explicit metric, was presented by Saar (1971a), who was inspired by the work of Isaacson (1968, see Section
8.4), but instead of Isaacson's covariant definition of averaging, he used the averaging by volume. He discussed the influence of averaged rapid fluctuations on a
slowly changing background metric. The metric is split into the background and the
fluctuation multiplied by a small parameter; then the Einstein equations are developed to second order in the small parameter and averaged by volume. The contribution of averaged fluctuations can be interpreted as a negative pressure in the
background. The result is that the cosmological expansion may proceed with
smaller velocity, and more time is available for structure formation. The paper does
not give details, for which the reader is referred to other papers by the same author
(Saar 1971b,c,d).

Nelson (1972) assumed the metric to decompose into smooth background and
small perturbations describing lumps. The averaging was done by integrating metric
components over volume, and Nelson emphatically disdained covariance. The background is then approximately equal to the average of the whole metric. The average
obeys "the usual set of cosmological equations" while the corrections to the average
obey equations "equivalent to instability equations. The large scale development of
such a Universe is therefore shown to be almost independent of the formation of
condensations provided the average of the energy-stress tensor is unaffected by the

condensations." This result seems to be opposite to the others. The condensations were assumed to be many and evenly distributed in space, each one having its radius large compared with the Schwarzschild radius. The averaging volume was assumed to contain many condensations.

Marochnik, Pelikhov and Vereshkov (1975a) calculated the influence of averaging on the Einstein equations. The metric, the density and the velocity field were assumed to be sums of an average (background) and a correction whose average is zero. The correction (called turbulence) and the background were assumed to obey:

$$\lambda_T^2 < L^2 \ll 1/\bar{R}, \tag{8.3.2}$$

where λ_T is the characteristic scale of the turbulence, L is the scale of averaging and $\bar{R}$ is the background curvature. The averaging was defined in terms of volume integrals. Marochnik *et al.* observed that corrections to energy and pressure resulting from averaging out small-scale inhomogeneities are not necessarily positive, and, formally, may be interpreted as antigravitation. Consequently, perturbations of a FLRW background may in principle result in a static solution (a nonexpanding Universe). In an evolutionary Universe, the corrections from averaged inhomogeneities may prevent the Big Bang singularity.

In a follow-up paper, the same authors (Marochnik, Pelikhov and Vereshkov 1975b) calculated explicitly the results of averaging on linearized perturbations of the FLRW models. The results are the following:

1. For the equation of state $p = \epsilon/3$ (obeying in the FLRW background), two subcases arise:
   (a) If the potential turbulence dominates over the vortical turbulence (in the terminology of Lifshitz 1946), then the perturbation causes an additional deceleration of the expansion and the initial singularity is avoided. As Marochnik *et al.* observe, this is only a formal result, since the linear approximation becomes invalid before the minimal radius is approached. However, this may indicate the behaviour of an exact solution.
   (b) If the vortical turbulence dominates over the potential one, then the expansion is accelerated and the singularity remains.
2. For other equations of state in the FLRW background, the turbulence would have significant effects on the evolution only when $p \geq \epsilon/3$. However, at those stages of evolution at which $p > \epsilon/3$, the linear approximation is invalid. Hence, Marochnik *et al.* concluded, the other equations of state need not be considered within this approximation scheme.

Marochnik *et al.* noted that their results in point 1 are in contradiction to those of Bialko (1973), who found that the main source of deviation from the Friedmann equation should be the gravitational waves. The contradiction was not discussed further, however.

Beginning with the two papers described above, a whole series of more detailed studies was published. The paper by Krymsky, Marochnik, Naselsky and Pelikhov

(1978) is not directly relevant for this book, and is mentioned for completeness only. In it, the authors studied the spectra of classical and quantum short-wave turbulences.

Marochnik, Naselsky and Pelikhov (1980) studied the influence of long-wave turbulence on the expansion of the background. Only those modes of turbulence were considered which remain finite when approaching the singularity, and it was shown that they influence the expansion in the same way as the perturbative solution of Lifshitz and Khalatnikov (1963) predicted.

Marochnik (1980a) wrote out the Marochnik–Pelikhov–Vereshkov (1975) equations up to the second approximation in the turbulent perturbation. In the next paper, the author (Marochnik 1980b) solved these equations and studied the quantitative effects of the perturbations on the background. No pronounced qualitative effects were found, and the same may be said about the last paper in the series (Marochnik 1981).

In two other papers by the same author (Marochnik 1980c, d), the influence of averaged-out small-scale inhomogeneities on the evolution of the FLRW background was discussed again. Marochnik calculated the changes in the most important cosmological parameters resulting from such an influence. The changes affect: (1) the transition moment between the hadron and the lepton eras (by a factor of up to 1.4); (2) the temperature in the transition moment (by the factor 0.88); and (3) the helium abundance (may remain unchanged if the energy-density of short-wave fluctuations is smaller than $1.5\bar{\epsilon}$, where $\bar{\epsilon}$ is the large-scale average energy density).

The papers by Marochnik and coworkers in fact have already initiated a detailed program in physical cosmology which takes small-scale inhomogeneities into account.

Zotov and Stoeger (1992) considered the effect of averaging over spatial volumes on the Einstein equations. They compared an exact FLRW model with one where galaxies are superposed on the FLRW background with a constant number density, each galaxy being represented by a Schwarzschild metric. Averaging the metric components simply results in changing the background FLRW model with the scale factor $R(t)$ to another FLRW model with the scale factor $S(t)$, where $S^2(t)=R^2(t)(1-K)$, $K=NV_1/V_2$, N is the number density of galaxies, V_1 is the averaging volume, and V_2 is the volume per galaxy in space. Thus, adding matter results in squeezing the volume of space, if the effect is calculated by means of average values of metric components. If the effect is calculated by substituting the new average density in the Einstein equations, then the volume of space is squeezed, too, but this time the dependence of $R(t)$ on the density parameter σ_0 is more complicated, and is given by the following parametric equations:

$$R_h(t)=H_0\sigma_0(1-2\sigma_0)^{-3/2}(\cosh 2\psi-1), \qquad (8.3.3)$$

$$ct=H_0\sigma_0(1-2\sigma_0)^{-3/2}(\sinh 2\psi-2\psi). \qquad (8.3.4)$$

Hence, averaging does indeed produce compensating terms in the Einstein equations, but the terms are neither calculated nor discussed in the paper.

8.4 Exact covariant averaging

The pioneer in this field was Isaacson (1968a,b). Although he considered only the vacuum Einstein equations, the idea and the result apply to cosmology as well, and the papers inspired some of the later activity. In the first paper, Isaacson assumed that the metric of a gravitational wave spacetime can be split into a low frequency background and a high frequency wave. By definition, the frequency is high when the wavelength is small compared with the curvature radius. The Einstein equations were then linearized with respect to the high frequency correction, and the waves were shown to obey a covariant generalization of the equation of massless spin 2 fields in a flat background and to travel on null geodesics of the background. The amplitude, frequency and polarization of the waves are modified by the background curvature.

In the second paper (Isaacson 1968b) the author continued the considerations of the preceding paper to a higher order of approximation. The corrections of the first nonlinear order to the vacuum Einstein equations were shown to provide a term simulating a stress-energy tensor that can be interpreted as the effective energy of the gravitational waves. This tensor was then used to define the total energy and momentum carried by the waves off to infinity.

The energy-momentum tensor of the gravitational waves arises from averaging the metric. The average was defined by parallel-transporting the tensors from the point x' to the representative point x along the geodesic joining x' to x, and then integrating the resulting object with respect to x' with a weighting function:

$$\langle T_{\mu\nu}(x)\rangle = \int_{\text{all space}} G_\mu{}^{\alpha'}(x,x')G_\nu{}^{\beta'}(x,x')T_{\alpha'\beta'}(x')f(x,x')\mathrm{d}_4x', \qquad (8.4.1)$$

where $G_\alpha{}^{\alpha'}$ are the propagators of parallel displacement, and $f(x,x')$ is the weighting function. The weighting function was not defined explicitly, but was only described through its desired properties:

(1) $f(x, x') \rightarrow 0$ when the distance d between x and x' obeys $\lambda \ll d \ll L$, λ is the wavelength of the high frequency wave, L is the wavelength of the low frequency background.
(2) $\int f(x,x')\mathrm{d}_4x' = 1$.
   all space

This definition of averaging is covariant, even though it is applied in the paper only within a perturbative scheme.

Matzner's (1968a,b) work was an improvement over that of Isaacson (1968) in that Matzner's method of identifying the background of a wave does not require the assumption that the wave is of high frequency. In the first paper, the author proposed another definition of averaging the geometry: a metric is Lie-dragged along a specific vector field to a chosen point, averaged there over all the points, and then Lie-dragged back. He also proposed:

$$\xi^{(\alpha;\beta)}{}_{;\beta} + \lambda \xi^{\alpha} = 0 \qquad\qquad (8.4.2)$$

as the definition of the vector field along which the metric is dragged; eq. (8.4.2) is a generalization of the Killing equations. The definition was then applied to the $t = \text{const}$ sections of the Taub–NUT space.[55] The averaged metric has the FLRW algebraic form, but is not related to the original metric in any easily interpretable way (for example, the volume of the averaged space is different from the volume at the beginning).

In the second paper (Matzner 1968b) the author proposed a measure of symmetry in a Riemannian manifold with a positive-definite metric. The measure was defined as the minimum value of the functional:

$$\lambda[\xi] = \left(\int \xi^{(\alpha;\beta)} \xi_{(\alpha;\beta)} \, dV \right) / \left(\int \xi^{\mu} \xi_{\mu} \, dV \right), \qquad\qquad (8.4.3)$$

where ξ^{α} is a vector field and the integral is taken over the whole space, $dV = g^{1/2} d^n x$. If the space is compact or the integral over the boundary at infinity is zero, then the minima of $\lambda[\xi]$ obey (8.4.2); on a positive-definite manifold, $\lambda = 0$ in (8.4.2) implies Killing equations. This definition was then applied to a few kinds of spacetimes and it was shown that the parameters λ defined by (8.4.2) can be interpreted as the averaged energy-density and averaged stresses of gravitational waves.

Rosen (1980) showed that averaging the metric results in corrections to the energy-momentum tensor. The form of the corrections was calculated explicitly under two assumptions: (1) The perturbed metric is conformally equivalent to the background metric. (2) The averaged energy-momentum tensor has the algebraic form of a perfect fluid. The author observed that fluctuations in the metric always accelerate the expansion in a FLRW background (they increase $\ddot{R}/R$). However, the subject was treated axiomatically and no definition of averaging was proposed. Rosen called his averages "with respect to the statistical ensemble".

The paper by Carfora and Marzuoli (1984b) was written under the direct influence of the ideas of Ellis (1984). Carfora and Marzuoli considered a continuous family of spacetimes, each being a Cauchy development of initial data on a closed spacelike hypersurface, such that the Ricci tensor of the hypersurface, R_{ab}, is positive-definite. The family (flow) of three-metrics $h_{ab}(\beta)$ on the initial hypersurfaces is defined by:

$$\frac{\partial}{\partial \beta} h_{ab}(\beta) = \tfrac{2}{3} \langle R(\beta) \rangle_{\beta} h_{ab}(\beta) - 2R_{ab}(\beta), \qquad\qquad (8.4.4)$$

where β is the parameter labelling the family and $\langle R(\beta) \rangle_{\beta}$ is the scalar curvature corresponding to $h_{ab}(\beta)$ averaged over the three-manifold $M(\beta)$:

$$\langle R(\beta) \rangle_{\beta} = [V(\beta)]^{-1} \int_{M(\beta)} R(\beta) \, dV(\beta), \qquad\qquad (8.4.5)$$

---

[55] The procedure was thus applied to a space of finite volume and may not be applicable otherwise.

$V(\beta)$ being the volume. The family (8.4.4) was defined by Hamilton (1982), and it has the following properties:

1. $V(\beta)$ is in fact independent of β.
2. Any symmetries of $h_{ab}(\beta_0)$ are inherited by all $h_{ab}(\beta)$ with $\beta \geq \beta_0$.
3. The limiting metric $\bar{h}_{ab} = \lim_{\beta \to \infty} h_{ab}(\beta)$ has positive constant curvature.

Suppose h_{ab} is the metric on the inhomogeneous manifold M on which we want to average out the inhomogeneities. Equation (8.4.4), with the initial condition $h_{ab}(0) = h_{ab}$, then defines a smooth family of deformations of the initial manifold M, gradually deforming it into a three-space of constant curvature and of the same volume as the initial manifold. For each β, the space-average of matter-density, $\langle \epsilon(\beta) \rangle_\beta$, may be defined analogously to (8.4.5), and then:

$$\bar{\epsilon} = \lim_{\beta \to \infty} \langle \epsilon(\beta) \rangle_\beta \tag{8.4.6}$$

is the matter-density in the limiting FLRW model. The main result of the paper is that $\bar{\epsilon}$ is not equal to the space-average of the matter-density in the original space, $\langle \epsilon \rangle_0$, but:

$$\bar{\epsilon} = (1 + \Sigma^2)^{-1} [\langle \epsilon \rangle_0 + \tfrac{1}{2} \langle \tilde{K}_{ab} \tilde{K}^{ab} \rangle_0 + \tfrac{1}{2} (\eta + \Sigma^2) \bar{R}], \tag{8.4.7}$$

where $\tilde{K}_{ab}$ is the traceless part of the second fundamental form K_{ab} of the initial manifold $M(0)$, $k \stackrel{\text{def}}{=} K^a{}_a$ and:

$$\Sigma = (\langle k^2 \rangle_0 - \langle k \rangle_0^2)^{1/2} / \langle k \rangle_0, \tag{8.4.8}$$

$$\bar{R} = \lim_{\beta \to \infty} \langle R(\beta) \rangle_\beta, \tag{8.4.9}$$

$$\eta = (\bar{R} - \langle R \rangle_0) / \bar{R}. \tag{8.4.10}$$

Carfora and Marzuoli were thus the first to confirm by an exact and covariant derivation the result predicted by Shirokov and Fisher (1962, see Sections 8.1 and 8.3).

A somewhat more elementary account of the same work is given in another paper (Carfora and Marzuoli 1984a).

Hemmerich (1987) criticized the Carfora–Marzuoli approach rather strongly by raising two points:

1. The spacetime metric in the limit $\beta \to \infty$ depends not only on $h_{ab}(\infty)$, but also on the foliation, that is, on the lapse and shift functions in the ADM formalism, and the resulting limiting four-metric is nonunique; in particular, it may not be FLRW at all.
2. The Carfora–Marzuoli equations, (8.4.7) among them, may be obtained directly from the ADM formalism without employing the Hamilton flow (8.4.4).

A solution to the first problem is proposed in the paper: the lapse and shift can be limited so that $g_{\mu\nu}(\infty)$ is the FLRW geometry. The second objection can be answered by pointing out that (8.4.4) provides not only a relation between the average values of scalars on $M(0)$ and on $M(\infty)$, but also a mapping of points of $M(0)$ into points of $M(\infty)$ – an element lacking in other schemes. Moreover, using (8.4.4) one can consider middle-scale smoothing, from $\beta=0$ to $\beta=\beta_0<\infty$, in addition to global smoothing from $\beta=0$ to $\beta\rightarrow\infty$ (see Ellis 1984 for more remarks on the problem of smoothing scales). The real disadvantages seem to be the assumptions that all $M(\beta)$ are closed and that their Ricci tensors are positive-definite. Another valid point was raised by Zotov and Stoeger (1992, see the preceding section): Carfora and Marzuoli used the standard form of the Einstein equations at each stage of smoothing, while the effect of smoothing on those equations was the effect under investigation.

The (currently, but not yet absolutely) last word on the subject is the paper by Zalaletdinov (1992). The author introduced an axiomatic description of averaging of tensors. An average of the tensor field $P^{\alpha\cdots}_{\beta\cdots}$ over a region $\Sigma\subset\mathcal{M}$ of the manifold $\mathcal{M}$ at a point $x\in\Sigma$ is defined as:

$$\bar{P}^{\alpha\cdots}_{\beta\cdots}(x)=\frac{1}{V_\Sigma}\int_\Sigma P^{\mu'\cdots}_{\nu'\cdots}(x')\mathcal{A}^{-1\alpha}{}_{\mu'}(x,x')\bullet\bullet\bullet\mathcal{A}^{\nu'}{}_{\beta}(x,x')\bullet\bullet\bullet\,\mathrm{d}\Omega', \qquad (8.4.11)$$

where:

$$V_\Sigma=\int_\Sigma(-g)^{1/2}\mathrm{d}_4x, \quad \mathrm{d}\Omega'=[-g(x')]^{1/2}\mathrm{d}_4x'. \qquad (8.4.12)$$

The averaging bivector $\mathcal{A}^{\alpha'}{}_{\beta}(x,x')$ is defined through its postulated properties:

(I) $\displaystyle\lim_{x\rightarrow x'}\mathcal{A}^{\alpha'}{}_{\beta}(x,x')=\delta^{\alpha'}{}_{\beta}$, $\qquad\qquad\qquad\qquad\qquad\qquad\qquad$ (8.4.13)

(II) In the limit of vanishing curvature, $\mathcal{A}^{\alpha'}{}_{\beta}=\delta^{\alpha'}{}_{\beta}$.

Property (I) implies the existence of $\mathcal{A}^{-1}$ with the properties:

$$\mathcal{A}^{\alpha'}{}_{\beta}\mathcal{A}^{-1\beta}{}_{\gamma'}=\delta^{\alpha'}{}_{\gamma'}, \quad \mathcal{A}^{\alpha'}{}_{\beta}\mathcal{A}^{-1\gamma}{}_{\alpha'}=\delta^{\gamma}{}_{\beta}. \qquad (8.4.14)$$

In addition to $\mathcal{A}$, Zalaletdinov defined the coordination bivector $\mathcal{W}^{\alpha'}{}_{\beta}(x,x')$ mapping the averaging region from x to x'. The $\mathcal{W}^{\alpha'}{}_{\beta}$ is assumed to have properties (I) and (II) and in addition:

(III) $\qquad\qquad\qquad\qquad\mathcal{W}^{\alpha'}{}_{\beta;\alpha'}=0,$ $\qquad\qquad\qquad\qquad\qquad$ (8.4.15)

(IV) $\qquad\qquad\mathcal{W}^{\alpha'}{}_{[\beta,\gamma]}+\mathcal{W}^{\alpha'}{}_{[\beta,\mu']}\mathcal{W}^{\mu'}{}_{\gamma]}=0.$ $\qquad\qquad\qquad$ (8.4.16)

Particularly simple formulae result if one demands that $\mathcal{W}^{\alpha'}{}_{\beta}=\mathcal{A}^{\alpha'}{}_{\beta}$.

Starting from these axioms, Zalaletdinov showed that the averaging has desirable properties, for example, the average of the Christoffel symbols has the properties of an affine connection, and the average of the Riemann tensor has all the algebraic properties of a curvature tensor. In consequence, the manifold acquires a second connection and a second curvature (see below). The Einstein equations in macro-scale acquire a geometric source term in addition to the energy-momentum tensor, and the equations of motion in the macroscale acquire another geometric source term.

Equations (8.4.13)–(8.4.16) defining the $W^{\alpha'}_{\ \beta}$ and $A^{\alpha'}_{\ \beta}$ are not mutually contradictory because Zalaletdinov showed that a class of solutions exists. The details of their derivation are, however, published only in a preprint (see paper for the reference).

Zalaletdinov deduced the following form of the averaged Einstein equations:

$$\bar{g}^{\alpha\beta}M_{\beta\gamma}-\tfrac{1}{2}\delta^{\alpha}_{\ \gamma}\bar{g}^{\mu\nu}M_{\mu\nu}=\kappa\bar{T}^{\alpha}_{\ \gamma}+(Z^{\alpha}_{\ \mu\nu\gamma}+\tfrac{1}{2}\delta^{\alpha}_{\ \gamma}Q_{\mu\nu})\bar{g}^{\mu\nu}, \tag{8.4.17}$$

where $\bar{g}^{\alpha\beta}$ and $\bar{T}^{\alpha\gamma}$ are the metric tensor and the microscopic energy-momentum tensor, respectively, averaged by the rule (8.4.11), and the other quantities are determined through the following chain of definitions:

1. The object $\mathscr{F}^{\alpha}_{\ \beta\gamma}$ is defined in terms of $\mathscr{W}^{\alpha'}_{\ \beta}$ as:

$$\mathscr{F}^{\alpha}_{\ \beta\gamma}=\mathscr{W}^{-1\alpha}_{\ \ \rho'}(\mathscr{W}^{\rho'}_{\ \beta;\gamma}+\mathscr{W}^{\rho'}_{\ \beta;\sigma'}\mathscr{W}^{\sigma'}_{\ \gamma}), \tag{8.4.18}$$

and then the corresponding affine connection is $\bar{\mathscr{F}}^{\alpha}_{\ \beta\gamma}$, that is, $\mathscr{F}^{\alpha}_{\ \beta\gamma}$ averaged by (8.4.11). This connection is taken to be the Levi–Civita connection of the averaged manifold.

2. The second curvature tensor $M^{\alpha}_{\ \beta\gamma\delta}$ is calculated from $\bar{\mathscr{F}}^{\alpha}_{\ \beta\gamma}$ in the usual way, and obeys the following equation:

$$M^{\alpha}_{\ \beta\gamma\delta}=\bar{R}^{\alpha}_{\ \beta\gamma\delta}+2\langle\mathscr{F}^{\mu}_{\ \beta[\gamma}\mathscr{F}^{\alpha}_{\ |\mu|\delta]}\rangle-2\bar{\mathscr{F}}^{\mu}_{\ \beta[\gamma}\bar{\mathscr{F}}^{\alpha}_{\ |\mu|\delta]}, \tag{8.4.19}$$

where $\bar{R}^{\alpha}_{\ \beta\gamma\delta}$ is the average of the microscopic curvature (Riemann) tensor calculated by (8.4.11). The tensor $\bar{R}^{\alpha}_{\ \beta\gamma\delta}$ is taken to be non-Riemannian; $M^{\alpha}_{\ \beta\gamma\delta}$ is taken to be Riemannian. The angle brackets $\langle\rangle$ denote the averaging (8.4.11) applied to the whole expression between them, and the index between the vertical strokes $\|$ does not participate in the antisymmetrization. Equation (8.4.19) results from averaging Cartan's second structure equation by the rule (8.4.11). The tensor $M_{\beta\gamma}$ in (8.4.17) is defined by:

$$M_{\beta\gamma}=M^{\rho}_{\ \beta\rho\gamma}. \tag{8.4.20}$$

3. The correlation tensor is defined by:

$$Z^{\alpha}_{\ \beta[\gamma}{}^{\mu}_{\ |\mu|\delta]}=\langle\mathscr{F}^{\alpha}_{\ \beta[\gamma}\mathscr{F}^{\mu}_{\ |\mu|\delta]}\rangle-\bar{\mathscr{F}}^{\alpha}_{\ \beta[\gamma}\bar{\mathscr{F}}^{\mu}_{\ |\mu|\delta]}. \tag{8.4.21}$$

The quantity $Z^{\alpha}_{\ \mu\nu\gamma}$ present in (8.4.17) results from the above by:

$$Z^{\alpha}_{\ \mu\nu\gamma}=2Z^{\alpha}_{\ \mu[\rho}{}^{\rho}_{\ |\nu|\gamma]}, \tag{8.4.22}$$

and the quantity $Q_{\mu\nu}$ is:

$$Q_{\mu\nu}=Q^{\rho}_{\ \mu\rho\nu}, \tag{8.4.23}$$

where:

$$Q^{\alpha}_{\ \beta\gamma\delta}=2Z^{\mu}_{\ \beta[\gamma}{}^{\alpha}_{\ |\mu|\delta]}. \tag{8.4.24}$$

One might object that the author did not provide any explicit prescription to calculate the averaged metric with a given scale and with the averaging and coordination bivectors determined along the averaging procedure. However, such a prescription should not be expected. It would require the microscopic metric to be calculated explicitly, and this is just not realistic. The explicit microscopic description of matter is not specified in the Lorentz theory in electrodynamics either. The correlation tensor $Z^\alpha{}_{\beta[\gamma}{}^\mu{}_{|\nu|\delta]}$ should be inferred from within the macroscopic theory. The differential equations defining Z are given in the paper (eqs. 14-16), but they are not quoted here because of their complexity.

In the subsequent paper (Zalaletdinov 1993) the author presented the same consideration translated into Cartan's exterior form calculus. The author then applied his averaging procedure to a vacuum metric. It turned out that the macroscopic metric (constructed from the Levi–Civita connection $\tilde{\mathcal{F}}^\alpha{}_{\beta\gamma}$, see the remark after (8.4.18)) obeys macroscopic (macrovacuum) equations that are no longer Ricci-flat. The correlation tensor serves as a source, and the macro-Ricci tensor represents the dynamical effect of averaged-out small-scale fluctuations in geometry. Moreover, if the microscopic vacuum metric represents a high-frequency wave superposed on a low-frequency background (macroscopic) metric, then the induced macroscopic energy-momentum tensor becomes the same as Isaacson's (1968a,b) energy-stress tensor for gravitational waves. Thus, in this theory there exists the energy-momentum tensor for macroscopic gravitational fields.

Short descriptions of the same procedure are contained in two other papers by the same author (Zalaletdinov 1994a,b).

The papers by Khiet (1971 and 1984) must be mentioned here for completeness. In the first one (Khiet 1971) the author briefly recalled the argument by Shirokov and Fisher (1962, see our Section 8.1) and studied spherically symmetric perturbations of the FLRW geometries, without, however, reaching any qualitative conclusion. In the other paper (Khiet 1984) the author announced a study of the results of averaging microscopic gravitational equations leading to macroscopic equations. However, no definition of an averaging procedure was provided, and the correction term $C_{\alpha\beta}(g)$ in the macroscopic equations $G_{\alpha\beta} + C_{\alpha\beta} = \kappa T_{\alpha\beta}$ was guessed on the basis of covariance requirements. As a result, Khiet obtained new field equations (nonlinear in curvature), and then studied exact solutions of these equations for a general FLRW metric. The paper is interesting as one more evidence of growing appreciation of the problem of averaging out microscopic inhomogeneities, but the equations considered in it are not in fact averaged Einstein equations, but field equations of a new metric gravitation theory.

The problem of averaging is connected with approximating a hypothetical fine-scale cosmological model by a large-scale model. An independent approach to this problem was presented by Spero and Baierlein (1977 and 1978). The authors proposed to define an approximate symmetry of an inhomogeneous model by "best fitting" a Bianchi-type model to it. In the first paper (Spero and Baierlein 1977), the "best fit" was defined in terms of a minimum of a functional with respect to variations of a triad of orthonormal vectors in the given spacetime and variations of the set of structure constants to be found.[56] The resulting Bianchi type is not always unique, but it will be unique when it is any of the types {IX, VIII, VII$_h$ or VI$_h$}. This shows that these types are generic. The classification is dependent on the slicing and not necessarily preserved in time. The approximant is not guaranteed to obey the Einstein's equations; an example is given in the 1978 paper of when it does not.

In the other paper (Spero and Baierlein 1978) the authors applied the definition of approximate symmetry from the preceding paper to two slicings of the Gowdy (1974) solutions, providing approximants of Bianchi types I and VI$_0$, and to the Kantowski–Sachs metric. The latter's symmetry group is not simply transitive, so the result was not obvious, the approximant turned out to be of Bianchi type I.

Another related paper is that by Stoeger, Ellis and Hellaby (1987). The authors proposed the following criterion of continuous homogeneity of the Universe: the mean mass-density in a sphere of volume V_L centered at the point $\vec{r}$ is:

$$\bar{\epsilon}_L(\vec{r}) = \frac{1}{V_L} \int_{V_L} \epsilon(\vec{r}') \mathrm{d}_3 r', \qquad (8.4.25)$$

where V_L is assumed to be small enough so that curvature inside V_L does not have to be taken into account. Then we say that the density distribution is spatially homogeneous on average at the level ε on scales larger than L_c if and only if there exists an $\varepsilon \ll 1$ and an L_c such that:

$$|\bar{\epsilon}_{L_1}(\vec{r}_1) - \bar{\epsilon}_{L_2}(\vec{r}_2)| < \varepsilon \bar{\epsilon}_{L_1}(\vec{r}_1) \qquad (8.4.26)$$

for all $\vec{r}_1$, $\vec{r}_2$ and all L_1, $L_2 \geq L_c$. The same criterion may be applied to other scalars. This criterion is in principle falsifiable by observations. Stoeger *et al.* discussed the virtues of this definition. Among other things, they showed that if galaxies are randomly distributed in space according to the Poisson distribution, then without further assumptions such a distribution is not in agreement with (8.4.26) and cannot be described by a FLRW model in any quantitative sense. Stoeger *et al.* proposed that for the observed Universe one should take $L_c = 200$ Mpc and $\varepsilon \leq 0.01$.

For completeness, let us note a paper in which the author proposed an approach to averaging dual to the one described in this section. Mahanta (1984) postulated that the Einstein equations as we know them are field equations of an already averaged (macroscopic) theory of gravitation. They should result by averaging over high frequencies of microscopic equations, similar in form to the Einstein equations. The micro-metric should be conformally flat and apply in the interior of elementary particles. Details of the averaging procedure were not discussed in the paper. The resulting equations of macro-gravity have the property that their vacuum solution is necessarily the Minkowski metric, so gravitational waves cannot exist in this theory.

---

[56] The formulae are not quoted here because they are rather large, and the subject is not directly related to this book. However, the paper is written very much in the spirit of this book and is strongly recommended as further reading.

9

Comments

9.1 Afterthoughts

In addition to justifying the two messages from Section 1.1, the main purpose of this book was to draw together and bring to the readers' attention all those fine and illuminating contributions to relativistic cosmology that do not fit into the "standard model", while still not negating its applicability as a first approximation. Hopefully, the book has proved that these contributions are not phantasies of isolated individuals, but that, taken together, they do tell us about interesting processes that might be going on in the Universe – processes that we would not even suspect using the FLRW models alone. The results presented highlight several problems in the current system of education and research in physics.

The first problem is the highly dogmatic approach of astronomers towards the "standard model" and other "standards". It seems that the hypothetical, provisional character of the assumptions that lead to the FLRW models has not been given sufficient emphasis in astronomy courses. As a result, the homogeneity and isotropy of the Universe are treated by many (most?) astronomers as a revealed truth, never to be questioned. This author has sometimes experienced outright aggression during seminars and conference talks, while presenting the various ideas that now appear in this book. Physicists do not seem to suffer from this problem, but astronomers would be well advised to treat the "current knowledge" in a more relaxed way, especially in view of the still highly unsatisfactory reliability and precision of most observations and the numerous changes in the apparently well-established results.

The other problem is the way in which relativity courses are taught. Almost every newly published book places its main emphasis on novel mathematical methods in deriving results known for decades. This is useful for newcomers into the field, but in this way an impression is created that little progress is occurring in relativity as applied to physics, while journals are overflowing with new ideas. Somehow we should become accustomed to the idea that *even relativity* is developing, and the new results should find their way into relativity courses. This book will perhaps fill the gap in relativistic cosmology, but relativity is advancing in other directions as well.

The third problem is the attitude of relativists, and of physicists in general, toward publications. Once upon a time it was obvious that papers are published in order that other people can read them, that reading is an important part of research, and that publishing results that have already been published before is of no use. It seems that now the cynical view has prevailed that papers are published in order that they can be entered in lists of publications and citation indexes, that writing papers is more exciting than reading them, and that it does not matter whether the result is new so long as it was obtained "independently". This is obviously the point of view of a trade union of scientists, and it contradicts the traditional definition of science. The urge to "publish or perish" has several negative side-effects. The same simple results are published again and again, independently of course, while important fine developments go unnoticed for a long time. Solving this problem may take much time and effort, but meanwhile relativists might do well to adopt the view that our field of research is no longer new, and that a result which extends our old course is not necessarily a new discovery. It is rather funny to read yet another author declaring that he or she is presenting a "new solution", while it is obvious from the paper that (s)he did not make the slightest effort to survey the literature and make sure that the solution is really new. I am myself partly guilty of this sin, and this is why I felt I had the right (and maybe duty) to say the foregoing. It would help if the editors and referees of physics journals were more critical when accepting papers for publication. Readers of this book might have noticed how many rediscoveries have been published in journals that have the reputation of being the best and most prestigious. Part of their reputation should stem from access to better informed referees, or at least this is what we all believe.

Appendix A

Dictionary of abbreviations and other symbols used in the text and in the diagrams

A_2 the two-dimensional Abelian group with spacelike orbits.

CF   conformal flatness.

ECI   embedding class 1.

$F_{\mu\nu}$ the electromagnetic field.

FLRW   Friedmann (1922 and 1924)–Lemaître (1927 and 1931)–Robertson (1929 and 1933)–Walker (1935); the models.

G_3/S_2 a three-dimensional group acting on two-dimensional orbits.

G–W   Goode and Wainwright (1982b); the representation of the Szekeres models.

HS   hyperbolic symmetry.

k the spatial curvature index of the limiting FLRW model.

K–Q   Kustaanheimo and Qvist (1948); the class of solutions and the equation.

K–S   Kantowski and Sachs (1966); the model and the symmetry group.

L–T   Lemaître (1933a)–Tolman (1934); the model.

nuq   nonunique. In some cases, the spatial homogeneity alone does not uniquely single out a subcase and further assumptions are needed to split the limit into definite solutions. The label "nuq" marks such limits.

P?   the Petrov type is unknown.

PS   plane symmetry.

q^α the heat-flow vector.

σ the shear.

SH   spatial homogeneity.

SS   spherical symmetry.

SSW   the subcase of the Szafron–Wainwright (1977) solution defined in the second paragraph after eq. (2.8.4). It appeared so frequently in other papers that it had to be given a name, see Figures 2.1 and 2.2.

$\dot{u}^\alpha$ the acceleration vector.

WIM   Wainwright, Ince and Marshman (1979); the algorithm for generating scalar field solutions out of vacuum solutions.

Appendix B

Why should one consider inhomogeneous models of the Universe?

For most physicists, sufficient reason to consider the Universe as being inhomogeneous is the intellectual challenge to explore the unknown. A mathematical argument in favour of inhomogeneous models was given by Tavakol and Ellis (1988). The authors demonstrated on examples that including new parameters (or changing values of parameters) in a set of ordinary differential equations may drastically change the behaviour of solutions, for example, from periodic to chaotic. This shows that the set of cosmological models is most probably structurally unstable. Tavakol and Ellis emphasized the importance of studying models without symmetry.

Several physicists were already able to see in the 1930s that the "cosmological principle" was not a summary of any kind of knowledge, but just a working hypothesis leading to the simplest models that were yet acceptable. A sample of early reservations about the cosmological principle is given in Appendix C. Unfortunately, the astronomical community was largely unconvinced. The discoveries of large inhomogeneities in the 1970s and 1980s were met with surprise. By the present day, those same discoveries could be sufficient argument for considering inhomogeneous models, and the idea seems to be gaining still wider acceptance (see an interesting proposal by Melott, 1990, to measure the topology of large-scale matter distribution in the Universe). However, over the years, astrophysical and philosophical arguments have been given by several authors in favour of generalizations of the FLRW models, and they are briefly recalled here. The collection does not pretend to be complete.

"... it is evident that some preponderating tendency for inhomogeneities to disappear with time would have to be demonstrated, before such models could be used with confidence to obtain extrapolated conclusions as to the behavior of the universe in very distant regions or over exceedingly long periods of time ... The result of the investigation will be to emphasize the possible dangers of drawing conclusions as to the actual universe from long range extrapolations made on the basis of a homogeneous model." (Tolman 1934, "such models" means the FLRW models here.)

McCrea (1939) presented several arguments for considering anisotropic and inhomogeneous models. The author quoted Shapley (1938) for observations that revealed anisotropic distribution of galaxies. Then he invoked the result by Milne, that any cloud of particles having different velocities and initially occupying a finite volume of space will disperse (if the particles interact weakly enough) and after some time will obey the Hubble law in the first approximation. From this McCrea concluded that in order to infer the distribution of matter in the Universe from the observed motion of galaxies, a higher approximation than the Hubble law is needed. Then he derived the following formula for redshift as a function of distance:

$$z=(-\sigma_{\alpha\beta}n^{\alpha}n^{\beta}+\tfrac{1}{3}\theta+n^{\alpha}\dot{u}_{\alpha})_{o}\delta l, \qquad (B.1)$$

where z is the observed redshift, $\sigma_{\alpha\beta}$, θ and $\dot{u}_{\alpha}$ have their usual meanings, n^{α} is the unit spacelike vector anticollinear with the projection of the tangent vector to the light ray k^{α} onto the three-space locally orthogonal to the velocity field:

$$n^{\alpha}=u^{\alpha}-k^{\alpha}/(k^{\mu}u_{\mu}), \qquad (B.2)$$

δl is the distance between the observer and the source of the light, and the subscript "o" indicates that the quantity is evaluated at the observer's position. The notation in (B.1) and the terminology were adapted to the present standards, but McCrea's formula is equivalent to (B.1). Equation (B.1) was derived in this form by Ehlers (1961). It is a linear approximation, valid only for small distances. The exact formula is:

$$1+z=(u_{\alpha}k^{\alpha})_{s}/(u_{\beta}k^{\beta})_{o}, \qquad (B.3)$$

where the subscript "s" indicates the evaluation at the source. McCrea referred (B.3) to Kermack, McCrea and Whittaker (1933). He observed that it is almost a trivial consequence of his result that to first approximation the redshift will be proportional to distance, but the coefficient of proportionality will be, even in this approximation, in general different for each direction. Then McCrea considered the Einstein equations for dust with the metric assumed in the form $ds^{2}=dt^{2}-h_{kl}dx^{k}dx^{l}$, $k, l=1, 2, 3$ and derived the second-order generalization of (B.1). He pointed out that comparison of local observations with his formula should allow us to infer about redshifts, distances and matter-density in the neighbourhood of the observer. Finally, he emphasized that his relations do not enable the observers to make this inference in reality; this is only a connection "in principle".[57]

"... cosmological solutions suffer from the disadvantage that the spatial part of spacetime is supposed to be homogeneous and isotropic. Therefore it is often difficult, owing to the lack of independent variables, to disentangle the causes of various effects" (Bondi 1947).

"It is pointed out that observations to date do not exclude the possible presence of anisotropies and inhomogeneities whose dynamical effects are comparable to the dynamical effects of expansion." (Kristian and Sachs 1966.) The authors introduced a formalism for comparing the predictions of inhomogeneous cosmological models with astronomical observations. Although quoted from time to time, the paper does

---

[57] Note that (B.1) does not depend on rotation. Rotation contributes to higher approximations only. Equation (B.1) is sometimes used to estimate rotation from observations, but this is done for the Bianchi-type spatially homogeneous models in which the rotation is coupled to shear and expansion through common arbitrary constants. Such coupling is a consequence of the assumed symmetry and is only one more piece of evidence for the Bianchi models being non-generic; conclusions drawn from it do not apply to the observed Universe.

not seem to be appreciated as highly as it deserves. See another quotation in Appendix C.

Kristian (1967) provided an indirect argument for considering inhomogeneous models. He made an attempt to measure the distortion of images of clusters of galaxies caused by conformal curvature of the Universe. Though the attempt was unsuccessful (in Kristian's own words), the effect is in principle very important because, being zero in the conformally flat models, it can allow us to detect departures from the FLRW geometry of the real Universe. The underlying principle is this: in a galaxy cluster, the angular distribution of galaxies of each shape should be random, but in a Universe that is not conformally flat the observed images of galaxies would have anisotropic distributions, and the anisotropy would be determined by the magnetic part of the Weyl tensor.

Several arguments in favour of inhomogeneous models were presented by Collins (1979). In the paper, the author proposed to investigate spacetimes with intrinsic symmetries (see our Section 4.9 on the Stephani Universe, 1967a, which is an example).

Spatially homogeneous models "may not be sufficiently general for problems where generality considerations are of prime importance (e.g. in the study of singularities)" (Collins and Szafron 1979a). The singularity "could differ greatly from the type found in FRW models" (MacCallum 1979).

The spatially homogeneous models "seem incapable of explaining . . . homogenization and isotropization of the Universe" and ". . . do not provide a suitable background for the formation of galaxies from small inhomogeneous perturbations" (Collins and Szafron 1979a). "Statistical fluctuations in FRW models cannot collapse fast enough to form the observed galaxies. This suggests that there must be real inhomogeneities at all stages in the universe. Moreover, some perturbations of FRW models are decaying modes which would have been more important in the past." (MacCallum 1979). More on this point can be found in the paper by Bonnor (1956, see our Section 3.2).

Ellis (1980) presented a general discussion of relations between cosmological models and cosmological observations. He discussed the following topics: (1) What quantities can be estimated from observations and how?; (2) How could observations imply that our Universe is FLRW?; (3) The practical limitations on observations. This last point is discussed in great detail, and the discussion highlights the possibility of, and the need of, introducing more general models, even though it explicitly concentrates on the difficulties of verifying the FLRW relations between observable quantities. The lack of well-defined criteria for acceptance or rejection of the FLRW models is emphasized in particular. Ellis explained that the isotropy of the microwave background radiation does not prove the isotropy of either the matter distribution or the spacetime. Finally, Ellis pointed out that instead of trying to deduce a model from observations, astronomers *assume* the FLRW models on the basis of the philosophically motivated "cosmological principle", and then only compare the results of the observations with the assumed models.

Wesson (1980, his section 11.4) presented several astrophysical arguments for studying inhomogeneous cosmological models, together with extensive references to observations supporting the arguments. For doubters, that book is worth reading in full.

"But this approximation[58] is a very crude one. The Einstein field equations are not linear, so that the disturbances in the field produced by the various stars cannot be just added, or averaged in any way." (Dirac 1981).

Fabbri and Melchiorri (1981) reported the observational discovery of quadrupole anisotropy in the microwave background radiation on a large angular scale ($\approx 80°$). Starting from this, they argued that the Universe will become "extremely irregular" in the future or perhaps it is already irregular today "at sufficiently large scales". This calls for the use of more general models than FLRW. Fabbri and Melchiorri stated that the arguments for the FLRW models are mainly philosophical: the "beauty and simplicity" of the isotropic Universe and "a remnant of the Ptolemaic view of the Cosmos". The paper is interesting as a very rare (especially at that time) call from within the observational community for the consideration of more general models.

Some more astrophysical arguments for studying inhomogeneous cosmological models were also given by MacCallum (1985).

"Although the cosmological principle is accepted as a fundamental corollary for the general picture of the Universe, it might be considered as a conjecture which cannot be proved by observations . . . It is sure that the scale of inhomogeneity is not of the order of ≈ 10 Mpc; it is practically sure that it is larger than ≈ 100 Mpc; it is highly probable that it is not smaller than ≈ 1000 Mpc." (Meszaros and Vanysek 1988.) The paper contains more elaborate arguments in favour of inhomogeneity.

The paper by van den Bergh (1990) can be listed here, too, although the author concentrated on emphasizing the difficulties of the currently accepted models without openly calling for inhomogeneous models. This is a voice from within the astronomical community, and it says, in short, that current cosmology has the following problems:

1. Large amounts of matter are revealed through the non-Keplerian motions of stars in galaxies and of galaxies themselves, but that dark matter is not directly visible.
2. Streaming motions of galaxies with velocities $v \leq 5000$ km/s towards invisible attractors are seen on giant scales.
3. Honeycomblike large-scale structures are observed in the distribution of galaxies, with some evidence of spatial periodicity.
4. Evidence of the presence of carbon, nitrogen and oxygen in quasars implies that quasars must contain stars and density concentrations of large magnitude. If quasars are accreting black holes, then they must be immersed in massive over-dense "fuel reservoirs". On the other hand, the very exact isotropy of the microwave background

---

[58] "This approximation" here means replacing the actual matter distribution in the Universe by a smoothed-out one (note by A.K.).

radiation implies isotropy of matter distribution at the last scattering.[59] It is a problem how such density contrasts might have evolved from such a homogeneous initial state in the time implied by the standard model.

Arp and van Flandern (1992) reviewed the astronomical observations that contradict the Hubble law. This article is not directly related to our subject, but several objections raised by the authors can be answered by invoking the L–T or Szekeres models with a nonsimultaneous Big Bang. For example, Arp and van Flandern argue that the Universe is not old enough to allow the density concentrations being formed out of an initially homogeneous distribution. This difficulty is solved if the Universe emerges from the Big Bang already inhomogeneous. The same concerns the existence of galaxies that are visibly younger than others – the nonsimultaneous Big Bang can solve this problem. What Arp and van Flandern call "a number of mini bangs" is realized in the L–T and Szekeres models as a nonsimultaneous Big Bang. Also their final suggestion of "creation of matter in the universe over an interval of time" is easily implemented in the L–T and Szekeres models.

"Recent deep surveys . . . have revealed superstructures . . . These results cast shadows on the long-cherished concept of a homogeneous isotropic universe . . . they are starting to seriously confront the standard model . . . they suggest that the clustering shows a periodic pattern . . . it is fundamentally self-inconsistent to use the homogeneous-isotropic Friedmann relation between redshifts and comoving distance." (Kurki–Suonio and Liang 1992).

Finally, let us note that there exists direct observational evidence that our Universe is not faithfully described by the FLRW models within the current observational error. In the Minkowski spacetime, two events situated on the same (future or past) light half-cone can only be in null or spacelike relation, that is, no timelike curve can join any two points of the same light half-cone. The conformal mappings preserve the light cones and the timelike/null/spacelike relations among events. Consequently, the light cones in conformally flat spacetimes have the same property, and this includes the FLRW models. However, in a gravitational lens system, consisting of the source, the lens and an observer, every light cone of an emission event develops a fold behind the lens: a generic observer (situated off any symmetry axis) receives light rays from the source along different optical paths that reach him or her at different times, although they left the source simultaneously. This means that the world-line of such an observer pierces the future light half-cone of every emission event in two (or more) points, that is, two events on that light half-cone *can* be joined by a timelike curve. Hence, a Universe that contains a gravitational lens cannot be conformally flat. Gravitational lenses are observed in our Universe at distances that are cosmologically significant, that is, at scales at which the cosmological models are supposed to apply. Consequently, our Universe is not FLRW within the limits set by observation.

---

[59] See, however, the remarks in Section 3.7 on this point.

Appendix C

Memorable statements about the cosmological principle

This appendix is meant to pay justice to those authors who could see earlier than others that the FLRW models are an oversimplification of Nature. It contains a short (and very possibly incomplete) selection of quotations.

> ... the grounds on which homogeneity is generally assumed appearing to be those of convenience rather than generality . . . We must categorically dissent from the extreme idea . . . that homogeneity is included in the definition of the universe . . . We take it to be perfectly conceivable that an increase of telescopic power may reveal a variation of material density with distance, and the denial of this possibility . . . seems to us to be inconsistent with the fundamental principles of science. We hold that the assumption of spatial homogeneity is . . . a working hypothesis, valid so long as it does not conflict with observation or with theoretical probability, and justifiable during that time as a restriction on arbitrary speculation . . . we have no grounds for supposing that the part of the universe which is observed is typical of the whole . . . the phenomena we embody in our models may be purely local characteristics . . . while, in the present state of knowledge, a spatially homogeneous universe has greater claims to consideration than any other, such claims . . . have no *a priori* justification. (*Dingle 1933*)

> The foregoing results demonstrate the lack of existence of any general kind of gravitational action which would necessarily lead to the disappearance of inhomogeneities in cosmological models . . . the phenomena of the actual universe will be affected by a more drastic kind of inhomogeneities than those here considered[60] . . . it is at least evident from the results obtained, that we must proceed with caution in applying to the actual universe any *wide* extrapolations – either spatial or temporal – of results deduced from strictly homogeneous models . . . it would appear wise . . . to envisage the possibility that regions of the universe beyond the range of our present telescopes might be contracting rather than expanding and contain matter with a density and stage of evolutionary development quite different from those with which we are familiar. It would also appear wise not to draw too definite conclusions from the behavior of homogeneous models as to a supposed initial state of the whole universe. (*Tolman 1934*)

[60] By this is meant the L–T model (note by A.K.).

It is often claimed that the universe in the large must be isotropic or homo-geneous. Certainly this view has immense aesthetic and philosophical appeal, but is it strongly supported by current observations? Unfortunately, it is not . . . observations neither confirm nor deny the "cosmological principle" that the universe is isotropic and homogeneous, or even homogeneous, and . . . measurements at the present time cannot prove, but can only disprove, that particular models represent the actual structure of the universe . . . global theoretical models that are inhomogeneous should be looked for. (*Kristian and Sachs 1966*)

. . . it is at least conceivable that the big bang is still going on in isolated parts of the universe . . . it would be very artificial to require that the big bang went off simultaneously in the past of each matter world line. (*Eardley, Liang and Sachs 1972*)

Appendix D

How was this review compiled?

Every review article or book raises the obvious question about its completeness. In order to give the readers an idea about the degree of completeness of this review, the method used to compile the bibliography is described briefly below. These were the essential steps:

1. The author has been interested in the subject since about 1980. Until 1988, when systematic compilation was begun, I studied every newly published research or review article on inhomogeneous cosmological models, and followed each reference whose context of citation suggested that it might contain more of the relevant material. The latest publications included are those that reached my hands in September 1994.

2. I looked through the subject indexes to all volumes of *Physics Abstracts*, beginning with the 1915 volume, and studied the sections on cosmology, general relativity, gravitation, gravitational collapse and spacetime configurations. Whenever any keyword of title or abstract suggested that the paper might be relevant for the review, I added the reference to the list of papers to look up. The last index so surveyed was Part I of the 1994 volume.

3. While reading or looking through the papers, I added every reference that seemed relevant to the list.

Stage 3 produced more than 1000 references in addition to about 1000 found in stage 2 (but about two-thirds of the total number of papers were discarded according to the criteria listed in Section 1.1). This means that half of this review is just the accumulated knowledge of the authors of the papers quoted. At the same time, this implies that the method adopted in stage 2 is not 100% reliable. Several authors concealed parts of their message under misleading titles or abstracts, and looking through a large number of lengthy indexes poses an extreme challenge to one's ability to concentrate one's attention. I cannot guarantee that no relevant papers were omitted. Still, it seems that no other review came closer to completeness.

While compiling this review, I came across many solutions that turned out to have no FLRW limit, but verifying this required elaborate reasoning and calculation. Those solutions have not been included, as explained in Section 1.1. This is the main reason why many papers having "inhomogeneous cosmological models" in their titles are not mentioned in this text.

Appendix E

Historical milestones

This appendix contains a list of papers which, in the opinion of this author, played a crucial role in the development of inhomogeneous cosmological models. It must be stressed that, except for a few, the papers listed below have never been properly appreciated, and many of them are virtually unknown even today. The list is thus a call for historical justice (based on a personal assessment by this author) rather than a presentation of development of the field.

Lemaître (1933a) – the pioneering paper, and probably the most underappreciated one. The author introduced the Lemaître–Tolman model, and in addition presented or solved a few problems commonly associated now with names and papers younger by a whole generation. Examples: the definition of mass for a spherically symmetric perfect fluid, a proof that the Schwarzschild horizon is not a singularity (by a coordinate transformation to a system of freely falling observers), a preliminary statement of a singularity theorem illustrated by a Bianchi I model.

McVittie (1933) – presented a superposition of the Schwarzschild and FLRW metrics which is a perfect fluid solution. A remarkably bold and early entry, but the solution has still not been satisfactorily interpreted.

Dingle (1933) – a preliminary investigation of spherically symmetric shearfree perfect fluid solutions, later completed by Kustaanheimo and Qvist (1948). The paper is remarkable for the author's strong criticism of the cosmological principle and an explicit call for inhomogeneous models (see Appendix C).

Tolman (1934) – the first extended investigation of the Lemaître (1933a) solution. The author showed that the solution implies the instability of the Einstein and FLRW models against the formation of condensations and rarefactions (which today would be called voids). Tolman cited Lemaître. All the contemporary authors who call this solution "the Tolman model" evidently did not read Tolman's paper.

Sen (1934) – the first explicit statement that voids should form in an inhomogeneous Universe: "the [Einstein and FLRW] models are unstable for initial rarefaction".

McCrea (1939) – the first elaborate call for inhomogeneous models and the first draft of a theoretical formalism to describe such models and compare them with observations.

Einstein and Straus (1945) – the first formally correct investigation of the influence of the expansion of the Universe on planetary orbits. Though formally correct, the investigation unfortunately produced an incorrect result because the model it used was rather peculiar and unstable.

Bondi (1947) – a thorough in-depth investigation of geometrical and physical properties of the Lemaître–Tolman model. The author was the first to observe several physically interesting possibilities offered by the model, but the paper was ignored for a long time. Several of Bondi's observations were rediscovered independently and picked up much later by other authors.

Kustaanheimo and Qvist (1948) – reduced the Einstein equations for a spherically symmetric metric with a shearfree expanding perfect fluid source to a single ordinary differential equation. Their class includes the McVittie (1933) solution.

Bonnor (1956) – showed, using the L–T model, that galaxies could not have originated as statistical fluctuations in an initially homogeneous matter distribution (because, according to the view that is still considered valid now, the Universe is much too young to have allowed enough time for the process to succeed).

Ehlers (1961) – a self-contained exposition of relativistic hydrodynamics and thermodynamics in curved space, with several ideas of McCrea (1939, see above) incorporated. A recent English translation is available (Ehlers 1993).

Shirokov and Fisher (1962) – first pointed out that an "averaged" cosmological model should obey modified field equations, with a "polarization term" arising from averaging out small-scale inhomogeneities.

Kristian and Sachs (1966) – a self-contained theory of observations in an inhomogeneous Universe.

Stephani (1967a) – the most general conformally flat solution of the Einstein equations with a perfect fluid source; an important link in the chain of generalizations of the FLRW models.

Ellis (1971) – foundations of relativistic cosmology with arbitrary geometry, based on the exposition of Ehlers (1961, see above).

Barnes (1973) – a complete set of solutions of the Einstein equations with a shearfree, nonrotating and expanding perfect fluid source; the set includes the solutions of Kustaanheimo and Qvist (1948) and Stephani (1967a).

Szekeres (1975a) – still the most sophisticated dust solutions generalizing the FLRW models; they include the $\Lambda = 0$ subcase of the L–T model.

Szafron (1977) – generalization of the Szekeres (1975a) solutions to nonzero pressure, it includes the L–T model in full generality.

Goode and Wainwright (1982b) – a reformulation of the Szekeres (1975a) solutions showing the interconnection between increasing and decreasing modes of perturbation of the FLRW background on the one hand, and the initial conditions (nonsimultaneous Big Bang + initial inhomogeneities) on the other. The authors established a link between the Szekeres models and the perturbative methods.

Ellis (1984) – the first apparently successful call for approaching the relativistic cosmology on a broader basis than the FLRW models.

Gautreau (1984) – the first quantitative estimate of the influence of cosmic expansion on planetary orbits; based on the L–T model.

Sato (1984) – an elaborate theory of evolution of voids in the Universe, based on the L–T model.

Panek (1992) – a thorough investigation of the influence of inhomogeneities in matter distribution on the anisotropies in the cosmic microwave background radiation, based on the L–T model. Several authors argued that the isotropy of the radiation actually proves the homogeneity of the Universe. However, no quantitative estimates existed of the amount of anisotropy that inhomogeneities in matter distribution would induce in the background radiation. Panek's paper provided the first such estimate which in fact proves the earlier arguments to be void.

Zalaletdinov (1992) – the first covariant theory of averaging out small-scale inhomogeneities in the metric.

Notes added in proof

Note 1 (to Section 1.2, before eq. (1.2.11))

The form (1.2.5) of the energy-momentum tensor guarantees that the transport of energy occurs only by means of mass-flow. This property is a necessary condition for a continuous medium to be a perfect fluid, but it is not a sufficient condition. A portion of a single-component perfect fluid must also obey the appropriate laws of thermodynamics, and among them the Gibbs–Duhem equation:

$$dU + p\,dV = T\,dS, \tag{N.1}$$

where U is the internal energy, V is the volume, T is the temperature and S is the entropy. In cosmology, this equation should apply when the quantities in it are referred to one particle of the medium, that is, $U = \epsilon/n$, $V = 1/n$, $S = s/n$, where s is the entropy density, and n is the particle number density that obeys:

$$(nu^{\alpha})_{;\alpha} = 0. \tag{N.2}$$

Equation (N.2) merely defines the function n (actually, this is a large family of functions), and is no limitation. However, eq. (N.1) can be solved for T and S only if its left-hand side has an integrating factor. Most of the cosmological solutions considered in the literature so far have symmetry groups with two- or three-dimensional orbits. In such solutions, the functions ϵ, n and p depend on one or two coordinates only, and eq. (N.1) is then no limitation either. However, for "perfect fluid" solutions with a one-dimensional symmetry or with no symmetry at all, the differential form $[d(\epsilon/n) + p\,d(1/n)]$ does not automatically have an integrating factor. In those solutions, eq. (N.1) introduces additional limitations.

Bona and Coll (1988) were the first to recognize this problem, and Coll and Ferrando (1989) discussed it at a general level. Of the solutions described in this book, only those by Szafron (1977) and Stephani (1967a) (see Sections 2.1 and 4.9 respectively) pose nontrivial problems for thermodynamical interpretation; these have been discussed in a few recent papers (see Notes 2 and 4 below). The other solutions either have a sufficiently high symmetry to obey (N.1) identically, or else they have constant pressure ($p=0$ in particular), and so have trivial thermodynamics.

When eq. (N.1) can be imposed without introducing new symmetries, this does not yet guarantee that the thermodynamical functions will obey any equation of state that is familiar from laboratory thermodynamics. This problem has not yet been investigated at all.

Note 2 (to the end of Section 2.2)

The Szafron solutions are an example to which the discussion of thermodynamical interpretation (see Note 1 above) applies. A preliminary study was done by Quevedo and Sussman (1995a) and the problem was solved by Krasiński, Quevedo and Sussman (1995). In this last paper, it was shown that in the $\beta' \neq 0$ subfamily eqs. (N.1) and (N.2) necessarily either make the thermodynamics trivial (because p=const) or else imply a symmetry group with at least two-dimensional orbits. Hence, the general $\beta' \neq 0$ Szafron (1977) solutions require an interpretation in terms of a more complicated source than a single-component perfect fluid; it could be a mixture in which chemical reactions occur, or a mixture of two fluids, such as was first introduced by Letelier (1980) and mentioned in our Section 6.6.

In the $\beta' = 0$ subfamily, there is a subset of solutions that obey (N.1) and (N.2) while still having no symmetry. This subset is defined by rather complicated formulae, so we refer the reader to the paper (Krasiński *et al.* 1995). As mentioned in Note 1 above, the problem of the explicit form of an equation of state remains unsolved.

Note 3 (to Section 2.9, after eq. (2.9.10))

The solution 3 by Stephani (1987, our eqs. (2.9.10)–(2.9.11)) has a deeper-hidden error in it; it is not in fact a perfect fluid solution, and the correction made in (2.9.10) does not remove the problem. Hint: a perfect fluid solution with p=const must have zero acceleration, but the acceleration in (2.9.10) will not be zero unless either all $f^m=0$ (then $\omega=0$) or $\Lambda=0$. The existence of this problem was confirmed by H. Stephani, and this author (A.K.) is very grateful to him for his assistance. The error disappears in the limits $\Lambda=0$ and $\omega=0$, so all the solutions considered in this book that result from (2.9.10) by taking these limits are correct.

Note 4 (to Section 4.9, third paragraph after eq. (4.9.6))

The Stephani solution (4.9.1)–(4.9.5) is another example to which the discussion from Note 1 applies. Its source can be interpreted as a single-component perfect fluid only if the functions are specialized so that the spacetime acquires a G_3/S_2 symmetry group, or a still higher symmetry (the six-dimensional FLRW symmetry, for example). This was first shown by Bona and Coll (1988), and those authors found an explicit formula that defines the G_3/S_2-symmetric subcases. Since the G_3/S_2 symmetry is disguised in a rather elaborate way in this case, Quevedo and Sussman (1995b) mistakenly denied the result of Bona and Coll and claimed that Bona and Coll's subcase has only a one-dimensional symmetry. The dispute was settled in the subsequent paper (Krasiński, Quevedo and Sussman 1995), in which Bona and Coll's result was confirmed by a different method. In all three papers thermodynamical properties of the subcases with higher symmetry were discussed, though not in much detail.

Bibliography

(The symbols in parentheses at the end of each entry refer to the section numbers, appendices, figure numbers and box numbers.)

Adams, P. J., Hellings, R. W., Zimmerman, R. L. (1985), *Astrophys. J.* **288**, 14 (3.9).

Adams, P. J., Hellings, R. W., Zimmerman, R. L. (1987), *Astrophys. J.* **318**, 1 (3.9).

Adams, P. J., Hellings, R. W., Zimmerman, R. L., Farhoosh, H., Levine, D. I., Zeldich, S. (1982), *Astrophys. J.* **253**, 1 (3.9).

Agnew, A. F., Goode, S. W. (1994), *Class. Q. Grav.* **11**, 1725 (6.7; Fig. 6.1).

Aguirregabiria, J. M., Feinstein, A., Ibañez, J. (1993), *Phys. Rev.* **D48**, 4669 (7.1; Fig. 7.1).

Alencar, P. S. C., Letelier, P. S. (1986), in: *Proceedings of the 4th Marcel Grossman Meeting on General Relativity*. Edited by R. Ruffini. Elsevier, Amsterdam, p. 885 (6.4; Fig. 6.1).

Alexander, D., Green, R. M., Emslie, A. G. (1989), *Mon. Not. Roy. Astr. Soc.* **237**, 93 (4.6, 4.11).

Ambartsumyan, V. A. (1964), *Raport 13 Conseil de Physique Solvay*, Brussels (3.2).

Argueso, F., Sanz, J. L. (1985), *J. Math. Phys.* **26**, 3118 (4.9).

Arnau, J. V., Fullana, M., Monreal, L., Saez, D. (1993), *Astrophys. J.* **402**, 359 (3.7).

Arnau, J. V., Fullana, M., Saez, D. (1994), *Mon. Not. Roy. Astr. Soc.* **268**, L17 (3.7).

Arp, H. C., van Flandern, T. (1992), *Phys. Lett.* **A164**, 263 (B).

Bailyn, M. (1973), *Phys. Rev.* **D8**, 1036 (2.14; Box 2.4.4).

Bali, R., Singh, G., Tyagi, A. (1987), *Astrophys. Space Sci.* **139**, 365 (1.2, 7.2; Fig. 7.1).

Bampi, F., Cianci, R. (1980), *Int. J. Theor. Phys.* **19**, 211 (1.3.1).

Banerjee, A. (1967), *Proc. Phys. Soc.* **91**, 794 (2.12, 3.6; Box 2.4.2).

Banerjee, A. (1971), *Progr. Theor. Phys.* **46**, 1625 (4.6).

Banerjee, A. (1972), *J. Phys.* **A5**, 1305 (4.10; Box 4.1.4).

Banerjee, A. (1975), *J. Phys.* **A8**, 281 (3.6).

Banerjee, A., Banerji, S. (1976), *Acta Phys. Polon.* **B7**, 389 (4.6; Box 4.1.3).

Banerjee, A., Chakrabarty, N., Dutta Choudhury, S. B. (1976), *Acta Phys. Polon.* **B7**, 675 (4.12; Box 4.1.5).

Banerjee, A., Chakravarty, N. (1979), *Acta Phys. Polon.* **B10**, 3 (4.2.2, 4.3, 4.9; Fig. 4.1; Box 4.1.1).

Banerjee, A., Chakravorty, N., Dutta Choudhury, S. B. (1975), *Nuovo Cimento* **B29**, 357 (4.12; Box 4.1.5).

Banerjee, A., De, U. K. (1975), *Acta Phys. Polon.* **B6**, 335 (4.13; Fig. 4.2).

Banerjee, A., Dutta Choudhury, S. B. (1977), *Progr. Theor. Phys.* **57**, 1070 (4.12).

Banerjee, A., Dutta Choudhury, S. B., Bhui, B. K. (1989), *Phys. Rev.* **D40**, 670 (4.14; Fig. 4.3).

Banerjee, A., Dutta Choudhury, S. B., Bhui, B. K. (1990), *Pramana* **34**, 397 (4.14; Fig. 4.3).

Banerjee, A., Sil, A., Chatterjee, S. (1994), *Astrophys. J.* **422**, 681 (2.16; Fig. 2.5).

Bardeen, J. M. (1968), *Bull. Am. Phys. Soc.* **13**, 41 (2.14).

Barnes, A. (1970), *J. Phys.* **A3**, 653 (2.4.2, 2.12, 3.1, 3.4, 3.6, 3.8, 3.9).

Barnes, A. (1973), *Gen. Rel. Grav.* **4**, 105 (4.1, 4.2, 4.2.1, 4.2.2, 4.3, 4.4, 4.8, 4.9, 4.11, E; Fig. 4.1; Boxes 4.1.1, 4.1.2).

Barnes, A. (1974), *Gen. Rel. Grav.* **5**, 147 (2.7; Fig. 2.1).

Barnes, A. (1984), in: *Classical General Relativity.* Edited by W. B. Bonnor, J. N. Islam and M. A. H. MacCallum. Cambridge University Press, p. 15 (4.1).

Barnes, A., Rowlingson, R. R. (1989), *Class. Q. Grav.* **6**, 949 (2.2, 2.3).

Barnes, A., Whitrow, G. J. (1970), *Mon. Not. Roy. Astr. Soc.* **148**, 193 (2.12; Box 2.4.2).

Barrow, J. D., Grøn, Ø. (1986), *Phys. Lett.* **B182**, 25 (7.1).

Barrow, J. D., Silk, J. (1981), *Astrophys. J.* **250**, 432 (2.4.1).

Barrow, J. D., Stein–Schabes, J. A. (1984), *Phys. Lett.* **A103**, 315 (2.4.1, 2.4.2, 2.8, 2.10.3, 2.15; Figs. 2.1, 2.2, 2.4, 2.5).
Bartnik, R. (1988), *Commun. Math. Phys.* **117**, 615 (3.8).
Bayin, S. S. (1979), *Phys. Rev.* **D19**, 2838 (5.7; Fig. 5.1).
Bayin, S. S. (1986), *Astrophys. J.* **303**, 101 (4.8; Fig. 4.1).
Baysal, H. (1992), *Astrophys. Space Sci.* **196**, 345 (2.10.1, 2.10.3; Fig. 2.2).
Bekenstein, J. D. (1971), *Phys. Rev.* **D4**, 2185 (3.5).
Belinskii, V. A. (1979), *ZhETF* **77**, 1239 [*Sov. Phys. JETP* **50**, 623 (1979)] (6.4, 6.7; Fig. 6.1).
Belinskii, V., Ruffini, R. (1980), *Phys. Lett.* **B89**, 195 (7.8).
Berger, B. K., Eardley, D. M., Olson, D. W. (1977), *Phys. Rev.* **D16**, 3086 (2.2).
Bergmann, O. (1981), *Phys. Lett.* **A82**, 383 (1.3.2, 4.14; Fig. 4.3).
Bialko, A. V. (1973), *ZhETF* **65**, 849 [*Sov. Phys. JETP* **38**, 421 (1974)] (8.2, 8.3).
Bičak, J., Griffiths, J. B. (1994), *Phys. Rev.* **D49**, 900 (6.7; Fig. 6.1).
Bildhauer, S., Futamase, T. (1992), *Gen. Rel. Grav.* **23**, 1251 (8.2).
Bogoyavlenskii, O. I. (1977), *ZhETF* **73**, 1201 [*Sov. Phys. JETP* **46**, 633 (1977)] (2.13).
Bokhari, A. H. (1992), *Int. J. Theor. Phys.* **31**, 2087 (4.14).
Bona, C. (1985), in: *Proceedings of the Relativity Meeting 1984, Universidad de Valencia.* Edited by M. Portilla, J. L. Sanz, R. Lapiedra. Universidad de Valencia, Santander, p. 11 (4.9).
Bona, C. (1988), *J. Math. Phys.* **29**, 2462 (3.8).
Bona, C., Coll, B. (1985), *C. R. Acad. Sci.* Paris I **301**, 613 (4.9).
Bona, C., Coll, B. (1988), *Gen. Rel. Grav.* **20**, 297 (4.9; Notes 1, 4).
Bona, C., Stela, J. (1987), *Phys. Rev.* **D36**, 2915 (2.13; Fig. 2.4).
Bona, C., Stela, J., Palou, P. (1987a), *Gen. Rel. Grav.* **19**, 179 (2.13; Fig. 2.4).
Bona, C., Stela, J., Palou, P. (1987b), *J. Math. Phys.* **28**, 654 (2.8, 2.13; Figs. 2.1, 2.4).
Bondarenko, N. P., Kobushkin, P. K. (1972), *Doklady ANSSSR* **202**, 558 [*Sov. Phys. Doklady* **17**, 48 (1972)] (1.1, 4.11, 4.14; Figs. 4.1, 4.3).
Bondi, H. (1947), *Mon. Not. Roy. Astr. Soc.* **107**, 410 (2.12, 2.13, 3.1, 3.4, 3.8, B, E).
Bondi, H. (1967), *Nature* **215**, 838 (4.10; Box 4.1.4).
Bondi, H. (1969), *Mon. Not. Roy. Astr. Soc.* **142**, 333 (4.10).
Bonnor, W. B. (1956), *Z. Astrophysik* **39**, 143 (3.2, B, E).
Bonnor, W. B. (1972), *Mon. Not. Roy. Astr. Soc.* **159**, 261 (3.8).
Bonnor, W. B. (1974), *Mon. Not. Roy. Astr. Soc.* **167**, 55 (3.8).
Bonnor, W. B. (1976a), *Commun. Math. Phys.* **51**, 191 (2.4.2).
Bonnor, W. B. (1976b), *Nature* **263**, 301 (2.4.2).
Bonnor, W. B. (1985a), *Class. Q. Grav.* **2**, 781 (3.8).
Bonnor, W. B. (1985b), *Mon. Not. Roy. Astr. Soc.* **217**, 597 (2.16, 3.8).
Bonnor, W. B. (1985c), *Phys. Lett.* **A112**, 26 (2.5).
Bonnor, W. B. (1986), *Class. Q. Grav.* **3**, 495 (2.5).
Bonnor, W. B. (1987), *Phys. Lett.* **A122**, 305 (4.14).
Bonnor, W. B., Chamorro, A. (1990), *Astrophys. J.* **361**, 21 (3.1).
Bonnor, W. B., Chamorro, A. (1991), *Astrophys. J.* **378**, 461 (3.1).
Bonnor, W. B., de Oliveira, A. K. G., Santos, N. O. (1989), *Phys. Rep.* **181** no. 5, 269 (4.14; Fig. 4.3).
Bonnor, W. B., Ellis, G. F. R. (1986), *Mon. Not. Roy. Astr. Soc.* **218**, 605 (2.5, 2.13, 3.8).
Bonnor, W. B., Faulkes, M. C. (1967), *Mon. Not. Roy. Astr. Soc.* **137**, 239 (4.11; Fig. 4.1).
Bonnor, W. B., Pugh, D. J. R. (1987), *South Afr. J. Phys.* **10**, 169 (2.5, 3.8).
Bonnor, W. B., Sulaiman, A. H., Tomimura, N. (1977), *Gen. Rel. Grav.* **8**, 549 (2.1.1, 2.2).
Bonnor, W. B., Tomimura, N. (1976), *Mon. Not. Roy. Astr. Soc.* **175**, 85 (2.4.1, 2.6; Fig. 2.1).
Boyer, R. H., Lindquist, R. W. (1967), *J. Math. Phys.* **8**, 265 (5.3).
Brauer, U. (1991), *Class. Q. Grav.* **8**, 1283 (3.9).
Brauer, U., Malec, E. (1992a), *Class. Q. Grav.* **9**, 905 (3.9).
Brauer, U., Malec, E. (1992b), *Phys. Rev.* **D45**, 1836 (3.9).
Brauer, U., Malec, E., O'Murchadha, N. (1994), *Phys. Rev.* **D49**, 5601 (3.9).
Bray, M. (1986), *Rendiconti di Matematica e delle sue Applicazioni*, ser. VII, **6**, 171 (2.10.2, 2.10.3; Fig. 2.2).
Brezhnev, V. S. (1966a), in: *Problemy teorii gravitatsii i elementarnykh chastits* [*Problems of Gravitation Theory and Elementary Particle Theory*], 1st edition. Edited by K. P. Stanyukovich and G. A. Sokolnik. Atomizdat, Moskva, p. 152 (4.10; Box 4.1.4).

Brezhnev, V. S. (1966b), in: *Problemy teorii gravitatsii i elementarnykh chastits* [*Problems of Gravitation Theory and Elementary Particle Theory*], 1st edition. Edited by K. P. Stanyukovich and G. A. Sokolnik. Atomizdat, Moskva, p. 158 (4.6; Box 4.1.3).

Brezhnev, V. S., Ivanenko, D. D., Frolov, V. N. (1966), *Izv. VUZ Fiz.* **9** no. 6, 119 (4.6; Box 4.1.3).

Bronnikov, K. A. (1980), *J. Phys.* **A13**, 3455 (6.5; Fig. 6.1).

Bronnikov, K. A. (1983), *Gen. Rel. Grav.* **15**, 823 (2.14, 3.5; Fig. 2.4).

Bronnikov, K. A., Pavlov, N. V. (1979), in: *Diskusyonnye voprosy teorii otnositelnosti i gravitatsii* [*Controversial Questions of the Theory of Relativity and Gravitation*]. Nauka, Moskva, p. 59 (2.14, 3.5; Fig. 2.4).

Bruni, M., Matarrese, S., Pantano, O. (1995), *Astrophys. J.* **445**, 958 (2.5).

Burd, A. A., Coley, A. A. (1994), *Class. Q. Grav.* **11**, 83 (2.11.3; Fig. 2.3).

Burlankov, D. E. (1987), *ZhETF* **93**, 1921 [*Sov. Phys. JETP* **66**, 1095 (1987)] (2.14).

Burnett, G. A. (1991), *Phys. Rev.* **D43**, 1143 (3.9).

Burnett, G. A. (1993), *Phys. Rev.* **D48**, 5688 (3.8).

Cahill, M. E., McVittie, G. C. (1970a), *J. Math. Phys.* **11**, 1382 (4.3, 4.10; Boxes 4.1.1, 4.1.4).

Cahill, M. E., McVittie, G. C. (1970b), *J. Math. Phys.* **11**, 1392 (4.3).

Cahill, M. E., Taub, A. H. (1971), *Commun. Math. Phys.* **21**, 1 (2.13, 3.8; Box 2.4.3).

Carfora, M., Marzuoli, A. (1984a), in: *Atti del VI Convegno Nazionale di Relativita Generale e Fisica della Gravitazione* (Firenze 1984) (8.4).

Carfora, M., Marzuoli, A. (1984b), *Phys. Rev. Lett.* **53**, 2445 (8.4).

Carmeli, M., Charach, C. (1984), *Found Phys.* **14**, 963 (6.7).

Carmeli, M., Charach, C., Feinstein, A. (1983a), *Ann. Phys.* **150**, 392 (6.7; Fig. 6.1).

Carmeli, M., Charach, C., Feinstein, A. (1983b), *Phys. Lett.* **A96**, 1 (6.7).

Carmeli, M., Charach, C., Malin, S. (1981), *Phys. Rep.* **76** no. 2, 79 (6.7; Fig. 6.1).

Carminati, J. (1990), *Class. Q. Grav.* **7**, 1543 (2.5).

Carot, J., Ibanez, J. (1985), *J. Math. Phys.* **26**, 2282 (6.4).

Carr, B. J. (1990), in: *Proceedings of the Relativity Meeting 89: Recent developments in gravitation, Barcelona 1989.* Edited by E. Verdaguer, J. Garriga, J. Cespedes. World Scientific, Singapore, p. 121 (2.13).

Carr, B. J., Hawking, S. W. (1974), *Mon. Not. Roy. Astr. Soc.* **168**, 399 (2.12, 2.13; Boxes 2.4.2, 2.4.3).

Carr, B. J., Yahil, A. (1990), *Astrophys. J.* **360**, 330 (2.13; Box 2.4.3).

Carter, B., Henriksen, R. N. (1989), *Ann. Physique* **14** colloq. no. 1 (suppl. au no. 6), 47 [*Journees Relativistes, Tours 1989.* Edited by C. Barrabes] (2.13).

Chakravarty, N., Chatterjee, S. (1978), *Acta Phys. Polon.* **B9**, 777 (4.12; Box 4.1.5).

Chakravarty, N., Dutta Choudhury, S. B., Banerjee, A. (1976), *Austral. J. Phys.* **29**, 113 (4.6; Box 4.1.3).

Chamorro, A. (1991), *Astrophys. J.* **383**, 51 (3.1).

Chan, R., Lemos, J. P. S., Santos, N. O., Pacheco, J. A. de F. (1989), *Astrophys. J.* **342**, 976 (4.14).

Chandrasekhar, S., Xanthopoulos, B. C. (1985), *Proc. Roy. Soc. London* **A402**, 37 (6.3, 6.7; Fig. 6.1).

Charach, C., Malin, S. (1979), *Phys. Rev.* **D19**, 1058 (6.4, 6.5, 6.6, 6.7; Fig. 6.1).

Charach, C., Malin, S. (1980), *Phys. Rev.* **D21**, 3284 (6.5; Fig. 6.1).

Chatterjee, S. (1984), *Gen. Rel. Grav.* **16**, 381 (4.12; Box 4.1.5).

Chatterjee, S., Banerji, S. (1980), *Int. J. Theor. Phys.* **19**, 599 (2.14; Fig. 2.4).

Chi, L. K. (1987), *J. Math. Phys.* **28**, 1539 (2.13; Box 2.4.3).

Christodoulou, D. (1984a), *Commun. Math. Phys.* **93**, 171 (3.6).

Christodoulou, D. (1984b), in: *General Relativity and Gravitation.* Edited by B. Bertotti, F. de Felice and A. Pascolini. D. Reidel, Dordrecht, p. 27 (3.6).

Clarke, C. J. S. (1993), *Class. Q. Grav.* **10**, 1375 (3.6).

Clarke, C. J. S., O'Donnell, N. (1992), *Rendiconti del Seminario Matematico della Universita e Politecnico de Torino* **50**(1), 39 (3.6).

Coley, A. A., Czapor, S. R. (1992), *Class. Q. Grav.* **9**, 1787 (4.11; Fig. 4.1).

Coley, A. A., McManus, D. J. (1994), *Class. Q. Grav.* **11**, 1261 (1.3.2).

Coley, A. A., Tupper, B. O. J. (1983), *Phys. Lett.* **A95**, 357 (1.1).

Coley, A. A., Tupper, B. O. J. (1990a), *Class. Q. Grav.* **7**, 1961 (4.11; Fig. 4.1).

Coley, A. A., Tupper, B. O. J. (1990b), *Class. Q. Grav.* **7**, 2195 (4.11; Fig. 4.1).

Coll, B., Ferrando, J. J. (1989), *J. Math. Phys.* **30**, 2918 (Note 1).

Collins, C. B. (1977), *J. Math. Phys.* **18**, 2116 (2.11.1; Figs. 2.1, 2.3).

Collins, C. B. (1979), *Gen. Rel. Grav.* **10**, 925 (1.1, 2.6, 2.8, 2.9, 4.9, B).

Collins, C. B. (1985), *J. Math. Phys.* **26**, 2009 (4.2.2, 4.4).

Collins, C. B. (1986), *Can. J. Phys.* **64**, 191 (4.2.2, 4.4).

Collins, C. B. (1990), *Class. Q. Grav.* **7**, 1983 (2.13; Box 2.4.3).

Collins, C. B., Ellis, G. F. R. (1979), *Phys. Rep.* **56** no. 2, 65 (1.1).

Collins, C. B., Hawking, S. W. (1973), *Astrophys. J.* **180**, 317 (3.8).

Collins, C. B., Lang, J. M. (1987), *Class. Q. Grav.* **4**, 61 (6.7, 7.1; Figs. 6.1, 7.1).

Collins, C. B., Szafron, D. A. (1979a), *J. Math. Phys.* **20**, 2347 (1.1, 2.2, 2.6, B).

Collins, C. B., Szafron, D. A. (1979b), *J. Math. Phys.* **20**, 2362 (1.1).

Collins, C. B., Wainwright, J. (1983), *Phys. Rev.* **D27**, 1209 (4.1, 4.2.1, 4.2.2, 4.4, 4.12; Figs. 4.1, 4.2).

Cook, M. W. (1974), *Bull. Austral. Math. Soc.* **10**, 149 (4.9, 4.10; Box 4.1.4).

Cook, M. W. (1975), *Austral. J. Phys.* **28**, 413 (4.9, 4.10; Box 4.1.4).

Covarrubias, M. (1980), *J. Phys.* **A13**, 3023 (2.4.2).

Covarrubias, G. M. (1983), in: *Relativity, Cosmology, Topological Mass and Supergravity. Proceedings of the 4th SILARG Symposium on Gravity, Gauge Theories and Supergravity, Caracas 1982.* World Scientific, Singapore, p. 251 (2.4.2, 2.15; Figs. 2.4, 2.5).

Covarrubias, G. M. (1984), *Astrophys. Space Sci.* **103**, 401 (2.4.2).

Cruzate, J., Diaz, M., Gleiser, R., Pullin, J. (1988), *Class. Q. Grav.* **5**, 883 (7.8).

Dąbrowski, M. (1993), *J. Math. Phys.* **34**, 1447 (4.9).

Dandach, N. F., Mitskievic, N. V. (1984), *J. Phys.* **A17**, 2335 (2.16).

Dandash, N. F. (1985), *Izv. VUZ Fiz.* **28** no. 12, 33 [*Sov. Phys. J.* **28**, 977 (1985)] (2.16).

Das, K. C. (1981), *J. Math. Phys. Sci. (India)* **15**, 151 (2.1).

Date, T. H. (1973), *Indian J. Pure Appl. Math.* **4**, 612 (4.12; Box 4.1.5).

Datt, B. (1938), *Z. Physik* **108**, 314 (2.6, 2.7, 2.12, 2.13; Figs. 2.1, 2.2; Boxes 2.4.2, 2.4.3).

Datta, A. K. (1976), *Indian J. Phys.* **50**, 590 (2.14; Box 2.4.4).

Dautcourt, G. (1980), in: *9th International Conference on General Relativity and Gravitation*, Jena, p. 315 (3.9).

Dautcourt, G. (1983a), *Astron. Nachr.* **304**, 153 (3.9).

Dautcourt, G. (1983b), *J. Phys.* **A16**, 3507 (3.9).

Dautcourt, G. (1985), *Astron. Nachr.* **306**, 1 (3.9).

Davidson, W. (1988), *Class. Q. Grav.* **5**, 147 (7.1; Fig. 7.1).

Davidson, W. (1992), *Gen. Rel. Grav.* **24**, 179 (7.1; Fig. 7.1).

De, U. K. (1968), *J. Phys.* **A1**, 645 (2.14).

De, U. K. (1971), *Indian J. Phys.* **45**, 487 (4.13; Fig. 4.2).

De, U. K., Ray, D. (1983), *J. Math. Phys.* **24**, 610 (2.14; Fig. 2.4).

Demiański, M. (1972), *Phys. Lett.* **A42**, 157 (5.3, 5.5; Fig. 5.1).

Demiański, M., Grishchuk, L. P. (1972), *Commun. Math. Phys.* **25**, 233 (2.9).

Demiański, M., Lasota, J. P. (1973), *Nature Phys. Sci.* **241**, 53 (3.4, 3.6, 3.9).

Deng, Y. (1989), *Gen. Rel. Grav.* **21**, 503 (4.14; Fig. 4.3).

Deng, Y., Mannheim, P. (1990), *Phys. Rev.* **D42**, 371 (4.14; Fig. 4.3).

Deng, Y., Mannheim, P. (1991), *Phys. Rev.* **D44**, 1722 (2.15; Fig. 2.5).

de Oliveira, A. K. G., Kolassis, C. A., Santos, N. O. (1988), *Mon. Not. Roy. Astr. Soc.* **231**, 1011 (4.14).

de Oliveira, A. K. G., Santos, N. O., Kolassis, C. A. (1985), *Mon. Not. Roy. Astr. Soc.* **216**, 1001 (4.14; Fig. 4.3).

de Souza, M. M. (1985), *Rev. Bras. Fis.* **15**, 379 (2.4.2).

de Vaucouleurs, G. (1970), *Science* **167**, 1203 (3.8).

Diaz, M., Gleiser, R. J., Gonzalez, G. I., Pullin, J. A. (1989), *Phys. Rev.* **D40**, 1033 (6.7).

Diaz, M. C., Gleiser, R. J., Pullin, J. A. (1987), *Class. Q. Grav.* **4**, L23 (7.8; Fig. 7.3).

Diaz, M. C., Gleiser, R. J., Pullin, J. A. (1988a), *Class. Q. Grav.* **5**, 641 (7.8; Fig. 7.3).

Diaz, M. C., Gleiser, R. J., Pullin, J. A. (1988b), *J. Math. Phys.* **29**, 169 (7.8; Fig. 7.3).

Diaz, M. C., Gleiser, R. J., Pullin, J. A. (1989), *Astrophys. J.* **339**, 1 (7.8).

Dingle, H. (1933), *Mon. Not. Roy. Astr. Soc.* **94**, 134 (4.1, 4.2.3, 4.3, C, E; Box 4.1.1).

Dirac, P. A. M. (1979a), in: *On the Path of A. Einstein.* Edited by B. Kursunoglu, A. Perlmutter, L. F. Scott. Plenum, New York, p. 1 (4.6, 4.12; Box 4.1.3).

Dirac, P. A. M. (1979b), *Proc. Roy. Soc. London* **A365**, 19 (4.6, 4.12; Box 4.1.3).

Dirac, P. A. M. (1981), in: *Gauge Theories, Massive Neutrinos and Proton Decay. Proceedings of Orbis Scientiae 1981*. Edited by B. Kursunoglu and A. Perlmutter. Plenum, New York and London, p. 1 (4.2, B).

Dodson, C. T. J. (1972), *Astrophys. J.* **172**, 1 (1.3.2, 2.13; Fig. 2.4).

Dwivedi, I. H., Joshi, P. S. (1992), *Class. Q. Grav.* **9**, L69 (3.6).

Dwivedi, I. H., Joshi, P. S. (1994), *Commun. Math. Phys.* **166**, 117 (3.6).

Dyer, C. C. (1979), *Mon. Not. Roy. Astr. Soc.* **189**, 189 (2.13; Box 2.4.3).

Dyer, C. C., McVittie, G. C., Oattes, L. M. (1987), *Gen. Rel. Grav.* **19**, 887 (4.3; Box 4.1.1).

Eardley, D. M. (1974a), *Commun. Math. Phys.* **37**, 287 (2.13).

Eardley, D. M. (1974b), *Phys. Rev. Lett.* **33**, 442 (3.8).

Eardley, D. M. (1979), *Gen. Rel. Grav.* **10**, 1033 (3.6).

Eardley, D. M. (1987), in: *Gravitation and Astrophysics (Cargese Lectures 1986)*. Edited by B. Carter and J. Hartle. Plenum, New York and London, p. 229 (3.6).

Eardley, D., Liang, E., Sachs, R. (1972), *J. Math. Phys.* **13**, 99 (2.4.2, 2.5, 2.12, 3.2, 3.6, 6.4, C; Box 2.4.1).

Eardley, D. M., Smarr, L. (1979), *Phys. Rev.* **D19**, 2239 (2.4.2, 3.6).

Eckart, C. (1940), *Phys. Rev.* **58**, 919 (1.2).

Ehlers, J. (1961), *Akad. Wiss. Lit. Mainz, Abhandl. Math. Naturw. Kl.,* Nr 11, 791 (1.2, B, E).

Ehlers, J. (1993), *Gen. Rel. Grav.* **25**, 1225 (1.2, E).

Einstein, A., Straus, E. G. (1945), *Rev. Mod. Phys.* **17**, 120 (2.13, 3.1, 3.3, E).

Einstein, A., Straus, E. G. (1946), *Rev. Mod. Phys.* **18**, 148 (2.13, 3.1, 3.3, 4.7).

Eisenstaedt, J. (1975a), *Phys. Rev.* **D11**, 2021 (4.4, 4.10; Boxes 4.1.2, 4.1.4).

Eisenstaedt, J. (1975b), *Phys. Rev.* **D12**, 1573 (4.4).

Eisenstaedt, J. (1976), *Ann. Inst. Poincare* **A24**, 179 (4.4, 4.6; Boxes 4.1.2, 4.1.3).

Eisenstaedt, J. (1977), *Astrophys. J.* **211**, 16 (4.4).

Ellis, G. F. R. (1967), *J. Math. Phys.* **8**, 1171 (2.6, 2.7, 2.8, 2.10.2, 2.10.3, 2.12, 2.14, 2.15, 3.6, 3.8; Figs. 2.1, 2.2, 2.4, 2.5).

Ellis, G. F. R. (1971), in: *Proceedings of the International School of Physics "Enrico Fermi", Course 47: General Relativity and Cosmology*. Edited by R. K. Sachs. Academic Press, New York and London, p. 104 (1.2, 1.3.2, 2.5, 4.1, E).

Ellis, G. F. R. (1980), *Ann. N. Y. Acad. Sci.* **336**, 130 (B).

Ellis, G. F. R. (1984), in: *General Relativity and Gravitation*. Edited by B. Bertotii, F. de Felice, A. Pascolini. D. Reidel, Dordrecht, p. 215 (8.1, 8.4, E).

Ellis, G. F. R. (1988), in: *Proceedings of the Second Canadian Conference on General Relativity and Relativistic Astrophysics*. Edited by A. A. Coley and B. O. J. Tupper. World Scientific, Singapore, p. 1 (8.1).

Ellis, G. F. R., Hellaby, C., Matravers, D. R. (1990), *Astrophys. J.* **364**, 400 (3.1).

Ellis, G. F. R., Maartens, R., Nel, S. D. (1978), *Mon. Not. Roy. Astr. Soc.* **184**, 439 (3.6).

Fabbri, R., Melchiorri, F. (1981), *Gen. Rel. Grav.* **13**, 201 (B).

Faulkes, M. C. (1969a), *Can. J. Phys.* **47**, 1989 (4.12; Box 4.1.5).

Faulkes, M. C. (1969b), *Progr. Theor. Phys.* **42**, 1139 (4.6; Box 4.1.3).

Feinstein, A., Griffiths, J. B. (1994), *Class. Q. Grav.* **11**, L109 (6.7; Fig. 6.1).

Feinstein, A., MacCallum, M. A. H., Senovilla, J. M. M. (1989), *Class. Q. Grav.* **6**, L217 (6.3).

Feng, L. L., Mo, H. J., Ruffini, R. (1991), *Astron. Astrophys.* **243**, 283 (2.12; Box 2.4.2).

Fennelly, A. J. (1977), *Mon. Not. Roy. Astr. Soc.* **181**, 121 (3.8).

Ferrando, J. J., Morales, J. A., Portilla, M. (1990a), *Gen. Rel. Grav.* **22**, 1021 (6.6).

Ferrando, J. J., Morales, J. A., Portilla, M. (1990b), in: *Proceedings of the Relativity Meeting, Barcelona 1989 [Recent Developments in Gravitation]*. Edited by E. Verdaguer, J. Garriga, J. Cespedes. World Scientific, Singapore, p. 356 (6.6).

Friedmann, A. A. (1922), *Z. Physik* **10**, 377 (1.1, A; all diagrams).

Friedmann, A. A. (1924), *Z. Physik* **21**, 326 (1.1, A; all diagrams).

Futamase, T. (1988), *Phys. Rev. Lett.* **61**, 2175 (8.2).

Futamase, T. (1989), *Mon. Not. Roy. Astr. Soc.* **237**, 187 (8.2).

Futamase, T. (1993), *Progr. Theor. Phys.* **89**, 581 (8.2).

Gautreau, R. (1984), *Phys. Rev.* **D29**, 198 (1.1, 2.13, 3.3, 3.8, 4.6, 4.7, E; Fig. 2.4; Box 2.4.3).

Geroch, R. (1967), *J. Math. Phys.* **8**, 782 (4.9).

Geroch, R. (1969), *Commun. Math. Phys.* **13**, 180 (1.3.1).

Gertsenshtein, M. E. (1966), *ZhETF* **51**, 129 [*Sov. Phys. JETP* **24**, 87 (1967)] (3.8).

Gertsenshtein, M. E. (1977), *Izv. VUZ Fiz.* **20** no. 7, 90 [*Sov. Phys. J.* **20**, 912 (1977)] (3.8).

Gertsenshtein, M. E., Stanyukovich, K. P. (1974), in: *Problemy teorii gravitatsii i elementarnykh chastits* [*Problems of Gravitation Theory and Elementary Particle Theory*], 5th issue. Edited by K.P. Stanyukovich. Atomizdat, Moskva, p. 162 (3.5).

Ginzburg, V. L., Ozernoi, L. M. (1964), *ZhETF* **47**, 1030 [*Sov. Phys. JETP* **20**, 689 (1965)] (3.8).

Glass, E. N. (1979), *J. Math. Phys.* **20**, 1508 (4.3, 4.4, 4.10; Boxes 4.1.1, 4.1.2, 4.1.4).

Glass, E. N. (1981), *Phys. Lett.* **A86**, 351 (4.14; Fig. 4.3).

Glass, E. N., Mashhoon, B. (1976), *Astrophys. J.* **205**, 570 (4.6; Box 4.1.3).

Gleiser, R. J. (1984), *Gen. Rel. Grav.* **16**, 1039 (2.4.2, 3.6).

Gleiser, R. J., Diaz, M. C., Grosso, R. D. (1988a), *Astron. Nachr.* **309**, 239 (6.7).

Gleiser, R. J., Diaz, M. C., Grosso, R. D. (1988b), *Class. Q. Grav.* **5**, 989 (6.7).

Goethals, M. (1975), *Ann. Soc. Sci. Bruxelles* **89**, 50 (2.11.1; Fig. 2.3).

Goicoechea, L. J., Martin–Mirones, J. M. (1987), *Astron. Astrophys.* **186**, 22 (3.8).

Gold, T. (1973), *Nature* **242**, 24 (3.9).

Goode, S. W. (1986), *Class. Q. Grav.* **3**, 1247 (2.10.1; Fig. 2.2).

Goode, S. W., Wainwright, J. (1982a), *Mon. Not. Roy. Astr. Soc.* **198**, 83 (2.5, 2.8; Fig. 2.1).

Goode, S. W., Wainwright, J. (1982b), *Phys. Rev.* **D26**, 3315 (1.3.4, 2.4.2, 2.5, 2.8, 2.10.1, 2.12, 2.13, 3.8, A, E).

Gorelik, G. E. (1972), *Vestn. Mosk. Univ. Fiz. Astr.* **13** no. 6, 727 [*Moscow Univ. Phys. Bull.* **27** no. 6, 86 (1972)] (2.14).

Gorini, V., Grillo, G., Pelizza, M. (1989), *Phys. Lett.* **A135**, 154 (3.6).

Gorini, V., Grillo, G., Pelizza, M. (1990), *Mod. Phys. Lett.* **A5**, 719 (3.8).

Götz, G. (1988), *Gen. Rel. Grav.* **20**, 23 (7.1; Fig. 7.1).

Gowdy, R. H. (1974), *Ann. Phys.* **83**, 203 (6.4, 8.4).

Grammenos, T., Kolassis, C. (1992), *Phys. Lett.* **A169**, 5 (4.14).

Graves, J. C., Brill, D. R. (1960), *Phys. Rev.* **120**, 1507 (2.14).

Gribkov, I. V., Soloviev, L. S. (1987), *Doklady ANSSSR* **296**, 566 [*Sov. Phys. Doklady* **32**, 723 (1987)] (3.8).

Griffiths, J. B. (1973), *Gen. Rel. Grav.* **4**, 361 (5.2).

Griffiths, J. B. (1974), *Gen. Rel. Grav.* **5**, 453 (5.2).

Griffiths, J. B. (1993a), *Class. Q. Grav.* **10**, 975 (6.7; Fig. 6.1).

Griffiths, J. B. (1993b), *J. Math. Phys.* **34**, 4064 (6.7).

Griffiths, J. B., Ashby, P. C. (1993), *Phys. Lett.* **A184**, 12 (6.3).

Griffiths, J. B., Newing, R. A. (1974), *Gen. Rel. Grav.* **5**, 345 (5.2).

Grillo, G. (1991), *Class. Q. Grav.* **8**, 739 (3.6).

Grishchuk, L. P. (1967), *Astron. Zh.* **44**, 1097 [*Sov. Astr. A. J.* **11**, 881 (1968)] (1.3.2).

Grøn, Ø (1985), *Phys. Rev.* **D32**, 1586 (7.1).

Gupta, P. S. (1959), *Ann. Physik* **2**, 421 (4.10; Box 4.1.4).

Gupta, P. S. (1962), *Nuovo Cimento* **26**, 379 (4.10).

Gurovich, V. T. (1966), *Doklady ANSSSR* **169**, 62 [*Sov. Phys. Doklady* **11**, 569 (1967)] (2.13; Box 2.4.3).

Gutman, I. I., Bespalko, R. M. (1967), in: *Sovremennye Problemy Gravitatsii. Sbornik Trudov II Sovetskoy Gravitatsyonnoy Konferentsii* [*Contemporary Problems of Gravitation. The Collection of Proceedings of the 2nd Soviet Conference on Gravitation*], Tbilisi, April 1965. Edited by D. D. Ivanenko, M. M. Miriyanashvili, V. S. Kiriya, A. B. Kereselidze. Izdatelstvo Tbilisskogo Universiteta, p. 201 (4.11).

Haantjes, J. (1937), *Koninklijke Nederlandsche Akademie van Wetenschappen Proceedings* **40**, 700 (4.2).

Haantjes, J. (1940), *Koninklijke Nederlandsche Akademie van Wetenschappen Proceedings* **43**, 1288 (4.2).

Hamilton, R. S. (1982), *J. Diff. Geom.* **17**, 255 (8.4).

Hamoui, A. (1969), *Ann. Inst. Poincare* **A10**, 195 (2.14, 3.5; Fig. 2.4).

Hellaby, C. (1987), *Class. Q. Grav.* **4**, 635 (2.12, 3.1, 3.8).

Hellaby, C. (1988), *Gen. Rel. Grav.* **20**, 1203 (3.8).

Hellaby, C. (1994), *Phys. Rev.* **D49**, 6484 (3.6).

Hellaby, C., Lake, K. (1984), *Astrophys. J.* **282**, 1 (3.6).

Hellaby, C., Lake, K. (1985a), *Astrophys. J.* **290**, 381 (3.1, 3.8).
Hellaby, C., Lake, K. (1985b), *Astrophys. J.* **294**, 702 (3.6).
Hellaby, C., Lake, K. (1985c), *Astrophys. J.* **300**, 461 (3.1, 3.8).
Hemmerich, A. (1987), *Astron. Astrophys.* **185**, 1 (8.4).
Henriksen, R. N. (1989), *Mon. Not. Roy. Astr. Soc.* **240**, 917 (2.13).
Henriksen, R. N., de Robertis, M. (1980), *Astrophys. J.* **241**, 54 (2.12, 3.1, 3.8; Box 2.4.2).
Henriksen, R. N., Emslie, A. G., Wesson, P. S. (1983), *Phys. Rev.* **D27**, 1219 (4.6, 4.11; Fig. 4.1; Box 4.1.3).
Henriksen, R. N., Patel, K. (1991), *Gen. Rel. Grav.* **23**, 527 (2.13; Box 2.4.3).
Henriksen, R. N., Wesson, P. S. (1978a), *Astrophys. Space Sci.* **53**, 429 (2.13; Box 2.4.3).
Henriksen, R. N., Wesson, P. S. (1978b), *Astrophys. Space Sci.* **53**, 445 (2.13).
Hewitt, C. G., Wainwright, J., Glaum, M. (1991), *Class. Q. Grav.* **8**, 1505 (2.13).
Hewitt, C. G., Wainwright, J., Goode, S. W. (1988), *Class. Q. Grav.* **5**, 1313 (2.13).
Hoffman, G. L., Salpeter, E. E., Wasserman, I. (1983), *Astrophys. J.* **268**, 527 (3.1).
Hogan, P. A. (1990), *Astrophys. J.* **360**, 315 (4.3, 4.7; Box 4.1.1).
Horsky, J., Lorenc, P., Novotny, J. (1977), *Phys. Lett.* **A63**, 79 (2.12, 3.4; Box 2.4.1).
Ibanez, J., Verdaguer, E. (1986), *Astrophys. J.* **306**, 401 (6.7, 7.8; Figs. 6.1, 7.3).
Isaacson, R. A. (1968a), *Phys. Rev.* **166**, 1263 (8.3, 8.4).
Isaacson, R. A. (1968b), *Phys. Rev.* **166**, 1272 (8.3, 8.4).
Israelit, M., Rosen, N. (1992), *Astrophys. J.* **400**, 21 (4.6; Box 4.1.3).
Ivanenko, D. D., Brezhnev, V. S., Frolov, B. N. (1967), in: *Sovremennye Problemy Gravitatsii. Sbornik Trudov II Sovetskoy Gravitatsyonnoy Konferentsii* [*Contemporary Problems of Gravitation. The Collection of Proceedings of the 2nd Soviet Conference on Gravitation*], Tbilisi 1965. Edited by D. D. Ivanenko, M. M. Miriyanashvili, V. S. Kiriya, A. B. Kereselidze. Izdatelstvo Tbilisskogo Universiteta, Tbilisi, p. 186 (4.6, 4.11; Fig. 4.1; Box 4.1.3).
Ivanenko, D. D., Krechet, V. G., Lapchinskii, V. G. (1973), *Izv. VUZ Fiz.* **16** no. 12, 63 [*Sov. Phys. J.* **16**, 1675 (1973)] (1.1, 2.14; Box 2.4.4).
Ivanov, G. G. (1980), *Izv. VUZ Fiz.* **23** no. 12, 22 [*Sov. Phys. J.* **23**, 1002 (1980)] (4.9; Fig. 4.1).
Jantzen, R. T. (1980), *Nuovo Cimento* **B59**, 287 (6.7).
Järnefelt, G. (1940a), *Ann. Acad. Soc. Sci. Fennicae* **A55** no. 3, 3 (3.3, 4.7).
Järnefelt, G. (1940b), *Arkiv Mat. Astr. Fys.* **27** no. 15, 1 (4.7).
Järnefelt, G. (1942), *Ann. Acad. Sci. Fennicae* **A** no. 12 (3.3, 4.7).
Jiang, S. (1992), *J. Math. Phys.* **33**, 3503 (4.14; Fig. 4.3).
Joshi, P. S. (1993), *Global Aspects in Gravitation and Cosmology.* Clarendon Press, Oxford, pp. 242–255 (3.6).
Joshi, P. S., Dwivedi, I. H. (1993), *Phys. Rev.* **D47**, 5357 (3.6).
Just, K. (1956), *Z. Physik* **145**, 235 (2.13; Box 2.4.3).
Just, K. (1960), *Z. Astrophysik* **49**, 19 (3.8).
Just, K., Kraus, K. (1962), *Z. Astrophysik* **55**, 127 (3.8).
Kantowski, R. (1969), *Astrophys. J.* **155**, 1023 (3.2).
Kantowski, R., Sachs, R. K. (1966), *J. Math. Phys.* **7**, 443 (1.1, 1.3.2, 1.4, 2.6, 2.7, 2.9, 2.11.1, 2.13, 3.8, 4.7, 4.11, 8.4, A; Figs. 2.1, 2.2, 2.3).
Karmarkar, K. R. (1948), *Proc. Indian Acad. Sci.* **27**, 56 (4.10).
Kasai, M. (1992), *Phys. Rev. Lett.* **69**, 2330 (2.5).
Kasai, M. (1993), *Phys. Rev.* **D47**, 3214 (2.5).
Kermack, W. O., McCrea, W. H., Whittaker, E. T. (1933), *Proc. Roy. Soc. Edinburgh* **53**, 31 (B).
Kerr, R. P. (1963), *Phys. Rev. Lett.* **11**, 237 (5.3, 5.6; Fig. 5.1).
Khiet, V. T. (1971), *C. R. Acad. Sci. Paris* **A272**, 509 (8.4).
Khiet, V. T. (1984), *Ukr. Fiz. Zh.* **29**, 969 (8.4).
Khlestkov, Yu. A. (1975), *ZhETF* **68**, 387 [*Sov. Phys. JETP* **41**, 188 (1975)] (2.14; Box 2.4.4).
Kitamura, S. (1994), *Class. Q. Grav.* **11**, 195 (4.11).
Kitchingham, D. W. (1984), *Class. Q. Grav.* **1**, 677 (6.7).
Kitchingham, D. W. (1986), *Class. Q. Grav.* **3**, 133 (6.7).
Knutsen, H. (1982), *Phys. Scripta* **26**, 365 (4.6).
Knutsen, H. (1983a), *Ann. Inst. Poincare* **A39**, 101 (4.5).
Knutsen, H. (1983b), *J. Math. Phys.* **24**, 2188 (4.4).

Knutsen, H. (1983c), *Phys. Scripta* **28**, 357 (4.10).
Knutsen, H. (1983d), PhD Thesis (unpublished). University of Oslo, Institute of Theoretical Astrophysics (1.1, Acknowl.).
Knutsen, H. (1984), *Gen. Rel. Grav.* **16**, 777 (4.6; Box 4.1.3).
Knutsen, H. (1985a), *Gen. Rel. Grav.* **17**, 1121 (4.6; Box 4.1.3).
Knutsen, H. (1985b), *Phys. Scripta* **31**, 305 (4.6; Box 4.1.3).
Knutsen, H. (1985c), *Phys. Scripta* **32**, 568 (4.8; Fig. 4.1).
Knutsen, H. (1987a), *Int. J. Theor. Phys.* **26**, 895 (4.6).
Knutsen, H. (1987b), *Phys. Scripta* **35**, 238 (4.8).
Knutsen, H., Stabell, R. (1979), *Ann. Inst. Poincare* **A31**, 339 (4.5).
Kolassis, C. A., Santos, N. O., Tsoubelis, D. (1988), *Astrophys. J.* **327**, 755 (4.14; Fig. 4.3).
Kolesnikov, S. M., Stanyukovich, K. P. (1965), *Priklad. Math. Mekh* **29**, 716 [*Appl. Math. Mech.* **29**, 848 (1965)] (3.8).
Kolesnikov, S. M., Stanyukovich, K. P. (1966), in: *Problemy teorii gravitatsii i elementarnykh chastits* [*Problems of Gravitation Theory and Elementary Particle Theory*], 1st issue. Edited by K.P. Stanyukovich and G. A. Sokolnik. Atomizdat, Moskva, p. 135 (3.8).
Kompaneets, A. S., Chernov, A. S. (1964), *ZhETF* **47**, 1939 [*Sov. Phys. JETP* **20**, 1303 (1965)] (1.3.2, 2.1.1, 2.6, 2.7, 2.9, 2.11.1, 2.11.2, 2.11.3, 2.11.4, 2.13, 4.11, 7.7; Figs. 2.1, 2.3, 7.3).
Koppar, S. S., Patel, L. K. (1988), *J. Math. Phys.* **29**, 182 (5.6; Fig. 5.1).
Korkina, M. P. (1991), *Ukr. Fiz. Zh.* **36**, 647 (3.8).
Korkina, M. P., Chernyi, L. M. (1975), *Izv. VUZ Fiz.* **18** no. 1, 23 [*Sov. Phys. J.* **18**, 17 (1975)] (3.8).
Korkina, M. P., Chernyi, L. M. (1976), *Ukr. Fiz. Zh.* **21**, 916 (2.14; Fig. 2.4).
Korkina, M. P., Martinenko, V. G. (1975a), *Ukr. Fiz. Zh.* **20**, 626 (2.6, 2.7, 2.11.1, 2.11.3, 2.11.4; Figs. 2.1, 2.3).
Korkina, M. P., Martynenko, V. G. (1975b), *Ukr. Fiz. Zh.* **20**, 2044 (2.11.4).
Korkina, M. P., Martynenko, V. G. (1976), *Ukr. Fiz. Zh.* **21**, 1191 (2.11.1; Fig. 2.3).
Korotkii, V. G., Krechet, V. G. (1988), *Izv. VUZ Fiz.* **31** no. 3, 48 [*Sov. Phys. J.* **31**, 214 (1988)] (2.9).
Kramer, E., Neugebauer, G., Stephani, H. (1972), *Fortschr. Physik* **20**, 1 (4.9).
Kramer, D., Stephani, H., Herlt, E., MacCallum, M. A. H. (1980), *Exact Solutions of Einstein Field Equations*. Cambridge University Press 1980 (1.1, 1.2, 1.3.2, 2.7, 4.9, 4.10, 4.11, 5.2, 5.3).
Krasiński, A. (1974), *Acta Phys. Polon.* **B5**, 411 (4.4).
Krasiński, A. (1981), *Gen. Rel. Grav.* **13**, 1021 (1.3.2, 4.9).
Krasiński, A. (1983), *Gen. Rel. Grav.* **15**, 673 (4.9).
Krasiński, A. (1984a), in: *The Big Bang and Georges Lemaître*. Edited by A. Berger. D. Reidel, Dordrecht, p. 63 (4.9).
Krasiński, A. (1984b), in: *Proceedings of the Sir Arthur Eddington Centenary Symposium. Vol. I: Relativistic Astrophysics and Cosmology*. Edited by V. de Sabbata and T. M. Karade. World Scientific, Singapore, p. 45 (4.1, 4.2, 4.3; Box 4.1.1).
Krasiński, A. (1985), in: *Proceedings of the 4th Marcel Grossman Meeting on General Relativity*. Edited by R. Ruffini. Elsevier, Amsterdam, p. 989 (4.2; Fig. 4.1).
Krasiński, A. (1986), in: *Gravitational Collapse and Relativity. Proceedings of Yamada Conference XIV*. Edited by H. Sato and T. Nakamura. World Scientific, Singapore, p. 500 (4.2, 4.2.1, 4.2.2; Fig. 4.1).
Krasiński, A. (1989), *J. Math. Phys.* **30**, 433 (4.1, 4.2, 4.2.1, 4.2.2, 4.3, 4.9; Fig. 4.1).
Krasiński, A. (1991), *Rep. Math. Phys.* **29**, 337 (4.4).
Krasiński, A. (1993), *Gen. Rel. Grav.* **25**, 165 (1.1).
Krasiński, A., Perkowski, M. (1981), *Gen. Rel. Grav.* **13**, 67 (1.1).
Krasiński, A., Plebański, J. (1980), *Rep. Math. Phys.* **17**, 217 (2.16, 4.3, 4.7).
Krasiński, A., Quevedo, H., Sussman, R. A. (1995), *On the thermodynamical interpretation of perfect fluid solutions of the Einstein equations with no symmetry*, preprint (Notes 2, 4).
Krechet, V. G., Ponomarev, V. N., Barvinskii, A. O. (1977), *Izv. VUZ Fiz.* **20** no. 9, 7 [*Sov. Phys. J.* **20**, 1123 (1977)] (2.14; Box 2.4.4).
Krishna Rao, J. (1973), *Gen. Rel. Grav.* **4**, 351 (4.3; Box 4.1.1).
Krishna Rao, J. (1990), *Pramana* **34**, 423 (2.12; Box 2.4.2).
Krishna Rao, J., Annapurna, M. (1986), *Pramana* **27**, 637 (2.12; Box 2.4.2).
Kristian, J. (1967), *Astrophys. J.* **147**, 864 (B).

Kristian, J., Sachs, R. K. (1966), *Astrophys. J.* **143**, 379 (B, C, E).

Królak, A., Czyrka, A., Gaber, J., Rudnicki, W. (1994), in: *Proceedings of the Cornelius Lanczos International Centenary Conference.* Edited by J. D. Brown, M. T. Chu, D. C. Ellison and R. J. Plemmons. Society for Industrial and Applied Mathematics, Philadelphia, 1994, p. 518. (2.4.2, 3.6).

Krori, K. D., Nandy, D. (1984), *J. Math. Phys.* **25**, 2515 (6.6; Fig. 6.1).

Krymsky, A. M., Marochnik, L. S., Naselsky, P. D., Pelikhov, N. V. (1978), *Astrophys. Space Sci.* **55**, 325 (8.3).

Kuchowicz, B. (1973), *Int. J. Theor. Phys.* **7**, 259 (4.10).

Kumar, M. M. (1969), *Nuovo Cimento* **A63**, 559 (4.11; Fig. 4.1).

Kundt, W., Trümper, M. (1962), *Akad. Wiss. Lit. Mainz, Abhandl. Math. Naturw. Kl.* Nr 12, p. 965 (7.3).

Kurki–Suonio, H., Liang, E. (1992), *Astrophys. J.* **390**, 5 (3.2, B).

Kustaanheimo, P. (1947), *Societas Scientiarum Fennicae Commentationes Physico-Mathematicae* **XIII** no. 12, 1 (4.4, 4.6, 4.12; Fig. 4.1; Box 4.1.2).

Kustaanheimo, P., Qvist, B. (1948), *Societas Scientiarum Fennicae Commentationes Physico-Mathematicae* **XIII** no. 16, 1 (4.1, 4.2.3, 4.3, 4.4, 4.5, 4.6, 4.8, 4.9, 4.10, 4.12, 4.14, A, E; Fig. 4.1; Boxes 4.1.1, 4.1.3, 4.1.4, 4.1.5; Fig. 4.3).

Lake, K. (1984), *Phys. Rev.* **D29**, 771 (3.8).

Lake, K. (1992), *Astrophys. J. Lett.* **401**, L1 (2.6).

Lake, K., Nelson, L. A. (1980), *Phys. Rev.* **D22**, 1266 (3.5).

Lake, K., Pim, R. (1985), *Astrophys. J.* **298**, 439 (3.1, 3.3).

Lake, K., Roeder, R. C. (1978), *Phys. Rev.* **D17**, 1935 (3.8).

Lal, K. B., Singh, T. (1973), *Tensor N. S.* **27**, 211 (6.4, 6.5; Fig. 6.1).

Lambas, D. G., Lamberti, W., Hamity, V. H. (1987), in: *Relativity, Supersymmetry and Cosmology. Proceedings of the 5th Simposio Latino–Americano de Relatividad y Gravitacion (SILARG 5) held at Bariloche, Argentina 1985.* Edited by O. Bressan, M. Castagnino and V. Hamity. World Scientific, Singapore, p. 271 (4.3).

Landau, L. D., Lifshitz, E. M. (1948), *Teoria pola.* Second, revised edition. OGIZ, Moskva, exercise 5 to section 96 (2.12; Box 2.4.2).

Landau, L. D., Lifshitz, E. M. (1962), *Teoria pola.* 4th edition. Gosudarstvennoye Izdatelstvo Fiziko-Matematicheskoy Literatury, Moskva, p. 345 (2.12; Box 2.4.2).

Lapedes, A. S. (1977), *Phys. Rev.* **D15**, 946 (6.4).

Laserra, E. (1985), *Meccanica* **20**, 267 (3.8).

Lawitzky, G. (1980), *Gen. Rel. Grav.* **12**, 903 (2.5).

Leibovitz, C. (1971), *Phys. Rev.* **D4**, 2949 (2.13; Fig. 2.4).

Lemaître, G. (1927), *Ann. Soc. Sci. Bruxelles* **A47**, 19 (1.1, A; all diagrams).

Lemaître, G. (1931), *Mon. Not. Roy. Astr. Soc.* **91**, 483 (1.1, A; all diagrams).

Lemaître, G. (1933a), *Ann. Soc. Sci. Bruxelles* **A53**, 51 (1.1, 1.4, 2.1, 2.6, 2.12, 2.13, 3.9, A, E; Fig. 2.4).

Lemaître, G. (1933b), *C. R. Acad. Sci. Paris* **196**, 903 (2.12, 2.13, 2.15; Figs. 2.4, 2.5).

Lemaître, G. (1933c), *C. R. Acad. Sci. Paris* **196**, 1085 (3.2).

Lemaître, G. (1934), *Proc. Nat. Acad. Sci. USA* **20**, 12 (3.2).

Lemos, J. P. S. (1991a), *Phys. Lett.* **A158**, 279 (3.6).

Lemos, J. P. S. (1991b), in: *Proceedings of the 7th Latin American Symposium on Relativity and Gravitation (SILARG 7).* Edited by J. C. Olivo, E. Nahmad–Achar, M. Rosenbaum, M. P. Ryan Jr., J. F. Urrutia, F. Zertuche. World Scientific, Singapore, p. 241 (3.6).

Lemos, J. P. S. (1992a), *Phys. Rev. Lett.* **68**, 1447 (3.6).

Lemos, J. P. S. (1992b), in: *Proceedings of the 6th Marcel Grossman Meeting.* Edited by H. Sato and T. Nakamura. World Scientific, Singapore, p. 1346 (3.6).

Lemos, J. P. S., Lynden–Bell, D. (1989), *Mon. Not. Roy. Astr. Soc.* **240**, 317 (3.8).

Letelier, P. S. (1975), *J. Math. Phys.* **16**, 1488 (6.4, 6.5, 6.7; Fig. 6.1).

Letelier, P. S. (1979), *J. Math. Phys.* **20**, 2078 (6.4; Fig. 6.1).

Letelier, P. S. (1980), *Phys. Rev.* **D22**, 807 (4.8, 6.6; Note 2).

Letelier, P. S. (1982), *Nuovo Cimento* **B69**, 145 (6.6; Fig. 6.1).

Letelier, P. S., Alencar, P. S. C. (1986), *Phys. Rev.* **D34**, 343 (6.6; Fig. 6.1).

Letelier, P. S., Machado, R. (1981), *J. Math. Phys.* **22**, 827 (6.6; Fig. 6.1).

Letelier, P. S., Tabensky, R. R. (1975a), *J. Math. Phys.* **16**, 8 (6.4).
Letelier, P. S., Tabensky, R. R. (1975b), *Nuovo Cimento* **B28**, 407 (6.4, 6.7; Fig. 6.1).
Letelier, P. S., Verdaguer, E. (1987), *J. Math. Phys.* **28**, 2431 (6.6; Fig. 6.1).
Li, J. (1992), *J. Math. Phys.* **33**, 3506 (6.4).
Li, J. Z., Liang, C. B. (1985), *Chin. Phys. Lett.* **2**, 23 (1.2).
Liang, E. P. T. (1974), *Phys. Rev.* **D10**, 447 (3.4).
Lie, S. (1912), *Vorlesungen über Differentialgleichungen.* Teubner, Leipzig und Berlin (4.2.3).
Lifshitz, E. M. (1946), *ZhETF* **16**, 587 (2.12, 8.3).
Lifshitz, E. M., Khalatnikov, I. M. (1963), *Usp. Fiz. Nauk* **80**, 391 (8.3).
Lightman, A., Press, W. H., Price, R. H., Teukolsky, S. A. (1975), *Problem Book in Relativity and Gravitation.* Princeton University Press, Princeton, pp. 98 and 460–461 (2.12; Box 2.4.2).
Lima, J. A. S. (1986), *Phys. Lett.* **A116**, 210 (2.10.1; Fig. 2.2).
Lima, J. A. S., Garcia Maia, M. R. (1985), *Phys. Lett.* **A110**, 366 (2.10.1, 2.10.3; Fig. 2.2).
Lima, J. A. S., Nobre, M. A. S. (1990), *Class. Q. Grav.* **7**, 399 (2.10.2; Fig. 2.2).
Lima, J. A. S., Tiomno, J. (1988), *Gen. Rel. Grav.* **20**, 1019 (2.8, 2.13; Figs. 2.1, 2.2).
Lima, J. A. S., Tiomno, J. (1989), *Class. Q. Grav.* **6**, L93 (2.8).
Lindquist, R. W., Schwartz, R. A., Misner, C. W. (1965), *Phys. Rev.* **B137**, 1364 (5.2).
Liu, H. (1990), *J. Math. Phys.* **31**, 2459 (3.8).
Liu, H. (1991), *J. Math. Phys.* **32**, 2279 (2.12; Box 2.4.2).
Lorenz, D. (1982), *J. Phys.* **A15**, 2997 (2.11.1; Fig. 2.3).
Lorenz, D. (1983), *J. Phys.* **A16**, 575 (2.11.1, 2.11.4, 2.11.6; Fig. 2.3).
Lorenz–Petzold, D. (1986), *J. Astrophys. Astron.* (India) **7**, 155 (4.9).
Lukacs, B., Meszaros, A. (1985), *Astrophys. Space Sci.* **114**, 211 (4.11; Fig. 4.1).
Lund, F. (1973a), *Phys. Rev.* **D8**, 3253 (3.8).
Lund, F. (1973b), *Phys. Rev.* **D8**, 4229 (3.8).
Lynden–Bell, D. (1987), in: *Gravitation and Astrophysics (Cargese Lectures 1986).* Edited by B. Carter and J. Hartle. Plenum, New York and London, p. 155 (2.12; Box 2.4.2).
Lynden–Bell, D., Lemos, J. P. S. (1988), *Mon. Not. Roy. Astr. Soc.* **233**, 197 (3.8).
Lyttleton, R. A., Bondi, H. (1959), *Proc. Roy. Soc. London* **A252**, 313 (4.10, 4.12).
MacCallum, M. A. H. (1973), in: *Cargese Lectures in Physics*, vol. 6. Edited by E. Schatzman. Gordon and Breach, New York, London and Paris, p. 61 (1.1).
MacCallum, M. A. H. (1979), in: *General Relativity, an Einstein Centenary Survey.* Edited by S. W. Hawking and W. Israel. Cambridge University Press, p. 533 (1.1, B).
MacCallum, M. A. H. (1984), in: *Exact Solutions of Einstein's equations, Techniques and Results.* Edited by C. Hoenselaers and W. Dietz. Springer, Berlin, p. 334 (1.1, 6.7).
MacCallum, M. A. H. (1985), in: *Observational and Theoretical Aspects of Relativistic Astrophysics and Cosmology.* Edited by J. L. Sanz and L. J. Goicoechea. World Scientific, Singapore, p. 183 (1.1, B).
Maeda, K., Sasaki, M., Sato, H. (1983), *Progr. Theor. Phys.* **69**, 89 (2.12, 3.1, 3.3, 3.7, 3.8; Box 2.4.2).
Maeda, K., Sato, H. (1983a), *Progr. Theor. Phys.* **70**, 772 (3.1, 3.3, 3.7, 3.8).
Maeda, K., Sato, H. (1983b), *Progr. Theor. Phys.* **70**, 1276 (3.1, 3.3, 3.7, 3.8).
Mahanta, M. N. (1984), *Ann. Physik* **41**, 357 (8.4).
Maharaj, S. D. (1988), *J. Math. Phys.* **29**, 1443 (2.13; Box 2.4.3).
Maharaj, S. D., Leach, P. G. L., Maartens, R. (1991), *Gen. Rel. Grav.* **23**, 261 (4.11; Fig. 4.1).
Maiti, S. R. (1982), *Phys. Rev.* **D25**, 2518 (4.14; Fig. 4.3).
Maiti, S. R. (1984), *Gen. Rel. Grav.* **16**, 297 (4.1).
Malec, E., O'Murchadha, N. (1993), *Phys. Rev.* **D47**, 1454 (3.9).
Mansouri, R. (1977), *Ann. Inst. Poincare* **A27**, 175 (4.3, 4.4; Box 4.1.1).
Mansouri, R. (1980), *Acta Phys. Polon.* **B11**, 193 (4.3, 4.4; Box 4.1.1).
Markov, M. A., Frolov, V. P. (1970), *Teor. Mat. Fiz.* **3**, 3 [*Theor. Math. Phys.* **3**, 301 (1970)] (2.14; Fig. 2.4; Box 2.4.4).
Markov, M. A., Frolov, V. P. (1972), *Teor. Mat. Fiz.* **13**, 41 [*Theor. Math. Phys.* **13**, 965 (1972)] (3.8).
Marochnik, L. S. (1980a), *Astron. Zh.* **57**, 903 [*Sov. Astr. A. J.* **24**, 518 (1981)] (8.3).
Marochnik, L. S. (1980b), *Astron. Zh.* **57**, 1129 [*Sov. Astr. A. J.* **24**, 651 (1980)] (8.3).
Marochnik, L. S. (1980c), *Astrophys. Space Sci.* **69**, 3 (8.3).
Marochnik, L. S. (1980d), *Astrophys. Space Sci.* **69**, 31 (8.3).
Marochnik, L. S. (1981), *Astron. Zh.* **58**, 15 [*Sov. Astr. A. J.* **25**, 8 (1981)] (8.3).

Marochnik, L. S., Naselsky, P. D., Pelikhov, N. V. (1980), *Astrophys. Space Sci.* **67**, 261 (8.3).
Marochnik, L. S., Pelikhov, N. V., Vereshkov, G. M. (1975a), *Astrophys. Space Sci.* **34**, 249 (8.3).
Marochnik, L. S., Pelikhov, N. V., Vereshkov, G. M. (1975b), *Astrophys. Space Sci.* **34**, 281 (8.3).
Martin, J., Senovilla, J. M. M. (1986a), *J. Math. Phys.* **27**, 265 (7.4, 7.7; Fig. 7.2).
Martin, J., Senovilla, J. M. M. (1986b), *J. Math. Phys.* **27**, 2209 (7.4, 7.7).
Martin–Pascual, F., Senovilla, J. M. M. (1988) *J. Math. Phys.* **29**, 937 (7.7; Fig. 7.3).
Martinez, E., Sanz, J. L. (1985), *J. Math. Phys.* **26**, 785 (4.9).
Mashhoon, B., Partovi, M. H. (1979), *Phys. Rev.* **D20**, 2455 (4.12; Box 4.1.5).
Mashhoon, B., Partovi, M. H. (1980), *Ann. Phys.* **130**, 99 (4.12; Box 4.1.5).
Mashhoon, B., Partovi, M. H. (1984), *Phys. Rev.* **D30**, 1839 (4.12; Box 4.1.5).
Mather, J. C., Bennett, C. L., Boggess, N. W., Hauser, M. G., Smoot, G. F., Wright, E. L. (1993), in: *General Relativity and Gravitation 1992. Proceedings of the 13th International Conference on General Relativity and Gravitation at Cordoba 1992.* Edited by R. J. Gleiser, C. N. Kozameh and O. M. Moreschi. Institute of Physics Publishing, Boston and Philadelphia, p. 151 (3.2, 3.7).
Matzner, R. A. (1968a), *J. Math. Phys.* **9**, 1063 (8.4).
Matzner, R. A. (1968b), *J. Math. Phys.* **9**, 1657 (8.4).
Matzner, R. A., Rosenbaum, M., Ryan, M. P. (1982), *J. Math. Phys.* **23**, 1984 (6.4).
Mavrides, S. (1976a), *C. R. Acad. Sci. Paris* **A282**, 451 (3.8).
Mavrides, S. (1976b), *Mon. Not. Roy. Astr. Soc.* **177**, 709 (3.8).
Mavrides, S., Tarantola, A. (1977), *Gen. Rel. Grav.* **8**, 665 (3.8).
McCrea, W. H. (1939), *Z. Astrophysik* **18**, 98 (B, E).
McIntosh, C. B. G. (1978), *Phys. Lett.* **A69**, 1 (6.7).
McManus, D. J., Coley, A. A. (1994), *Class. Q. Grav.* **11**, 2045 (1.3.2).
McVittie, G. C. (1933), *Mon. Not. Roy. Astr. Soc.* **93**, 325 (3.3, 4.1, 4.3, 4.4, 4.6, 4.7, 4.10, 4.12, 5.5, 5.6, E; Box 4.1.3).
McVittie, G. C. (1966), *Astrophys. J.* **143**, 682 (4.7).
McVittie, G. C. (1967a), *Ann. Inst. Poincare* **A6**, 1 (4.4, 4.5, 4.6, 4.8; Fig. 4.1; Box 4.1.3).
McVittie, G. C. (1967b), *Mem. Soc. Roy. Sci. Liege* **15**, 41 (4.4, 4.5, 4.6; Box 4.1.3).
McVittie, G. C. (1984), *Ann. Inst. Poincare* **A40**, 235 (4.5).
McVittie, G. C., Stabell, R. (1967), *Ann. Inst. Poincare* **A7**, 103 (4.11; Fig. 4.1).
McVittie, G. C., Stabell, R. (1968), *Ann. Inst. Poincare* **A9**, 371 (4.6, 4.11; Fig. 4.1; Box 4.1.3).
McVittie, G. C., Wiltshire, R. J. (1975), *Int. J. Theor. Phys.* **14**, 145 (2.11.1, 2.11.4; Fig. 2.3).
McVittie, G. C., Wiltshire, R. J. (1977), *Int. J. Theor. Phys.* **16**, 121 (7.5; Fig. 7.2).
Melott, A. L. (1990), *Phys. Rep.* **193** no. 1, 1 (B).
Meszaros, A. (1985), *Astrophys. Space Sci.* **108**, 415 (4.6).
Meszaros, A. (1986), *Acta Phys. Hung.* **60**, 75 (3.8).
Meszaros, A. (1991), *Mon. Not. Roy. Astr. Soc.* **253**, 619 (3.1).
Meszaros, A. (1993), *Astrophys. Space Sci.* **207**, 5 (3.1).
Meszaros, A., Vanysek, V. (1988), *Bull. Astron. Inst. Czech.* **39**, 185 (2.12, B; Box 2.4.2).
Michalski, H., Wainwright, J. (1975), *Gen. Rel. Grav.* **6**, 289 (1.2).
Mignani, R. (1978), *Lett. Nuovo Cimento* **23**, 349 (1.3.1).
Miller, B. D. (1976), *Astrophys. J.* **208**, 275 (3.8).
Mineur, H. (1933), *Ann. de l'Ecole Normale Superieure* ser. 3, **5**, 1 (5.2).
Misner, C. W. (1965), *Phys. Rev.* **B137**, 1360 (1.4).
Misner, C. W., Sharp, D. H. (1964), *Phys. Rev.* **B136**, 571 (1.4, 3.5).
Misner, C. W., Thorne, K. S., Wheeler, J. A. (1973), *Gravitation.* W. H. Freeman and Co, San Francisco, p. 859 (3.8).
Misra, R. M., Srivastava, D. C. (1973), *Phys. Rev.* **D8**, 1653 (4.10; Box 4.1.4).
Misra, R. M., Srivastava, D. C. (1974), *Phys. Rev.* **D9**, 844 (2.14; Box 2.4.4).
Mitskevich, N. V. (1969), *Fizicheskiye polya v obshchey teorii otnositelnosti* [*Physical Fields in the General Theory of Relativity*]. Nauka, Moskva, pp. 178–183 (8.3).
Mitskievič, N. V., Senin, Y. E. (1981), *Acta Phys. Polon.* **B12**, 541 (1.3.2, 2.8, 2.13, 7.2; Figs. 2.4, 7.1).
Modak, B. (1984a), *J. Astrophys. Astron.* (India) **5**, 317 (4.14; Fig. 4.3).
Modak, B. (1984b), *Pramana* **23**, 809 (2.15; Fig. 2.5).
Moffat, J. W., Tatarski, D. C. (1992), *Phys. Rev.* **D45**, 3512 (3.2).

Møller, C. (1975), *Det Kongelige Danske Videnskarbens Selskab Matematisk-fysiske Meddelelser* **39** no. 7 (2.13; Fig. 2.4).

Motta, D. C., Tomimura, N. (1990), *Astrophys. Space Sci.* **165**, 237 (2.10.3; Fig. 2.2).

Müller, J. (1969), *Wiss. Z. Friedrich Schiller Univ. Jena, Math. Naturw. Reihe* **18**, 169 (4.4; Fig. 4.1).

Nariai, H. (1967a), *Progr. Theor. Phys.* **38**, 92 (3.8, 4.3, 4.6, 4.14; Box 4.1.3; Fig. 4.3).

Nariai, H. (1967b), *Progr. Theor. Phys.* **38**, 740 (3.8, 4.6; Box 4.1.3).

Nariai, H. (1968), *Progr. Theor. Phys.* **40**, 1013 (4.3; Box 4.1.1).

Nariai, H., Tomita, K. (1968), *Progr. Theor. Phys.* **40**, 679 (4.10; Box 4.1.4).

Nariai, H., Tomita, K. (1971), *Suppl. Progr. Theor. Phys.* **49**, 83 (3.8).

Nariai, H., Tomita, K., Hayakawa, M. (1968), *Progr. Theor. Phys.* **39**, 601 (4.6).

Narlikar, V. V. (1936), *Phil. Mag.* **22**, 767 (7.5; Fig. 7.2).

Narlikar, V. V. (1947), *Curr. Sci.* **16**, 113 (4.2.3; Box 4.1.1).

Narlikar, V. V., Moghe, D. N. (1935), *Phil. Mag.* **20**, 1104 (7.5; Fig. 7.2).

Narlikar, V. V., Singh, K. P. (1950), *Phil. Mag.* **41**, 152 (4.11).

Narlikar, V. V., Vaidya, P. C. (1947), *Nature* **159**, 642 (5.2).

Nduka, A. (1979), *Acta Phys. Polon.* **B10**, 479 (4.12; Box 4.1.5).

Nduka, A. (1981), *Acta Phys. Polon.* **B12**, 833 (4.10, 4.12; Boxes 4.1.4, 4.1.5).

Neeman, Y., Tauber, G. (1967), *Astrophys. J.* **150**, 755 (3.2).

Nelson, A. H. (1972), *Mon. Not. Roy. Astr. Soc.* **158**, 159 (8.3).

Newman, E. T., Couch, E., Chinnapared, K., Exton, A., Prakash, A., Torrence, R. (1965), *J. Math. Phys.* **6**, 918 (5.3; Fig. 5.1).

Newman, R. P. A. C. (1986a), *Class. Q. Grav.* **3**, 527 (3.6).

Newman, R. P. A. C. (1986b), in: *Topological Properties and Global Structure of Spacetime.* Edited by P. G. Bergmann and V. de Sabbata. Plenum, New York, p. 153 (3.6).

Newman, R. P. A. C., Joshi, P. S. (1988), *Ann. Phys.* **182**, 112 (3.6).

Noerdlinger, P. D., Petrosian, V. (1971), *Astrophys. J.* **168**, 1 (3.3, 4.7).

Nolan, B. (1993), *J. Math. Phys.* **34**, 178 (4.3, 4.7, 4.10; Boxes 4.1.1, 4.1.4).

Noonan, T. W. (1984), *Gen. Rel. Grav.* **16**, 1103 (8.2).

Noonan, T. W. (1985), *Gen. Rel. Grav.* **17**, 535 (8.2).

Novikov, I. D. (1962a), *Vestn. Mosk. Univ.* no. 5, 90 (3.8).

Novikov, I. D. (1962b), *Vestn. Mosk. Univ.* no. 6, 66 (3.8, 3.9).

Novikov, I. D. (1963), *Astron. Zh.* **40**, 772 [*Sov. Astr. A. J.* **7**, 587 (1964)] (1.4, 2.6, 2.12; Box 2.4.2).

Novikov, I. D. (1964a), *Astron. Zh.* **41**, 1075 [*Sov. Astr. A. J.* **8**, 857 (1965)] (1.1, 3.2).

Novikov, I. D. (1964b), *Soobshcheniya GAISH* [*Communications of the State Sternberg Astronomical Institute*] **132**, 3 (3.8, 3.9).

Novikov, I. D. (1964c), *Soobshcheniya GAISH* [*Communications of the State Sternberg Astronomical Institute*] **132**, 43 (3.9).

Novikov, I. D. (1966), *Astron. Zh.* **43**, 911 [*Sov. Astr. A. J.* **10**, 731 (1967)] (2.14, 3.5).

Novikov, I. D. (1970), *ZhETF* **59**, 262 [*Sov. Phys. JETP* **32**, 142 (1971)] (2.14, 3.5).

Novotny, J., Horsky, J. (1979), *Scripta Fac. Sci. Nat. Univ. J. E. Purkynianae Brunensis, Physica* 2 **9**, 69 (2.12, 3.4; Box 2.4.1).

Nuñez, L., Rago, H., Aulestia, L. (1983), *Rev. Mex. Fis.* **30**, 83 (4.12; Box 4.1.5).

Obozov, V. I. (1979), *Izv. VUZ Fiz.* **22** no. 8, 71 [*Sov. Phys. J.* **22**, 869 (1979)] (4.9).

Occhionero, F., Santangelo, P., Vittorio, N. (1982), in: *Clustering in the Universe.* Edited by D. Gerbal and A. Mazure. Editions Frontieres, Gif sur Yvette, p. 103 (3.1, 3.3, 3.8).

Occhionero, F., Santangelo, P., Vittorio, N. (1983), *Astron. Astrophys.* **117**, 365 (3.1, 3.3, 3.8).

Occhionero, F., Vecchia–Scavalli, L., Vittorio, N. (1981a), *Astron. Astrophys.* **97**, 169 (3.1, 3.3, 3.8).

Occhionero, F., Vecchia–Scavalli, L., Vittorio, N. (1981b), *Astron. Astrophys.* **99**, L12 (3.1, 3.3, 3.8).

Occhionero, F., Vignato, A., Vittorio, N. (1978), *Astron. Astrophys.* **70**, 265 (3.1, 3.3, 3.8).

Oleson, M. (1971), *J. Math. Phys.* **12**, 666 (7.3, 7.4; Fig. 7.2).

Oliver, G., Verdaguer, E. (1989), *J. Math. Phys.* **30**, 442 (6.7; Fig. 6.1).

Olson, D. W. (1980), *Astrophys. J.* **236**, 335 (3.8).

Olson, D. W., Silk, J. (1979), *Astrophys. J.* **233**, 395 (3.1).

Omer, G. C. (1949), *Astrophys. J.* **109**, 164 (2.13, 3.8).

Omer, G. C. (1965), *Proc. Nat. Acad. Sci. USA* **53**, 1 (2.12).

Oppenheimer, J. R., Snyder, H. (1939), *Phys. Rev.* **56**, 455 (2.13, 3.4; Box 2.4.3).

Ori, A. (1990), *Class. Q. Grav.* **7**, 985 (2.14, 3.5; Fig. 2.4; Box 2.4.4).

Ori, A. (1991), *Phys. Rev.* **D44**, 2278 (2.14).

Ori, A., Piran, T. (1990), *Phys. Rev.* **D42**, 1068 (2.13; Box 2.4.3).

Pachner, J. (1966a), *Bull. Astron. Inst. Czech.* **17**, 105 and 108 (3.8).

Pachner, J. (1966b), *Phys. Rev.* **147**, 910 (3.8).

Pachner, J. (1967a), *Bull. Astron. Inst. Czech.* **18**, 219 (3.8).

Pachner, J. (1967b), *Mem. Soc. Roy. Sci. Liege* **15**, 45 (3.8).

Paczyński, B., Piran, T. (1990), *Astrophys. J.* **364**, 341 (3.7).

Paiva, F. M., Rebouças, M. J., MacCallum, M. A. H. (1993), *Class. Q. Grav.* **10**, 1165 (1.3.1).

Pandey, S. N., Gupta, Y. K., Sharma, S. P. (1983), *Indian J. Pure Appl. Math.* **14**, 79 (4.10; Box 4.1.4).

Panek, M. (1992), *Astrophys. J.* **388**, 225 (1.1, 3.7, E).

Papapetrou, A. (1974), *Lectures on General Relativity*. D.Reidel, Dordrecht, pp. 99–103 (2.12; Box 2.4.2).

Papapetrou, A. (1976a), *Ann. Inst. Poincare* **A24**, 165 (3.8).

Papapetrou, A. (1976b), *Ann. Inst. Poincare* **A24**, 171 (3.8).

Papapetrou, A. (1978), *Ann. Inst. Poincare* **A29**, 207 (3.4).

Papapetrou, A., Hamoui, A. (1967), *Ann. Inst. Poincare* **A6**, 343 (3.8).

Papini, G., Weiss, M. (1986), *Nuovo Cimento* **B91**, 31 (4.9).

Parenago, P. P. (1954), *A Course of Stellar Astronomy* (in Russian). Moskva (4.6).

Partovi, M. H., Mashhoon, B. (1984), *Astrophys. J.* **276**, 4 (3.9).

Patel, L. K., Koppar, S. S. (1987a), *Indian J. Pure Appl. Math.* **18**, 260 (5.4, 5.5, 5.6; Fig. 5.1).

Patel, L. K., Koppar, S. S. (1987b), *Phys. Lett.* **A121**, 267 (4.14; Fig. 4.3).

Patel, L. K., Koppar, S. S. (1988), *Acta Phys. Hung.* **64**, 353 (5.5, 5.6; Fig. 5.1).

Patel, L. K., Koppar, S. S., Pandya, N. R. (1990), *Acta Phys. Hung.* **67**, 135 (5.6).

Patel, L. K., Koppar, S. S., Yadav, S. R. (1989), *J. Math. Phys. Sci.* (India) **23**, 261 (5.6; Fig. 5.1).

Patel, L. K., Pandya, B. (1986), *Indian J. Pure Appl. Math.* **17**, 1224 (1.1).

Patel, L. K., Trivedi, H. B. (1982), *J. Astrophys. Astron.* (India) **3**, 63 (5.6; Fig. 5.1).

Patel, L. K., Yadav, S. R. (1987), *J. Math. Phys. Sci.* (India) **21**, 167 (5.6; Fig. 5.1).

Patel, L. K., Yadava, S. R. (1987), *Indian J. Pure Appl. Math.* **18**, 840 (5.6; Fig. 5.1).

Patel, R. B. (1969), *Curr. Sci.* **38**, 487 (5.7).

Pavlov, N. V. (1976), *Izv. VUZ Fiz.* **19** no. 4, 107 [*Sov. Phys. J.* **19**, 489 (1976)] (3.5).

Pavlov, N. V., Bronnikov, K. A. (1976), *Izv. VUZ Fiz.* **19** no. 7, 106 [*Sov. Phys. J.* **19**, 916 (1976)] (2.14, 3.5).

Peebles, P. J. E. (1967), *Astrophys. J.* **147**, 859 (2.12; Box 2.4.2).

Peebles, P. J. E. (1980), *The Large Scale Structure of the Universe*. Princeton University Press, Princeton, sec. 87 (3.8).

Penrose, R. (1979), in: *General Relativity, an Einstein Centenary Survey*. Edited by S. W. Hawking and W. Israel. Cambridge University Press, p. 581 (2.5).

Pim, R., Lake, K. (1986), *Astrophys. J.* **304**, 75 (3.1).

Pim, R., Lake, K. (1988), *Astrophys. J.* **330**, 625 (3.1).

Podurets, M. A. (1964), *Astron. Zh.* **41**, 28 [*Sov. Astr. A. J.* **8**, 19 (1964)] (1.4, 3.5).

Pollock, M. D., Caderni, N. (1980), *Mon. Not. Roy. Astr. Soc.* **190**, 509 (2.8; Fig. 2.1).

Polnarev, A. G. (1977), *Astrofizika* **13**, 375 [*Astrophysics* **13**, 203 (1977)] (3.4).

Ponce de Leon, J. (1986), *J. Math. Phys.* **27**, 271 (4.10; Box 4.1.4).

Ponce de Leon, J. (1988), *J. Math. Phys.* **29**, 2479 (2.11.3, 2.11.4; Fig. 2.3).

Ponce de Leon, J. (1991a), *J. Math. Phys.* **32**, 3546 (4.11; Fig. 4.1).

Ponce de Leon, J. (1991b), *Mon. Not. Roy. Astr. Soc.* **250**, 69 (2.12; Fig. 2.4).

Pullin, J. (1990), *Astrophys. Space Sci.* **164**, 309 (7.8).

Quevedo, H., Sussman, R. A. (1995a), *Class. Q. Grav.* **12**, 589 (Note 2).

Quevedo, H., Sussman, R. A. (1995b), *J. Math. Phys.* **36**, 1365 (Note 4).

Qvist, B. (1947), *Societas Scientiarum Fennicae Commentationes Physico-Mathematicae* **XIII** no. 11 (4.4, 4.6, 4.12; Box 4.1.2).

Raine, D. J., Thomas, E. G. (1981), *Mon. Not. Roy. Astr. Soc.* **195**, 649 (3.7).

Rao, J. R., Tiwari, R. N., Bhamra, K. S. (1974), *Ann. Phys.* **87**, 470 (6.7; Fig. 6.1).

Ray, D. (1976), *J. Math. Phys.* **17**, 1171 (6.4, 6.5; Fig. 6.1).

Ray, D. (1978), *Int. J. Theor. Phys.* **17**, 153 (4.11).

Ray, D. (1982), *Phys. Rev.* **D26**, 3752 (6.5).
Ray, J. R., Thompson, E. L. (1975), *J. Math. Phys.* **16**, 345 (1.2).
Ray, J. R., Zimmerman, J. C. (1977), *Nuovo Cimento* **B42**, 183 (2.11.7).
Raychaudhuri, A. K. (1952), *Phys. Rev.* **86**, 90 (4.2.3, 4.3).
Raychaudhuri, A. K. (1955), *Z. Astrophysik* **37**, 103 (4.10; Box 4.1.4).
Raychaudhuri, A. K. (1966), *Proc. Phys. Soc.* **88**, 545 (3.4).
Raychaudhuri, A. K. (1975), *Ann. Inst. Poincare* **A22**, 229 (2.14).
Raychaudhuri, A. K. (1979), *Theoretical Cosmology.* Clarendon Press, Oxford (2.5).
Raychaudhuri, A. K., De, U. K. (1970), *J. Phys.* **A3**, 263 (3.5).
Ribeiro, M. B. (1992a), *Astrophys. J.* **388**, 1 (3.8).
Ribeiro, M. B. (1992b), *Astrophys. J.* **395**, 29 (3.8).
Ribeiro, M. B. (1993a), *Astrophys. J.* **415**, 469 (3.8).
Ribeiro, M. B. (1993b), in: *Proceedings of the NATO Advanced Research Workshop on Deterministic Chaos in General Relativity.* Edited by D. W. Hobill *et al.*, in press (3.8).
Rindler, W., Suson, D. (1989), *Astron. Astrophys.* **218**, 15 (3.8).
Robertson, H. P. (1929), *Proc. Nat. Acad. Sci. USA* **15**, 822 (1.1, A; all diagrams).
Robertson, H. P. (1933), *Rev. Mod. Phys.* **5**, 62 (1.1, A; all diagrams).
Rosen, G. (1980), *Nuovo Cimento* **B57**, 125 (8.4).
Rosquist, K., Jantzen, R. (1988), *Phys. Rep.* **166** no. 2, 89 (1.1).
Roy, S. R., Bali, R. (1978a), *Indian J. Pure Appl. Math.* **9**, 871 (4.11).
Roy, S. R., Bali, R. (1978b), *Indian J. Pure Appl. Math.* **9**, 1236 (4.11; Fig. 4.1).
Roy, S. R., Banerjee, S. K. (1988), *Astrophys. Space Sci.* **150**, 213 (7.2; Fig. 7.1).
Roy, S. R., Prasad, A. (1989), *Progr. Math.* **23**, 153 (7.1; Fig. 7.1).
Roy, S. R., Prasad, A. (1991), *Astrophys. Space Sci.* **181**, 61 (7.1; Fig. 7.1).
Roy, S. R., Singh, J. P. (1982), *Indian J. Pure Appl. Math.* **13**, 1285 (2.15; Fig. 2.5).
Roy, S. R., Tiwari, O. P. (1983), *Indian J. Pure Appl. Math.* **14**, 233 (2.10.3, 2.15; Figs. 2.2, 2.5).
Ruban, V. A. (1968), *Pisma v Red. ZhETF* **8**, 669 [*Sov. Phys. JETP Lett.* **8**, 414 (1968)] (1.4, 2.6; Fig. 2.1).
Ruban, V. A. (1969), *ZhETF* **56**, 1914 [*Sov. Phys. JETP* **29**, 1027 (1969)] (1.4, 2.6, 2.7, 2.8, 2.10.2, 2.11.1; Figs. 2.1, 2.2, 2.3).
Ruban, V. A. (1972), in: *Tezisy dokladov 3y Sovetskoy Gravitatsyonnoy Konferentsii* [*Theses of Lectures of the 3rd Soviet Conference on Gravitation*]. Izdatelstvo Erevanskogo Universiteta, Erevan, p. 348 (2.10.2; Fig. 2.2).
Ruban, V. A. (1983), *ZhETF* **85**, 801 [*Sov. Phys. JETP* **58**, 463 (1983)] (1.4, 2.6, 2.10.2, 2.11.1, 2.11.3, 2.11.4; Figs. 2.1, 2.2, 2.3).
Ruiz, E., Senovilla, J. M. M. (1992), *Phys. Rev.* **D45**, 1995 (7.1; Fig. 7.1).
Ryan, M. P. (1972), *Ann. Phys.* **72**, 584 (3.9).
Ryżyk, I. M., Gradsztejn, I. S. (1964), *Tablice całek, sum, szeregów i iloczynów* [*Tables of Integrals, Sums, Series and Products*, in Polish, translated from Russian]. Państwowe Wydawnictwo Naukowe, Warszawa (5.3).
Saar, E. (1971a), *Eesti NSV Akadeemia Toimetised* [*Izv. Akad. Nauk. Eston. SSR*] **20**, 420 (8.3).
Saar, E. (1971b), *Tartu Astron. Obs. Publ.* **39**, 206 (8.3).
Saar, E. (1971c), *Tartu Astron. Obs. Publ.* **39**, 234 (8.3).
Saar, E. (1971d), *Tartu Astron. Obs. Publ.* **39**, 249 (8.3).
Saez, D., Arnau, J. V. (1990), in: *Proceedings of the Relativity Meeting 89: Recent Developments in Gravitation, Barcelona 1989.* Edited by E. Verdaguer, J. Garriga, J. Cespedes. World Scientific, Singapore, p. 145 (3.7).
Saez, D., Arnau, J. V., Fullana, M. J. (1993), *Mon. Not. Roy. Astr. Soc.* **263**, 681 (3.7).
Sakoto, M. (1977), *C. R. Acad. Sci. Paris* **A284**, 633 (2.12).
Samoilov, S. N. (1981), *Ukr. Fiz. Zh.* **26**, 672 (3.8).
Sannan, S. (1986), *J. Math. Phys.* **27**, 2592 (3.8).
Santos, N. O. (1985), *Mon. Not. Roy. Astr. Soc.* **216**, 403 (4.14).
Sanyal, A. K., Ray, D. (1984), *J. Math. Phys.* **25**, 1975 (4.14; Fig. 4.3).
Sato, H. (1982), *Progr. Theor. Phys.* **68**, 236 (3.1, 3.3, 3.7, 3.8).
Sato, H. (1984), in: *General Relativity and Gravitation.* Edited by B. Bertotti, F. de Felice, A. Pascolini. D. Reidel, Dordrecht, p. 289 (1.1, 2.12, 3.1, 3.3, 3.7, 3.8, E; Box 2.4.2).

Sato, H., Maeda, K. (1983), *Progr. Theor. Phys.* **70**, 119 (2.12, 3.1, 3.3, 3.7, 3.8; Box 2.4.2).

Schmidt, H. J. (1982), *Astron. Nachr.* **303**, 283 (2.6, 2.8; Fig. 2.1).

Sedov, L. I. (1959), *Similarity and Dimensional Methods in Mechanics.* Academic Press, New York, p. 105 (3.1).

Seifert, H. J. (1979), *Gen. Rel. Grav.* **10**, 1065 (3.6).

Sen, N. R. (1934), *Z. Astrophysik* **9**, 215 (3.1, 3.2, E).

Sen, N. R. (1935), *Z. Astrophysik* **10**, 291 (3.1).

Senin, Yu. E. (1982), in: *Problemy teorii gravitatsii i elementarnykh chastits [Problems of Gravitation Theory and Elementary Particle Theory]*, 13th issue. Edited by K. P. Stanyukovich. Energoizdat, Moskva, p. 107 (2.8, 2.13, 7.2; Fig. 2.1).

Senovilla, J. M. M. (1990), *Phys. Rev. Lett.* **64**, 2219 (7.1).

Senovilla, J. M. M., Sopuerta, C. F. (1994), *Class. Q. Grav.* **11**, 2073 (7.4; Fig. 7.2).

Shah, Y. P., Vaidya, P. C. (1967), *Ann. Inst. Poincare* **A6**, 219 (4.12, 4.13; Box 4.1.5).

Shah, Y. P., Vaidya, P. C. (1968), *Tensor N. S.* **19**, 191 (4.1, 4.12; Fig. 4.1; Box 4.1.5).

Shapley, H. (1938a), *Proc. Nat. Acad. Sci. USA* **24**, 148 (B).

Shapley, H. (1938b), *Proc. Nat. Acad. Sci. USA* **24**, 527 (B).

Shikin, I. S. (1966), *Doklady ANSSSR* **171**, 73 [*Sov. Phys. Doklady* **11**, 944 (1967)] (2.11.6; Fig. 2.3).

Shikin, I. S. (1967), *Doklady ANSSSR* **176**, 1048 [*Sov. Phys. Doklady* **12**, 950 (1968)] (2.11.5; Fig. 2.3).

Shikin, I. S. (1972a), *Commun. Math. Phys.* **26**, 24 (1.1, 2.10.2, 2.14; Fig. 2.4).

Shikin, I. S. (1972b), in: *Tezisy dokladov 3-y Sovetskoy Gravitatsyonnoy Konferentsii [Theses of Lectures of the 3rd Soviet Conference on Gravitation]*, Erevan 1972. Izdatelstvo Erevanskogo Universiteta, p. 178 (2.14).

Shikin, I. S. (1974), *ZhETF* **67**, 433 [*Sov. Phys. JETP* **40**, 215 (1975)] (2.14; Fig. 2.4).

Shikin, I. S. (1981), *ZhETF* **81**, 801 [*Sov. Phys. JETP* **54**, 427 (1981)] (7.1; Fig. 7.1).

Shirokov, M. F. (1967), in: *Sovremennye problemy gravitatsii. Sbornik trudov II Sovetskoy Gravitatsyonnoy Konferentsii [Contemporary Problems of Gravitation. The Collection of Proceedings of the 2nd Soviet Conference on Gravitation]*. Tbilisi University Publishing House p. 376 (8.3).

Shirokov, M. F., Fisher, I. Z. (1962), *Astron. Zh.* **39**, 899 [*Sov. Astr. A. J.* **6**, 699 (1963)] (8.1, 8.3, 8.4, E).

Shvetsova, N. A., Shvetsov, V. A. (1976), *Izv. VUZ Fiz.* **19** no. 7, 138 [*Sov. Phys. J.* **19**, 949 (1976)] (4.11; Fig. 4.1).

Sibgatullin, N. R. (1971), *Doklady ANSSSR* **200**, 308 [*Sov. Phys. Doklady* **16**, 697 (1972)] (8.2).

Siemieniec–Oziębło, G. (1983), *Acta Phys. Polon.* **B14**, 465 (5.7; Fig. 5.1).

Siemieniec–Oziębło, G., Klimek, Z. (1978), *Acta Phys. Polon.* **B9**, 79 (2.12).

Silk, J. (1977), *Astron. Astrophys.* **59**, 53 (3.8).

Singh, K. M., Bhamra, K. S. (1990), *Int. J. Theor. Phys.* **29**, 1015 (1.3.2, 5.6; Fig. 5.1).

Singh, K. P., Abdussattar (1983), *Proc. Indian Nat. Sci. Acad.* **A49**, 448 (3.5).

Singh, T. (1978), *Progr. Math.* [this journal is sometimes referred to as *Acad. Prog. Math.*] **12**, 27 (6.5; Fig. 6.1).

Singh, T., Rai, L. N., Yadav, R. B. S. (1979), *J. Sci. Res. Banaras Hindu Univ.* **30** no. 2, 9 (1979–80) (6.5; Fig. 6.1).

Singh, T., Yadav, R. B. S. (1978a), *Indian J. Pure Appl. Math.* **9**, 900 (6.4, 6.5; Fig. 6.1).

Singh, T., Yadav, R. B. S. (1978b), *Acta Phys. Acad. Sci. Hung.* **45**, 107 (6.4, 6.5; Fig. 6.1).

Som, M. M., Bedran, M. L., Vasconcellos–Vaidya, E. P. (1988), *Nuovo Cimento* **B102**, 573 (2.10.4; Fig. 2.2).

Som, M. M., Santos, N. O. (1981), *Phys. Lett.* **A87**, 89 (4.14; Fig. 4.3).

Spero, A., Baierlein, R. (1977), *J. Math. Phys.* **18**, 1330 (8.4).

Spero, A., Baierlein, R. (1978), *J. Math. Phys.* **19**, 1324 (8.4).

Spero, A., Szafron, D. A. (1978), *J. Math. Phys.* **19**, 1536 (2.2).

Srivastava, D. C. (1986), *Gen. Rel. Grav.* **18**, 1159 (4.12; Box 4.1.5).

Srivastava, D. C. (1987), *Class. Q. Grav.* **4**, 1093 (4.2.3, 4.5, 4.6, 4.8; Fig. 4.1; Box 4.1.3).

Srivastava, D. C. (1989), *Pramana* **32**, 741 (4.12; Box 4.1.5).

Srivastava, D. C. (1992), *Fortschr. Physik* **40**, 31 (4.10, 4.12; Boxes 4.1.4, 4.1.5).

Srivastava, D. C., Prasad, S. S. (1983), *Gen. Rel. Grav.* **15**, 65 (4.4, 4.12).

Srivastava, D. C., Prasad, S. S. (1991), *Class. Q. Grav.* **8**, 1001 (4.1, 4.2.2, 4.12, 4.13; Figs. 4.1, 4.2).

Stanyukovich, K. P. (1969), *Doklady ANSSSR* **186**, 809 [*Sov. Phys. Doklady* **14**, 547 (1969)] (3.8).

Stanyukovich, K. P., Sharshekeev, O. S. (1973), *Prikl. Math. Mekh.* **37**, 739 [*Appl. Math. Mech.* **37**, 697 (1973)] (3.8).

Stein–Schabes, J. A. (1985), *Phys. Rev.* **D31**, 1838 (3.5).

Stephani, H. (1967a), *Commun. Math. Phys.* **4**, 137 (1.1, 1.3.2, 2.7, 4.1, 4.2.3, 4.4, 4.7, 4.9, 4.10, 4.11, 4.12, 4.14, 7.5, B, E; Notes 1, 4; Figs. 4.1, 4.3).

Stephani, H. (1967b), *Commun. Math. Phys.* **5**, 337 (4.9).

Stephani, H. (1968), *Commun. Math. Phys.* **9**, 53 (2.7, 2.8, 2.9, 2.10.2; Figs. 2.1, 2.2).

Stephani, H. (1983), *J. Phys.* **A16**, 3529 (4.2.3, 4.6).

Stephani, H. (1987), *Class. Q. Grav.* **4**, 125 (2.1.1, 2.9; Note 3; Figs. 2.1, 2.2).

Stephani, H. (1988), *J. Math. Phys.* **29**, 1650 (6.1).

Stephani, H. (1990), *General Relativity. An Introduction to the Theory of the Gravitational Field.* Second edition, Cambridge University Press, 321 pp. (Preface).

Stephani, H., Wolf, T. (1985), in: *Galaxies, Axisymmetric Systems and Relativity. Essays presented to W. B. Bonnor on his 65th birthday.* Edited by M. A. H. MacCallum. Cambridge University Press, p. 275 (2.13, 7.5; Fig. 7.2).

Stoeger, W. R., Ellis, G. F. R., Hellaby, C. (1987), *Mon. Not. Roy. Astr. Soc.* **226**, 373 (8.4).

Stoeger, W. R., Ellis, G. F. R., Nel, S. D. (1992), *Class. Q. Grav.* **9**, 509 (2.12, 3.9).

Strobel, H. (1968), *Wiss. Z. Friedrich Schiller Univ. Jena, Math. Naturw. Reihe* **17**, 195 (4.10, 4.14; Box 4.1.4; Fig. 4.3).

Strobel, H. (1972), *Wiss. Z. Friedrich Schiller Univ. Jena, Math. Naturw. Reihe* **21**, 111 (4.14; Fig. 4.3).

Sussman, R. (1986), in: *Proceedings of the 4th Marcel Grossman Meeting on General Relativity.* Edited by R. Ruffini. Elsevier, Amsterdam, p. 867 (4.6).

Sussman, R. (1987), *J. Math. Phys.* **28**, 1118 (3.5, 4.6, 4.12; Boxes 4.1.3, 4.1.5).

Sussman, R. (1988a), *J. Math. Phys.* **29**, 945 (3.5, 4.4, 4.6, 4.12; Box 4.1.3).

Sussman, R. (1988b), *J. Math. Phys.* **29**, 1177 (3.5, 4.4, 4.6, 4.7, 4.12; Box 4.1.3).

Sussman, R. (1989a), *Gen. Rel. Grav.* **21**, 1281 (4.3, 4.10, 4.11; Fig. 4.1; Boxes 4.1.1, 4.1.4).

Sussman, R. (1989b), *Phys. Rev.* **D40**, 1364 (4.11).

Sussman, R. (1990), in: *Proceedings of the 3rd Canadian Conference on General Relativity and Relativistic Astrophysics.* Edited by A. Coley, F. Cooperstock and B. Tupper. World Scientific, Singapore, p. 40 (4.11).

Sussman, R. (1991), in: *Proceedings of the 7th Latin American Symposium on Relativity and Gravitation (SILARG 7).* Edited by J. C. Olivo, E. Nahmad–Achar, M. Rosenbaum, M. P. Ryan Jr., J. F. Urrutia, F. Zertuche. World Scientific, Singapore, p. 248 (4.11; Fig. 4.1).

Sussman, R. A. (1992), *Class. Q. Grav.* **9**, 1891 (2.13).

Sussman, R. A. (1993), *Class. Q. Grav.* **10**, 2675 (4.14).

Suto, Y., Sato, K., Sato, H. (1984a), *Progr. Theor. Phys.* **71**, 938 (3.1, 3.3).

Suto, Y., Sato, K., Sato, H. (1984b), *Progr. Theor. Phys.* **72**, 1137 (2.12, 3.1, 3.3).

Synge, J. L. (1934), *Proc. Nat. Acad. Sci. USA* **20**, 635 (2.12).

Szafron, D. A. (1977), *J. Math. Phys.* **18**, 1673 (2.1, 2.2, 2.3, 2.6, 2.8, 2.9, 2.10.1, 2.10.2, 2.10.3, 2.12, 2.13, 2.15, 4.11, E; Notes 1, 2; Figs. 2.1, 2.2, 2.4; Box 2.4.3; Fig. 2.5).

Szafron, D. A., Collins, C. B. (1979), *J. Math. Phys.* **20**, 2354 (2.2, 2.3, 2.12).

Szafron, D. A, Wainwright, J. (1977), *J. Math. Phys.* **18**, 1668 (2.8, 2.10.1, 2.10.3, 2.10.4, 2.13, A; Figs. 2.1, 2.2).

Szekeres, P. (1966a), *J. Math. Phys.* **7**, 751 (7.3).

Szekeres, P. (1966b), *Nuovo Cimento* **A43**, 1062 (2.7).

Szekeres, P. (1971), *Ann. Phys.* **64**, 599 (8.2).

Szekeres, P. (1972), *Proc. Astron. Soc. Australia* **2**, 110 (3.4).

Szekeres, P. (1975a), *Commun. Math. Phys.* **41**, 55 (1.1, 2.1, 2.4.1, 2.4.2, 2.5, 2.6, 2.7, 2.8, 2.12, 2.13, 3.6, 3.8, A, B, E; Figs. 2.1, 2.2, 2.4).

Szekeres, P. (1975b), *Phys. Rev.* **D12**, 2941 (2.4.2, 3.6).

Szekeres, P. (1980), in: *Gravitational Radiation, Collapsed Objects Exact Solutions.* Edited by C. Edwards. Springer (Lecture notes in physics, vol. 124), New York, p. 477 (2.4.2, 3.6).

Szekeres, P., Iyer, V. (1993), *Phys. Rev.* **D47**, 4362 (3.6).

Tabensky, R., Taub, A. H. (1973), *Commun. Math. Phys.* **29**, 61 (6.4, 6.5, 6.6, 6.7; Fig. 6.1).

Tabensky, R., Zamorano, N. (1975), *Int. J. Theor. Phys.* **13**, 1 (6.1, 6.4).

Taub, A. H. (1951), *Ann. Math.* **3**, 472 (2.12).

308 *Bibliography*

Taub, A. H. (1968a), in: *Abstracts, 5th International Conference on Gravitation and the Theory of Relativity*. Publishing House of the Tbilisi University, Tbilisi, p. 145 (4.3, 4.4, 4.5; Box 4.1.1).
Taub, A. H. (1968b), *Ann. Inst. Poincare* **A9**, 153 (4.3, 4.4, 4.5, 4.10; Fig. 4.1; Boxes 4.1.1, 4.1.2, 4.1.4).
Taub, A. H. (1972), in: *General Relativity (papers in honour of J. L. Synge)*. Edited by O'Raiffertaigh. Clarendon Press, Oxford, p. 133 (4.2.2).
Tavakol, R. K., Ellis, G. F. R. (1988), *Phys. Lett.* **A130**, 217 (B).
Thakurta, S. N. G. (1981), *Indian J. Phys.* **B55**, 304 (7.6; Fig. 7.2).
Thompson, I. H., Whitrow, G. J. (1967), *Mon. Not. Roy. Astr. Soc.* **136**, 207 (4.10; Box 4.1.4).
Thompson, I. H., Whitrow, G. J. (1968), *Mon. Not. Roy. Astr. Soc.* **139**, 499 (4.10).
Thorne, K. S. (1967), *Astrophys. J.* **148**, 51 (2.11.6; Fig. 2.3).
Tolman, R. C. (1934), *Proc. Nat. Acad. Sci. USA* **20**, 169 (1.1, 2.12, 2.13, 3.1, 3.2, A, B, C, E).
Tomimura, N. (1977), *Nuovo Cimento* **B42**, 1 (2.8; Fig. 2.1).
Tomimura, N. (1978), *Nuovo Cimento* **B44**, 372 (2.6, 2.12, 3.1; Fig. 2.1; Box 2.4.1).
Tomimura, N. (1981), *Astrophys. J.* **249**, 23 (2.4.1).
Tomimura, N., Motta, D. C. (1990a), *Astrophys. Space Sci.* **165**, 231 (2.10.3).
Tomimura, N., Motta, D. C. (1990b), *Astrophys. Space Sci.* **165**, 243 (2.10.3).
Tomimura, N., Waga, I. (1987), *Astrophys. J.* **317**, 52 (2.10.2; Fig. 2.2).
Tomita, K. (1969a), *Progr. Theor. Phys.* **42**, 9 (3.2).
Tomita, K. (1969b), *Progr. Theor. Phys.* **42**, 978 (3.2).
Tomita, K. (1975), *Progr. Theor. Phys.* **54**, 730 (2.12; Box 2.4.1).
Tomita, K. (1978), *Progr. Theor. Phys.* **59**, 1150 (6.3, 6.4, 6.5, 6.7; Fig. 6.1).
Tomita, K. (1992), *Astrophys. J.* **394**, 401 (3.8).
Tomita, K., Nariai, H. (1968), *Progr. Theor. Phys.* **40**, 1184 (4.10).
Trevese, D., Vignato, A. (1977), *Astrophys. Space Sci.* **49**, 229 (3.2).
Tupper, B. O. J. (1983), *Gen. Rel. Grav.* **15**, 47 (1.1).
Uggla, C. (1992), *Class. Q. Grav.* **9**, 2287 (7.1; Fig. 7.1).
Unnikrishnan, G. S. (1994), *Gen. Rel. Grav.* **26**, 655 (3.6).
Vaidya, P. C. (1943), *Curr. Sci.* **12**, 183 (3.6, 5.2).
Vaidya, P. C. (1951), *Proc. Indian Acad. Sci.* **A33**, 264 (3.6, 5.2).
Vaidya, P. C. (1953), *Nature* **171**, 260 (3.6, 4.14, 5.2).
Vaidya, P. C. (1966), *Astrophys. J.* **144**, 943 (5.7; Fig. 5.1).
Vaidya, P. C. (1968a), in: *Abstracts, 5th International Conference on Gravitation and the Theory of Relativity*. Publishing House of the Tbilisi University, Tbilisi, p. 10 (4.10).
Vaidya, P. C. (1968b), *Phys. Rev.* **174**, 1615 (4.10; Box 4.1.4).
Vaidya, P. C. (1976), *Curr. Sci.* **45**, 490 (5.3).
Vaidya, P. C. (1977), *Pramana* **8** 512 (5.4, 5.6, 7.6; Fig. 5.1).
Vaidya, P. C., Patel, L. K. (1985), *Math. Today* **3**, 41 (5.6; Fig. 5.1).
Vaidya, P. C., Patel, L. K., Bhatt, P. V. (1976), *Gen. Rel. Grav.* **7**, 701 (5.3, 5.5).
Vaidya, P. C., Shah, K. B. (1960), *Progr. Theor. Phys.* **24**, 111 (5.7; Fig. 5.1).
Vaidya, P. C., Shah, Y. P. (1967), *Curr. Sci.* **36**, 120 (4.12; Box 4.1.5).
Vajk, J. P., Eltgroth, P. G. (1970), *J. Math. Phys.* **11**, 2212 (2.11.1, 2.11.3, 2.11.6; Fig. 2.3).
Van den Bergh, N., Wils, P. (1985), *Gen. Rel. Grav.* **17**, 223 (2.11.1, 2.11.2; Fig. 2.3).
Van den Bergh, S. (1990), *J. Roy. Astr. Soc. Can.* **84**, 275 (B).
Verdaguer, E. (1985), in: *Observational and Theoretical Aspects of Relativistic Astrophysics and Cosmology*. Edited by J. L. Sanz and L. J. Goicoechea. World Scientific, Singapore p. 311 (6.7).
Verdaguer, E. (1993), *Phys. Rep.* **229** no. 1, 1 (6.1, 6.7, 7.8).
Verma, D. N., Roy, S. N. (1956), *Bull. Calcutta Math. Soc.* **48**, 129 (4.11; Fig. 4.1).
Vickers, P. A. (1973), *Ann. Inst. Poincare* **A18**, 137 (2.14, 3.5; Fig. 2.4; Box 2.4.4).
Wagh, R. V. (1955a), *J. Univ. Bombay* **24**, 1 (4.3, 4.6; Boxes 4.1.1, 4.1.3).
Wagh, R. V. (1955b), *J. Univ. Bombay* **24**, 5 (4.3).
Wagh, R. V. (1958), *J. Univ. Bombay* **26** no. 5, 16 (4.3; Box 4.1.1).
Wainwright, J. (1974), *Int. J. Theor. Phys.* **10**, 39 (7.7; Fig. 7.3).
Wainwright, J. (1977), *J. Math. Phys.* **18**, 672 (2.3).
Wainwright, J. (1979), *J. Phys.* **A12**, 2015 (1.1, 2.2, 4.1, 5.1, 6.1, 7.1, 7.3, 7.4, 7.5, 7.7, 7.8).
Wainwright, J. (1981), J. Phys. **A14**, 1131 (1.1, 2.2, 4.1, 5.1, 7.3, 7.4, 7.5, 7.7).
Wainwright, J. (1984), *Gen. Rel. Grav.* **16**, 657 (2.5).

Wainwright, J., Goode, S. W. (1980), *Phys. Rev.* **D22**, 1906 (7.1; Fig. 7.1).
Wainwright, J., Ince, W. C. W., Marshman, B. J. (1979), *Gen. Rel. Grav.* **10**, 259 (6.1, 6.7, 7.8, A; Fig. 6.1).
Wainwright, J., Yaremovicz, P. E. A. (1976a), *Gen. Rel. Grav.* **7**, 345 (1.2).
Wainwright, J., Yaremovicz, P. E. A. (1976b), *Gen. Rel. Grav.* **7**, 595 (1.2).
Walker, A. G. (1935), *Quart. J. Math. Oxford,* ser. 6, 81 (1.1, A; all diagrams).
Waugh, B., Lake, K. (1988), *Phys. Rev.* **D38**, 1315 (3.6).
Weber, E. (1984), *J. Math. Phys.* **25**, 3279 (2.11.1; Fig. 2.3).
Weber, E. (1985), *J. Math. Phys.* **26**, 1308 (2.11.1; Fig. 2.3).
Weber, E. (1986), *J. Math. Phys.* **27**, 1578 (2.11.1; Fig. 2.3).
Weinberg, S. (1972), *Gravitation and Cosmology. Wiley*, New York (2.5).
Wesson, P. S. (1978a), *Astrophys. Lett.* **19**, 121 (2.13).
Wesson, P. S. (1978b), *Astrophys. Space Sci.* **54**, 489 (2.13).
Wesson, P. S. (1978c), *Cosmology and Geophysics.* Adam Hilger Ltd, Techno House, Bristol, p. 199 (2.13).
Wesson, P. S. (1979), *Astrophys. J.* **228**, 647 (2.13).
Wesson, P. S. (1980), *Gravity, Particles and Astrophysics.* D. Reidel, Dordrecht, pp. 147–161 (B).
Wesson, P. S. (1984), *J. Math. Phys.* **25**, 3297 (2.13; Box 2.4.3).
Wesson, P. S. (1989), *Astrophys. J.* **336**, 58 (2.6; Fig. 2.1).
Wesson, P. S. (1990), *Can. J. Phys.* **68**, 824 (4.11).
Wesson, P. S., Ponce de Leon, J. (1989), *Phys. Rev.* **D39**, 420 (4.11; Fig. 4.1).
Wils, P. (1990), *Class. Q. Grav.* **7**, L43 (7.1; Fig. 7.1).
Wolf, T. (1986), *J. Math. Phys.* **27**, 2340 (7.5).
Woolley, M. L. (1973), *Commun. Math. Phys.* **31**, 75 (1.2).
Wu, Z. C. [also quoted as Chao] (1981), *Gen. Rel. Grav.* **13**, 625 (2.13; Box 2.4.3).
Wu, Z. C. (1982), Sci. Sinica **A25**, 737 (2.13).
Wyman, M. (1946), *Phys. Rev.* **70**, 396 (4.1, 4.2.2, 4.2.3, 4.3, 4.4, 4.5, 4.6, 4.10, 4.12; Fig. 4.1; Boxes 4.1.1, 4.1.3, 4.1.4, 4.1.5).
Wyman, M. (1976), *Canad. Math. Bull.* **19**, 343 (4.3; Box 4.1.1).
Wyman, M. (1978), *Austral. J. Phys.* **31**, 111 (4.6; Box 4.1.3).
Xanthopoulos, B. C., Zannias, T. (1992), *J. Math. Phys.* **33**, 1415 (2.11.1; Fig. 2.3).
Yodzis, P., Seifert, H. J., Müller zum Hagen, H. (1973), *Commun. Math. Phys.* **34**, 135 (2.4.2, 2.13, 3.6; Box 2.4.3).
York, J. W. Jr. (1972), *Phys. Rev. Lett.* **28**, 1082 (2.2).
Zakharov, A. V. (1987), *Izv. VUZ Fiz.* **30** no. 12, 20 [*Sov. Phys. J.* **30**, 1015 (1987)] (2.6, 2.12).
Zalaletdinov, R. M. (1992), *Gen. Rel. Grav.* **24**, 1015 (8.1, 8.4, E).
Zalaletdinov, R. M. (1993), *Gen. Rel. Grav.* **25**, 673 (8.4).
Zalaletdinov, R. M. (1994a), in: *Proceedings of the International Symposium on Experimental Gravitation.* Edited by M. Karim and A. Qadir. Institute of Physics Publishing, Bristol, p. A363 (8.4).
Zalaletdinov, R. M. (1994b), in: *Proceedings of the 7th Marcel Grossman Meeting on General Relativity.* Edited by M. Kaiser. World Scientific, Singapore (in press) (8.4).
Zecca, A. (1991), *Nuovo Cimento* **B106**, 413 (2.12; Fig. 2.4).
Zecca, A. (1993a), *Int. J. Theor. Phys.* **32**, 615 (3.8).
Zecca, A. (1993b), *Nuovo Cimento* **B108**, 403 (2.14).
Zeldovich, Ya. B. (1961), *ZhETF* **41**, 1609 [*Sov. Phys. JETP* **14**, 1143 (1962)] (6.1).
Zeldovich, Ya. B., Grishchuk, L. P. (1984), *Mon. Not. Roy. Astr. Soc.* **207**, 23p (3.8).
Zelmanov, A. L. (1959), in: *Trudy 6-go Soveshchanya po Voprosam Kosmogonii* [*Proceedings of the 6th Meeting on Questions of Cosmogony*]. Edited by D. A. Frank–Kamenetskii. Izdatelstvo Akademii Nauk SSSR, Moskva, p. 144 (1.2).
Zhu, S. (1983), in: *Proceedings of the 3rd Marcel Grossman Meeting on General Relativity.* Edited by Hu Ning. Science Press and North Holland Publishing Company, Amsterdam, p. 1391 (3.1).
Zotov, N. V., Stoeger, W. R. (1992), *Class. Q. Grav.* **9**, 1023 (8.3, 8.4).

Index